Biotechnology in Agriculture and Forestry

Springer

Berlin
Heidelberg
New York
Barcelona
Budapest
Hong Kong
London
Milan
Paris
Santa Clara
Singapore
Tokyo

Volumes already published

Volume 1: Trees I (1986)
Volume 2: Crops I (1986)
Volume 3: Potato (1987)
Volume 4: Medicinal and Aromatic Plants I (1988)
Volume 5: Trees II (1989)
Volume 6: Crops II (1988)
Volume 7: Medicinal and Aromatic Plants II (1989)
Volume 8: Plant Protoplasts and Genetic Engineering I (1989)
Volume 9: Plant Protoplasts and Genetic Engineering II (1989)
Volume 10: Legumes and Oilseed Crops I (1990)
Volume 11: Somaclonal Variation in Crop Improvement I (1990)
Volume 12: Haploids in Crop Improvement I (1990)
Volume 13: Wheat (1990)
Volume 14: Rice (1991)
Volume 15: Medicinal and Aromatic Plants III (1991)
Volume 16: Trees III (1991)
Volume 17: High-Tech and Micropropagation I (1991)
Volume 18: High-Tech and Micropropagation II (1992)
Volume 19: High-Tech and Micropropagation III (1992)
Volume 20: High-Tech and micropropagation IV (1992)
Volume 21: Medicinal and Aromatic Plants IV (1993)
Volume 22: Plant Protoplasts and Genetic Engineering III (1993)
Volume 23: Plant Protoplasts and Genetic Engineering IV (1993)
Volume 24: Medicinal and Aromatic Plants V (1993)
Volume 25: Maize (1994)
Volume 26: Medicinal and Aromatic Plants VI (1994)
Volume 27: Somatic Hybridization in Crop Improvement I (1994)
Volume 28: Medicinal and Aromatic Plants VII (1994)
Volume 29: Plant protoplasts and Genetic Engineering V (1994)
Volume 30: Somatic Embryogenesis and Synthetic Seed I (1995)
Volume 31: Somatic Embryogenesis and Synthetic Seed II (1995)
Volume 32: Cryopreservation of Plant Germplasm I (1995)
Volume 33: Medicinal and Aromatic Plants VIII (1995)
Volume 34: Plant Protoplasts and Genetic Engineering VI (1995)
Volume 35: Trees IV (1996)
Volume 36: Somaclonal Variation in Crop Improvement II (1996)
Volume 37: Medicinal and Aromatic Plants IX (1996)
Volume 38: Plant Protoplasts and Genetic Engineering VII (1996)
Volume 39: High-Tech and Micropropagation V (1997)
Volume 40: High-Tech and Micropropagation VI (1997)
Volume 41: Medicinal and Aromatic Plants X (1998)
Volume 42: Cotton (1998)

Volume in preparation

Volume 43: Medicinal and Aromatic Plants XI

Biotechnology in Agriculture and Forestry 42

Cotton

Edited by Y.P.S. Bajaj

With 118 Figures, 2 in Color, and 80 Tables

Springer

Professor Dr. Y.P.S. Bajaj
A-137
New Friends Colony
New Delhi 110065, India

ISSN 0934-943-X
ISBN 3-540-62728-6 Springer-Verlag Berlin Heidelberg New York

Library of Congress Cataloging-in-Publication Data. Cotton/edited by Y.P.S. Bajaj. p. cm. – (Biotechnology in agriculture and forestry; 42) Includes bibliographical references and index. ISBN 3-540-62728-6 (hardcover) 1. Cotton–Biotechnology. 2. Cotton–Micropropagation. I. Bajaj, Y.P.S., 1936– . II. Series. SB249.C79355 1998 633.5′123–dc21 97-45484 CIP

Cover design: *design & production* GmbH, Heidelberg

Typesetting: Best-set Typesetter Ltd., Hong Kong

SPIN: 10554001 31/3136 – 5 4 3 2 1 0 – Printed on acid-free paper

Dedicated to
Ruchi and Satvinder Bakshi

Preface

Cotton is one of the most important sources of fiber, oil, and seed meal. At present, it is being grown virtually all over the world. About 90% of the commercial cotton is obtained from *Gossypium hirsutum* (upland cotton). The genus *Gossypium* contains 50 species, both wild and cultivated. Though wild species have been a rich source of genetic variability, up until now little success has been achieved in utilizing this wide variability already existing in wild germplasm. Tremendous loss to cotton production and fiber quality is caused by various diseases and pests. Conventional methods of cotton breeding alone are not sufficient to cope with the situation, thus innovative approaches have to be incorporated.

The importance and prospects of biotechnology for the improvement of cotton were stressed by the author more than a decade ago in an article, *Biotechnology of Cotton Improvement* (Bajaj and Gill 1986). Since then there has been an upsurge of activity in this field, and much progress has been made, especially in the in vitro regeneration of plants and in the production of transgenic cotton showing resistance to insects, herbicides, etc.; field studies have also been conducted on such plants.

Taking the above-mentioned aspects into consideration, the present book on the biotechnology of cotton has been compiled. Twenty-three chapters on various aspects of in vitro manipulation and other biotechnological approaches to the improvement of cotton are arranged in six sections:

Section I. Biotechnology for the improvement of cotton, callus culture and somatic embryogenesis, protoplast culture, micropropagation and the conservation of germplasm
Section II. Interspecific hybridization, embryo/ovule culture, in vitro fertilization, RAPD, heterosis
Section III. Anther/pollen culture, parthenogenesis and haploid breeding
Section IV. Somaclonal variation, plants tolerant to salt, water stress, high temperature, and nematodes
Section V. Genetic transformation, transgenic cotton resistant to insects and herbicides, field trials, risk assessment studies, and genetic engineering of fiber.
Section VI. Secondary metabolites – gossypol and tannins

Special emphasis has been placed on interspecific hybridization, somaclonal variation, transgenic cotton resistant to insects and herbicides, and re-engineering of fiber. This book is of special interest to advanced students, teachers, and research workers in the field of cotton breeding, genetics, tissue culture, molecular biology, and plant biotechnology in general.

New Delhi, February 1998

Professor Dr. Y.P.S. BAJAJ
Series Editor

Contents

Section I Biotechnology for the Improvement of Cotton; Establishment of Callus Culture and Somatic Embryogenesis; Protoplast Culture; Micropropagation and the Conservation of Germplasm

I.4 Micropropagation and the Conservation of Germplasm of Cotton
Y.P.S. Bajaj (With 3 Figures)

Section II Interspecific Hybridization, Embryo/Ovule Culture; In Vitro Fertilization; RAPD; Heterosis

II.1 Interspecific Hybridization in Cotton Through Embryo Rescue and Culture
Y.P.S. Bajaj and M.S. Gill

II.2 In Vitro Fertilization in Cotton
C. Liu and Y.P.S. Bajaj (With 27 Figures)

II.3 In Vitro Culture of Cotton Ovules
J.E. Mellon

II.4 Use of Random Amplified Polymorphic DNA (RAPD) Markers to Assist Wide Hybridization in Cotton
G. Mergeai, I. Vroh Bi, J.P. Baudoin, and P. du Jardin (With 3 Figures)

II.5 Heterosis in Cotton
D.G. Roupakias, E. Gouli-Vavdinoudi, M. Koutsika-Sotiriou, S. Galanopoulou-Sendouka, and A.S. Mavromatis
(With 6 Figures)

Section III Anther/Pollen Culture; Parthenogenesis and Haploid Breeding

III.1 Anther Culture Studies and Pollen Embryogenesis in Cotton
Y.P.S. Bajaj and M.S. Gill (With 4 Figures)

III.2 In Vitro Culture of Isolated Pollen of Cotton
A.M. Touraev and Z.B. Shamina (With 4 Figures)

III.3 Parthenogenesis and Haploid Breeding in *Gossypium hirsutum*
S.Q. Zhou and D.Q. Qian (With 20 Figures)

Section IV Somaclonal Variation/Genetic Variability; Plants Tolerant to Salt and Water Stress, High Temperature, and Nematodes (for plants resistant to insects and herbicides see Chaps. V.2,3 this Volume)

Section V Genetic Transformation; Transgenic Cotton Resistant to Insects and Herbicides; Field Trials; Risk Assessment Studies; Gentic Engineering of Fiber

V.I Genetic Transformation of Cotton Through Particle Bombardment
D.E. McCabe, B.J. Martinell, and M.E. John (With 1 Figure)

V.2 Insect-Resistant Transgenic Cotton
J.C. Thomas (With 2 Figures)

V.3 Transgenic Cotton Resistant to the Herbicide 2,4-D
D.W. Ow and J.E. Quisenberry (With 4 Figures)

V.4 Effect of Transgenic Cotton Expressing the *Bacillus thuringiensis* var. *kurstaki* Endotoxin on Soil Microorganisms – Risk Assessment Studies
K.K. Donegan and R.J. Seidler (With 2 Figures)

V.5 Genetic Engineering of Cotton Fiber
M.E. JOHN (With 6 Figures)

Section VI Secondary Metabolites – Gossypol and Tannins

VI.1 Gossypol – A Unique Biologically Active Compound from the Cotton Plant
J.W. JAROSZEWSKI (With 6 Figures)

VI.2 Tannins in Cotton
K.E. LEGÉ (With 4 Figures)

List of Contributors

BAJAJ, Y.P.S., Biotechnology in Agriculture and Forestry, A-135 New Friends Colony, New Delhi 110065, India

BAUDOIN, J.P., Unité de Phytotechnie des Régions intertropicales, Faculté Universitaire des Sciences agronomiques, B-5030 Gembloux, Belgium

CAI, X.N., Institute of Agrobiological Genetics and Physiology, Jiangsu Academy of Agricultural Sciences, Nanjing, Jiangsu Province 210014, China

CREECH, R.G., Department of Plant and Soil Sciences, Mississippi State University, Mississippi State, Mississippi 39762, USA

DONEGAN, K.K., Dynamac Corporation, US Environmental Protection Agency, National Health and Environmental Effects Research Laboratory – Western Ecology Division, 200 SW 35th St., Corvallis, Oregon 97333, USA

DU JARDIN, P., Unité de Biologie Végétale, Faculté Universitaire des Sciences Agronomiques, B-5030 Gembloux, Belgium

GALANOPOULOU-SENDOUCA, S., Laboratory of Agronomy, School of Agriculture, University of Thessaly, 38334 Volos, Greece

GILL, M.S., Punjab Agricultural University, Ludhiana (Punjab), India

GORHAM, J., Centre for Arid Zone Studies and School of Biological Sciences, Memorial Building, University of Wales, Bangor, Gwynedd, Wales, LL57 2UW, UK

GOULI-VAVDINOUDI, E., Department of Genetics and Plant Breeding, School of Agriculture, Aristotelian University of Thessaloniki, 54006 Thessaloniki, Greece

JAROSZEWSKI, J.W., Department of Medicinal Chemistry, Royal Danish School of Pharmacy, DK-2100 Copenhagen, Denmark

Jenkins, J.N., USDA/ARS Crop Science Research Laboratory, Mississippi State, Mississippi 39762, USA

John, M.E., Agracetus, 8520 University Green, Middleton, Wisconsin 53562, USA

Koutsika-Sotiriou, M., Department of Genetics and Plant Breeding, School of Agriculture, Aristotelian University of Thessaloniki, 54006 Thessaloniki, Greece

Lawrence, G.W., Department of Entomology and Plant Pathology, Mississippi State University, Mississippi State 39762, USA

Legé, K.E., Sure Grow Seed Inc., Route 1, Box 310, Centre, Alabama 35960, USA

Leidi, E.O., Department of Plant Biology, Instituto de Recursos Naturales y Agrobiología de Sevilla, CSIC, Avda, Reina Mercedes 10, Apartado 1052 EP, 41012 Sevilla, Spain

Liu, C., Department of Agronomy, Huazhong Agricultural University, Wuhan 430070, China

Martinell, B.J., Agracetus, a unit of Monsanto, 8520 University Green, Middleton, Wisconsin 53562, USA

Mavromatis, A.S., Department of Genetics and Plant Breeding, School of Agriculture, Aristotelian University of Thessaloniki, 54006 Thessaloniki, Greece

McCabe, D.E., Agracetus, a unit of Monsanto, 8520 University Green, Middleton, Wisconsin 53562, USA

McCarty, J.C. Jr., USDA/ARS Crop Science Research Laboratory, Mississippi State, Mississippi 39762, USA

Mellon, J.E., US Department of Agriculture, Agricultural Research Service, Southern Regional Research Center, P.O. Box 19687, New Orleans, Louisiana 70179-0687, USA

Mergeai, G., Unité de Phytotechnie des Régions intertropicales, Faculté Universitaire des Sciences agronomiques, B-5030 Gembloux, Belgium

Ow, D.W., United States Department of Agriculture, Agricultural Research Service, Plant Gene Expression Center, 800 Buchanan Street, Albany, California 94710, USA

PEETERS, M.-C., Laboratory of Tropical Crop Improvement, Catholic University of Leuven, Kardinaal Mercierlaan 92, B-3001 Heverlee, Belgium

QIAN, D.Q., Institute of Genetics, Shenyang Agricultural University, Shenyang 110161, People's Republic of China

QUISENBERRY, J., United States Department of Agriculture, Agricultural Research Service, Cropping Systems Research Laboratory, Route 3, Box 213, Lubbock, Texas 79401, USA

ROUPAKIAS, D.G., Department of Genetics and Plant Breeding, School of Agriculture, Aristotelian University of Thessaloniki, 54006 Thessaloniki, Greece

SEIDLER, R.J., US Environmental Protection Agency, National Health and Environmental Effects Research Laboratory – Western Ecology Division, 200 SW 35th St., Corvallis, Oregon 97333, USA

SHAMINA, Z.B., Timiriazev Institute of Plant Physiology, Russian Academy of Sciences, Botanicheskaya ul., 35, Moscow 127276, Russia

SHANG, X., Department of Agronomy, Horticulture, and Entomology, Texas Tech University, Lubbock, Texas 79401, USA

SHE, J.M., Institute of Agrobiological Genetics and Physiology, Jiangsu Academy of Agricultural Sciences, Nanjing, Jiangsu Province 210014, China

SWENNEN, R.L., Laboratory of Tropical Crop Improvement. Faculty of Agricultural and Applied Biological Sciences, Catholic University of Leuven, 92 Kardinaal Mercierlaan, B-3000 Heverlee, Belgium

THOMAS, J.C., Department of Naturral Sciences, University of Michigan-Dearborn, 4901 Evergreen Road, Dearborn, Michigan 48128-1491, USA

TROLINDER, N.L., USDA-ARS, Cropping Systems Research Laboratory. Present address: BIOTEX, 599 E. 50th, Lubbock, Texas 79404, USA

TOURAEV, A.M., Vienna Biocenter, Institute of Microbiology and Genetics, Vienna University, Dr. Bohrgasse 9, A-1030 Vienna, Austria

VROH BI, I., Unité de Phytotechnie des Régions intertropicales, Faculté Universitaire des Sciences agronomiques, B-5030 Gembloux, Belgium

WU, J.Y., Institute of Agrobiological Genetics and Physiology, Jiangsu Academy of Agricultural Sciences, Nanjing, Jiangsu Province 210014, China

ZHOU, S.Q., Institute of Genetics, Shenyang Agricultural University, Shenyang 110161, People's Republic of China

Section I
Biotechnology for the Improvement of Cotton; Establishment of Callus Culture and Somatic Embryogenesis; Protoplast Culture; Micropropagation, and the Conservation of Germplasm

I.1 Biotechnology for the Improvement of Cotton

Y.P.S. Bajaj[1]

1 General Account

1.1 Importance and Distribution

Cotton belongs to the genus *Gossypium* (family Malvaceae). It is a multipurpose crop and produces lint (fiber), oil, seed meal, and hulls. Lint is the most important source of fiber currently being used in the textile industry. The oil content of different cultivated species varies from 16–25% of dry seed weight (Lawhon et al. 1977). Glandless cotton seed have been used for human consumption, e.g., a popped seed snack called TAMU nuts, a flour used for commercial production of a high-protein bread, and as a meat extender (Price and Smith 1984). Seed meal contains a high percentage (16–22%) of protein and is rich in essential amino acids like lycine, methionine, tryptophane, and other amino acids (El-Nockrashy et al. 1969; Beradi and Cherry 1980) and is primarily used as a fertilizer and cattle feed. Seed hulls are used for cattle feed and as a soil covering called mulch. Linters, which are short fibers hanging to the seed after ginning, are removed at the oil mill to become a valuable source of cellulose, for plastics, synthetic fibers, etc.

The cotton plant is native to the tropics and subtropics, but the annual types selected during domestication have allowed it to be grown in temperate regions with long warm summers. Today, cotton is grown virtually all over the world (Table 1). The cultivation of Old World diploid cotton is restricted to Asia and some parts of Africa. *G. barbadense* includes cultivars grown in the Nile River drainage area (Egyptian cotton), in Peru (Tanguis cotton), and in restricted areas of the United States (Pima cotton). About 90% of the commercial cotton is obtained from *G. hirsutum* (upland cotton), which is found in almost all the cotton-growing areas of the world (Lee 1984).

1.2 *Gossypium* Species

The genus *Gossypium* comprises 50 species (Fryxell 1992); including 4 cultivated species, two Old World diploid species (2n = 26), *G. aroboreum* and

[1] Biotechnology in Agriculture and Forestry, A-135 New Friends Colony, New Delhi 110065, India

Biotechnology in Agriculture and Forestry, Vol. 42
Cotton (ed. by Y.P.S. Bajaj)

Table 1. Some important cotton-growing areas/countries. (FAO 1993)

Region/country	Area harvested (1000 ha)	Yield (kg/ha)	Production (1000 ha)
World	30186	1592	49059
Egypt	372	2947	1096
NC America	5241	1787	9364
USA	5175	1780	9211
Brazil	1061	1074	1139
Asia	16608	1492	24781
China	5000	2256	11280
India	7485	861	6446
Pakistan	2718	1502	4083
Europe	398	2773	1103
Australia	239	3770	901

G. herbaceum, and two New World tetraploid (2n = 52) species, *G. hirsutum* and *G. barbadense*. These species have been categorized into different genomic groups (Tables 2, 3) on the basis of the cytology of the interspecific hybrids (Beasley 1942; Phillips and Strickland 1966; Edwards and Mirza 1979). The A genome contains only two diploid cultivated species (*G. arboreum* and *G. herbaceum*). In the B, C, D, and E genomes there are 4, 9, 12, and 4 species, respectively. *G. longicalyx* has been included in the F genome and the cultivated type *G. bickii* in the G genome. There are six tetraploid species, including the cultivated ones *G. hirsutum* and *G. barbadense*, all of which have a mixture of the A and D genomes.

The only cotton with spinnable lint that grows wild is *G. herbaceum* var. *africanum*, and this is probably the ancestor of all linted cottons in both the Old and New World. The New World cottons (tetraploids) are natural amphidiploids that contain a set of chromosomes from one of the two A genome species and a chromosome set from a species of the D genome group (Skovsted 1937; Beasley 1940; Harland 1940). Gerstel (1953) demonstrated that the A genome of the natural amphidiploid differs from *G. herbaceum* by two reciprocal translocations, thus indicating that *G. herbaceum* contributed the A genome to New World cotton. Cytogenetic studies (Phillips 1963) and seed electrophoresis of the known D genome species (Cherry et al. 1970) have shown that *G. raimondii* is most closely related to the contributor of the New World D genome.

1.3 Diseases and Pests

Tremendous loss is caused to cotton production and fiber quality by various diseases and pests. About 100 species of insects are known to be associated with cotton (Berger 1969). The damage caused by insects alone has been estimated to cost the consumers over 600 million US$ per year (Perlak et al. 1990). Sometimes, because of insects such as bollworms and jassids, or diseases such as bacterial blight, wilt, or root rot, the entire crop may be destroyed. Some of the important diseases and pests are listed in Table 4.

Table 2. Species of *Gossypium*. (Endrizzi et al. 1984; Percival and Kohel 1990)

Species	Distribution
1. Diploids (2n = 2x = 26)	
G. herbaceum L.	Old World cultigen
G. arboreum L.	Old World cultigen
G. anomalum Wawr. & Peyr.	Africa
G. triphyllum (Harv. & Sand.) Hochr.	Africa
G. capitis-uiridis Mauer	Cape Verde Islands
G. trifurcatum Vollesen	Africa
G. bricchettii (Ulbri.) Vollesen	Africa
G. benadirense Mattei	Africa
G. sturtianum J. H. Willis	Australia
G. nandewarense (Derera) Fryx	Australia
G. robinsonii F. Muell	Australia
G. australe F. Muell	Australia
G. costulatum Tod.	Australia
G. cunninghamii Tod.	Australia
G. nelsonii Fryx.	Australia
G. pilosum Fryx.	Australia
G. populifolium (Benth.) Tod.	Australia
G. pulchellum (C. A. Gardn.) Fryx.	Australia
G. thurberi Tod.	Mexico, USA (Arizona)
G. armourianum Kearn	Mexico
G. harknessii Brandg.	Mexico
G. davidsonii Kell.	Mexico
G. klotzschianum Anderss.	Galapagos Islands
G. aridum (Rose & Standl.) Skov.	Mexico
G. raimondii Ulbr.	Mexico
G. gossypioides (Ulbr.) Standl.	Mexico
G. lobatum Gentry	Mexico
G. trilobum (Moc. & Sess. ex DC.) Skov. emend. Kearn.	Mexico
G. laxum Phillips	Mexico
G. turneri Fryx.	Mexico
G. schwendimanii Fryx.	Mexico
G. stocksii Mast. ex Hook.	Arabia
G. somalense (Gurke) Hutch.	Arabia
G. areysianum (Defl.) Hutch.	Arabia
G. incanum (Schwartz) Hillc.	Arabia
G. longicalyx Hutch. & Lee	Africa
G. bickii Prokh.	Australia
2. Allotetraploids (2n = 4x = 52)	
G. hirsutum L.	New World cultigen
G. barbadense L.	New World cultigen
G. tomentosum Nutt ex Seem.	Hawaii
G. mustelinum Miers ex Watt	Brazil
G. darwinii Watt	Galapagos Isiands
G. lanceolatum Tod.	Mexico

Table 3. Some wild species of cotton possessing useful characters

Species	Genome	Useful characters
1. *G. anomalum* Wawrex Wawra & Peyr	B_1	Lint quality, resistance to black arm and jassid
2. *G. harkensii* Brendeg.	D_2^{-2}	Resistance to drought, source of cytoplasmic male sterility, resistance to spider mites
3. *G. raimondii* Ulbr.	D_5	Resistance to black arm and jassids, high density of seed hairs, and drought tolerance
4. *G. armourianum* Kearn.	D_2^{-1}	Resistance to black arm and bollworm, increased number of ovules per loculus
5. *G thurberi* Tod.	D_1	Resistance to bollworms and potentiality for imparting high fiber strength
6. *G. aridum* (Rose & Standl.) Skov.	D_4	Drought tolerance
7. *G. somalense* (Gurke) J.B. Hutch.	E_1	Resistance to bollworm
8. *G. stocksii* Mast. in Hook.	E_2	Resistance to drought
9. *G. tomentosum* Nutt. ex Seem.	AD	Resistance to drought and jassids, lint quality, and strength
10. *G. bickii* Prokh.	$2G_1$	Resistance to insects, gossypol-free seeds

Table 4. Some common diseases and pests of cotton

Disease/pest	Causal organism
1. Bacterial blight	*Xanthomonas malvacearum*
2. Ascochyt blight	*Ascochyta gossypii*
3. Verticillium wilt	*Verticillium alboatrum*
4. Root rot	*Phymatotrichum omnivorum*
5. Anthracnose	*Glomerella gossypii*
6. Rhizoctonia root rot	*Rhizoctonia solani*
7. Fusarium wilt	*Fusarium oxysporum* f. *vasinfectum*
8. Cotton rust	*Puccinia cabata*
9. Root knot nematode	*Meloidogyne incognita*
10. Boll weevil	*Anthonomous grandis*
11. Spotted bollworm	*Earias* spp.
12. American bollworm	*Heliothis* spp.
13. Pink bollworm	*Pectinophora gossypiella*
14. Leaf worm	*Alabama argillacea*
15. Aphid	*Aphis gossypii*
16. Thrips	*Thrips tabaci*
17. Flea hopper	*Psallus seriatus*
18. Tarnished and rapid plant bugs	*Lygus hesperus*
19. Jassid	*Amrasca biguttula biguttula*
20. Whitefly	*Bemisia tabaci*
21. Cabbage looper	*Trichoplusiani*

1.4 Objectives, and Conventional Methods of Cotton Breeding

The main objectives for the improvement of cotton are, (1) increased yield, (2) upgrading the quality of fiber, (3) early maturing types, (4) gossypol-free cotton seed, and (5) resistance to various insects, nematodes, and diseases.

Cotton is a cross-pollinated crop, the conventional breeding usually consisting of introduction, selection, and hybridization as in self-pollinated crops, although different procedures are usually practiced. Plant introductions have either been grown as introduced or improved by natural or artificial selection. Selection is required to maintain varietal purity because of the partial cross-pollination, as well as to aid in new variety development. Selection may be based on single plant selections, or mass selection may be practiced by bulking seed from open-pollinated or selfed plants selected on the basis of appearance. Crosses between different varieties or strains can be used to produce segregating populations from which selections can be isolated.

All the methods that have been used in cotton breeding are based on selection in variable populations. Hybridization and mutation breeding have often been used to introduce variability into populations. However, until now, not much success has been achieved in utilizing the wide variability existing in wild germplasm. Mutation breeding also has not been very successful.

1.5 Need to Incorporate Unconventional Methods

The prime requisite for the success of any crop improvement program is the presence of sufficient genetic variability for the traits that are to be improved. When the required variability is not present in a species, then it has to be

Table 5. Target genes for genetic engineering in cotton. (John and Stewart 1992)

Target genes	Potential application
Insecticidal genes	
Bacillus thuringiensis toxins	Control of *Helizoverpa zea*, *Heliothis virescens*, *Pectinophora gossypiella*
Protease inhibitors, lectins	Feeding deterrents
venoms, neuropeptides	Kill or paralyze feeding pests
Herbicidal genes	
5-Enolpyruvylshikimic acid 3-phosphate	Glyphosphate tolerance
Nitrilase	Bromoxynil tolerance
Acetolactate synthase	Sulfonylurea tolerance
2,4-Dichlorophenoxyacetate monooxygenase	2,4-D tolerance
Phosphinotricin acetyltransferase	Bialaphos tolerance
Environmental stress-resistance genes	
Superoxide dismutase	Free-radical quenching
Thermal and water stress-tolerance genes	Heat, cold, and drought tolerance
Fiber-modification genes	
Cotton genes	Modification of existing fiber properties
Other plant genes (extensins, peroxidase)	Modification of existing properties
Bacterial genes (e.g., hormone genes)	Modification of existing properties
Genes for hybrids	
Pollen-specific antisense genes	Production of male sterile plants
Cytotoxic genes (e.g., RNases)	Protection of proprietary seeds

induced by mutations or we have to look toward some related species for characters that are absent in the cultivated types. In the genus *Gossypium*, a number of wild species (Table 3) have useful attributes that can be transferred from the Old World diploids to the cultivated tetraploids. Success in this direction has been hampered because of various incompatibility barriers like abortion of embryo/endosperm. Even if interspecific hybrids are obtained, it is difficult to secure any recombinants. Thus, there is a strong need to resort to unconventional techniques that can ensure hybridization between different species so that hybrids can be produced in large numbers and utilized for growing a large number of selfed or backcrossed progenies. In this way, the chances of recovering useful recombinants may be increased. In addition, since New World cotton is tetraploid, the induction of mutations through culture of haploid cells and protoplasts would be ideal. Moreover, induction of genetic variability through somaclonal variation in cell cultures and genetic transformation, which has already given positive results, needs to be further incorporated into current cotton breeding programs. The potential applications of some target genes for genetic engineering in cotton are cited in Table 5.

2 Biotechnology Studies on Cotton

The importance of and prospects for biotechnology in the improvement of cotton were stressed by the author more than a decade ago in an article *Biotechnology of Cotton Improvement* (Bajaj and Gill 1986b). Since then, there has been an upsurge of activity in this field, and tremendous progress has been made, especially in the in-vitro regeneration of plants, and the production of transgenic cotton showing resistance to insects, herbicides, etc, and field studies have been conducted on such plants. In this chapter, the following aspects of tissue culture/biotechnology of cotton are discussed.

1. Establishment of callus cultures
2. Somatic embryogenesis and plant regeneration
3. Protoplast culture and regeneration of plants
4. Micropropagation
5. Storage and cryopreservation of germplasm
6. Interspecific hybridization through embryo culture and in vitro fertilization
7. In vitro culture of ovules
8. Haploid production through anther/pollen culture
9. Somaclonal variation/genetic variability for tolerance to high temperature, salt, and nematodes
10. Genetic transformation through *Agrobacterium*, and particle bombardment
11. Transgenic cotton plants resistant to insects and herbicides; induction of male sterility

12. Field trial/performance of transgenic plants
13. Risk assessment studies on transgenic plants
14. Genetic engineering of cotton fiber
15. Molecular marker – RFLP, RAPD, and genome mapping
16. Secondary metabolites – gossypol and tannins.

This introductory chapter, which is primarily based on information extracted from various chapters in this Volume, summarizes the state-of-the-art in various in vitro cultures/biotechnology studies on cotton, gives a review of the work, and the recent literature. At the end of the chapter references are also given for further reading, though some of these are not cited in the text, but concern the biotechnology of cotton.

2.1 Establishment of Callus Culture

Regeneration of complete plants from cells is a prerequisite for the induction of genetic variability and genetic manipulation studies. In cotton, callus capable of differentiation (organogenesis and somatic embryogenesis) has been established in a number of species of *Gossypium*, and the regenerated plants transferred to pots/field.

Callus cultures have been induced from excised ovaries and ovules, embryos (Gill and Bajaj 1987, 1991), anthers (Bajaj and Gill 1989), segments of young seedlings (Bajaj and Gill 1985), root (Kuo et al. 1989), leaves (Wu et al. 1994), and protoplasts (She et al. 1989) – for more references see Table 1 in Chapter I.2, this Volume.

Young flower buds denuded of petals and sepals cultured on MS medium (Murashige and Skoog 1962) containing 2,4-D proliferated to form callus in 2 weeks. Likewise, young ovules of various crosses (Bajaj and Gill 1985) cultured on MS + IAA (1.5 mg/l) + kinetin (0.5 mg/l + casein hydrolysate (250 mg/l) underwent profuse proliferation, and soft and friable callus was formed. The embryos of a cross *G. hirsutum* × *G. arboreum* cultured on Smith et al. (1977) medium proliferated to form callus which was creamy white, soft, and friable (Bajaj and Gill 1985). The excised anthers of *G. hirsutum* and *G. arboreum* cultured on MS supplemented with various concentrations and combinations of 2,4-D, NAA, BAP, kinetin, and coconut milk started to proliferate within 2 weeks and formed a mass of callus in 4 weeks (Bajaj 1982; Bajaj and Gill 1989).

Segments from in vitro-grown seedlings, i.e, hypocotyl, cotyledons, and mesocotyl, have been used for the induction of callus; in general, hypocotyl segments proved superior, and most regeneration studies have been conducted on hypocotyl-derived callus. The leaf segments and young leaves of *G. davidsonii* cultured on MS + zeatin (1 mg/l) + IAA (0.2 mg/l) + CH (500 mg/l) formed embryogenic callus (Wu et al. 1990). Suspension-derived protoplasts in cultures form cell colonies and masses of callus (She et al. 1989; Peeters et al. 1994). The calli obtained from most of the explant sources are capable of repeated subculturing.

2.2 Somatic Embryogenesis and Plant Regeneration

Induction of somatic embryogenesis has been a significant achievement in plant tissue culture. Its applications for mass production of plants, genetic transformation, and synthetic seed formation have been highlighted by the author (see Bajaj 1995) The phenomenon of somatic embryogenesis has been reported in about 300 species of plants. During the past 10 years, considerable work has been done on somatic embryogenesis in cotton (see Gawel and Robacker 1995), and it is a preferred method for plant regeneration. In most of the studies, hypocotyl-derived callus has been used for the induction of somatic embryogenesis. It was first reported in a wild species, *G. klotzschianum*, in 1979 by Price and Smith, followed by numerous reports on various American and Chinese cultivars of *G. hirsutum* (Firoozabady and DeBoer 1993; She et al. 1995).

In earlier studies, genotype was reported to play an important role in determining the induction of somatic embryogenesis in cotton (Shoemaker et al. 1986). For instance, Trolinder and Goodin (1987, 1988a,b) defined the culture parameters for somatic embryogenesis in Coker varieties, and suggested the importance of the genotype. Trolinder and Chen (1989) also observed that it is genotype-dependent. Genetic analysis of cotton somatic embryogenesis sustained that the trait was heritable (Gawel and Robacker 1995). Recent studies have shown that the induction of somatic embryogenesis is a widespread phenomenon, not restricted to a few Coker cvs. of *G. hirsutum*, but also reported in a number of Chinese cvs., and other species such as *G. davidsonii*, *G. raimondii*, and *G. arboreum* (see Chap. I.2, this Vol.).

Shang et al. (1991) observed that the starting pH of the culture medium affects long-term viability, growth, and differentiation of cells. The pH of 3 to 5 was best for cell viability, pH 3 to 4 enhanced cell elongation, and pH 4, 7, or 8 stimulated somatic embryogenesis. These somatic embryos mature in culture, germinate on sterile vermiculite (Firoozabady et al. 1987; Umbeck et al. 1987) or Gelrite (Finer 1988), and on transfer to greenhouse/field bloom and set bolls (Trolinder and Goodin 1987; Wu et al. 1990).

2.3 Protoplast Culture and Plant Regeneration

Isolated protoplasts have proved to be an excellent tool for DNA uptake/gene transfer, and somatic hybridization studies in a number of agricultural crops (see Bajaj 1994). In cotton, although protoplasts have been isolated by a number of workers (see She et al. 1995), the regeneration of complete plants is a comparatively recent development. The protoplasts released by enzymatic treatment of embryogenic cell suspension of *G. hirsutum* (cvs. Sumian 3118, Lumian 9554, Jinmian 4, and Coker 312) form cell colonies in 12–20 days in cultures under repeated division (Chen et al. 1989; She et al. 1989, 1993). The calli of yellow, compact pellet differentiated somatic embryos on MS-3 medium containing 0.1–0.5 NAA. The mature embryos eventually formed plants.

Recently, Peeters et al. (1994) used feeder layers for successful culture of hypocotyl callus-derived protoplasts of Coker 312. The protoplasts plated on cellulose nitrate filters and placed over feeder layers formed embryogenic callus from which plants were regenerated. The plating efficiency was 12.8% (for details see Chap. I.3, this Vol.).

2.4 Micropropagation

Cotton is a seed-propagated crop. The varieties suffer rapid deterioration due to natural crossing and through mechanical mixtures during ginning. The vegetative propagation by cuttings (Parmar et al. 1978) shows a slow rate of multiplication. Therefore, the need for suitable means of fast clonal propagation has been felt. Micropropagation of cotton (Bajaj and Gill 1992) can be employed for collection and maintenance of germplasm, rapid multiplication of seed of new varieties, production of clones for evaluation against diseases etc., and for propagating male sterile lines.

In cotton, successful micropropagation has been carried out by the culture of excised shoot tips and meristems (Bajaj and Gill 1986a), embryos and ovules (Gill and Bajaj 1984b, 1996) and somatic embryos (Shoemaker et al. 1986).

1. *Micropropation Through Meristem Culture.* Shoot tip and meristem culture are two of the simplest methods of in vitro propagation, and the plants obtained are true to type. The shoot tips and meristems of *G. arboreum*, *G. hirsutum*, and three interspecific hybrids, *G. arboreum* × *G. anomalum*, *G. arboreum* × *stocksii*, and *G. herbaceum* × *G. stocksii*, produced normal plants (Bajaj and Gill 1986a).

Altman et al. (1990) devised an in vitro method for collecting the germplasm of various wild species of *Gossypium*, especially those where viable seeds could not be obtained. The cuttings obtained from various species were surface sterilized in the field itself and cultured on the medium without any further precautions to assure asepsis. The shoots grew vigorously. Upon return to the laboratory, the cuttings were rinsed, treated with various chemicals, and transferred to soil. The overall response was 10% shoot formation from six *Gossypium* species.

Shoot apex and axillary bud culture has also been used to obtain plants of various Chinese cultivars (Wu and She 1990), and trasgenic plants of cotton (Gould et al. 1991).

2. *Micropropagation Through Embryo and Ovule Culture.* This technique can be employed to propagate species/genotypes from which viable seed cannot be obtained. Numerous studies have been conducted on this aspect by the author (Gill and Bajaj 1984a,b, 1986, 1987, 1991, 1996) on *G. hirsutum*, *G. arboreum*, *G. herbaceum*, *G. stocksii*, and *G. anomalum*, and various interspecific hybrids were obtained from 3 DAP ovules and 15 DAP embryos cultured on MS medium supplemented with growth regulators. In some cases,

multiple shoots were formed. On transfer to the pots/field these plants continued growth, and set flowers and bolls.

3. *Micropropagation Through Somatic Embryos (SEs).* Somatic embryogenesis is one of the fastest methods for large-scale propagation (see Bajaj 1995). In cotton, there are a number of studies on the induction, germination, and development of plants from somatic embryos, and the subject has been recently reviewed (Gawel and Robacker 1995). Shoemaker et al. (1986) reported that mature SEs of Coker 201 and 315 germinated on transfer to auxin-free medium. The SEs were also germinated on sterile vermiculite saturated with medium (Trolinder and Goodin 1987; Umbeck et al. 1987), and on Gelrite medium (Firoozabady et al. 1987). The percent germination varied, and the degree of fertility in the regenerated plants ranged from 15–100% (Trolinder and Goodin 1987; Finer 1988). Somatic embryos obtained from protoplast-derived callus also formed plants (She et al. 1995). Since SEs have been multiplied on a large scale in bioreactors in a number of plant species (Denchev et al. 1992), this method can be extended to cotton.

2.5 Storage and Cryopreservation of Germplasm

The conservation of germplasm of cotton is an important aspect. With recent progress in the biotechnology of cotton to induce genetic variability through in vitro culture and genetic transformation, methods need to be developed for storage of desirable cell lines/germplasm. In this regard, tissue cultures for short-, and medium-term storage, and cryopreservation for long-term conservation are the alternatives (Bajaj 1986).

Cotton germplasm is traditionally conserved through seed. The USDA maintains about 5100 seed accessions, which is one of the largest collections in the world (Percival and Kohel 1990). However, seeds are likely to suffer deterioration, thus alternative in vitro methods and cryopreservation of cell cultures have been experimented.

Shoot-tip cultures of cotton were maintained, without change of medium, for more than 6 months (Bajaj and Gill 1986a). This period can certainly be enhanced by growing cultures on minimal medium at low temperatures. Altman et al. (1990) developed a simple tissue culture method for the collection/storage of living vegetative cotton material from the field. Since the availability of viable seed cannot always be assured during botanical expeditions, this method may prove useful.

Plant-cell cultures are known to undergo genetic erosions on long-term storage. Recent advances in somaclonal variation and genetic transformation in cotton have resulted in cell lines which need to be stably conserved. Callus cultures and cell suspensions of *G. arboreum* and *G. hirsutum* frozen in liquid nitrogen ($-196\,^{\circ}$C) revived and resumed active growth (Bajaj 1982). It is envisaged that cryopreservation will be increasingly employed for the conservation of desirable cell lines, and also for the establishment of germplasm banks (Bajaj 1979b).

2.6 Interspecific Hybridization Through Embryo Rescue, and In Vitro Fertilization

2.6.1 Embryo Culture

Out of 50 species of *Gossypium*, only 4 are cultivated, the rest are wild species (Table 2), but a rich source of genetic variability (Table 3) for characters such as fiber quality and resistance to pests and diseases such as cotton bollworm, boll weevils, *Fusarium* and *Verticillium* wilts, bacterial blight, boll rot, etc. (Prentice 1972). In the past, numerous attempts have been made to transfer some of these characters by conventional methods of breeding, but the results have not been very encouraging due to failure of endosperm development and early degeneration of the embryo (Weaver 1958). Another important point in cotton improvement is the presence of gossypol in the seed, which makes the protein products and oil from cottonseed of little commercial use. However, these aspects can be taken care of by the use of embryo/ovule culture techniques.

Extensive studies have been conducted by the author (Gill and Bajaj 1984a,b, 1986, 1987, 1991, 1996) involving various wild and cultivated species, both diploid and tetraploid. The following crosses were successful; *G. arboreum* × *G. anomalum*, *G. arboreum* × *G. herbaceum*, *G. arboreum* × *G. stocksii*, *G. arboreum* × *G. hirsutum*, and *G. herbaceum* × *G. stocksii*. In all these crosses, hybrids were obtained by preventing the degeneration of embryos by treating the flowers/young bolls with growth regulators, and then culturing the rescued embryos/ovules.

The daily application of a mixture of GA (50 mg/l) + NAA (100 mg/l) to the flowers immediately after crossing considerably enhanced their retention and ensured further development. The base of the pedicel became thick and strong, and abscission was delayed. The response of the cultured hybrid embryos varied in different crosses. The plants were transferred to pots at the two- to three-leaf stage with well-developed roots. There was 80–90% survival of potted plants, which were then grown in the greenhouse and the field (Gill and Bajaj 1984b). The hybrids of *G. arboreum* × *G. anomalum* flowered throughout the year (Gill and Bajaj 1991). The *G. herbaceum* × *G. stocksii* (Gill and Bajaj 1996) hybrids, however, produced only few flowers in winter; *G. arboreum* × *G. stocksii* did not flower in winter at all, but resumed flowering in late spring.

The F_1 plants predominantly exhibited characters of the wild parent, as is expected, since wild plant characters are usually dominant (Mehboob Ali et al. 1964). Efforts to backcross the F_1 plants with their respective cultivated parent met with limited success. The *G. arboreum* × *G. anomalum* hybrid produced some seeds when pollinated with *G. arboreum* pollen or vice versa. The seeds set on the hybrid plant possessed strong and fine fiber. Second backcross generation seeds were obtained from all the plants after pollinating with *G. arboreum* pollen (Gill and Bajaj 1986).

Studies have also been successfully conducted on crosses between diploid and tetraploid species, i.e., *G. arboreum* × *G. hirsutum* (Gill and Bajaj 1987),

G. trilobum × *G. hirsutum* (Umbeck and Stewart 1985), *G. sturtianum* × *G. hirsutum* (Altman et al. 1987), etc. (for details on these crosses see Chap. II.1, this Vol.). Recently, a system of randomly amplified polymorphic DNA (RAPD) markers has been developed to facilitate the transfer of *G. sturtianum* genes into *G. hirsutum* genome by hybridization and backcrossing (Vroh et al. 1996).

2.6.2 In Vitro Fertilization

In vitro fertilization and pollination technique can be used in producing hybrids in incompatible crosses. This technique involves growing unpollinated flowers, ovaries, and ovules on synthetic media, and then sprinkling pollen over them, or on to the cut end of the style. In vitro fertilization has been successfully used in the production of seed in a number of species (see Bajaj 1979a; Zenkteler 1990; Kranz and Loerz 1993).

In cotton, Rafaat et al. (1984) fertilized *G. hirsutum* in vitro with pollen of *G. barbadense*, and obtained hybrids. Later, Liu et al. (1992) brought about in vitro fertilization between *G. hirsutum* and *G. arboreum*. They studied the effect of various media, temperatures, relative humidity, and shake culture, and obtained a few hybrid seed/plants. These plants had stronger vegetative growth than their parents, but were highly sterile. The shapes of leaves and flowers were somewhat intermediate to those of their parents (for details see Chap. II.2, this Vol.).

2.7 In Vitro Culture of Ovules – Studies on Fiber Development, Peroxidases, and Plant-Fungal Interaction

In addition to its use for interspecific hybridization, ovule culture has also been applied for understanding the development of fiber in vitro, study of peroxidases, and the interaction with the aflatoxin-producing fungus *Aspergillis flavus*.

1. *In Vitro Development of Fiber.* Cotton fiber is a single, elongated epidermal cell on the surface of the developing cotton ovule. It may undergo a 1000- to 3000-fold increase in length during its differentiation, and deposits a thick secondary cell wall of cellulose during the final stages of differentiation. The culture of ovule permits in vitro manipulation of fiber development, and the various factors affecting its development and differentiation. Numerous studies (see Chap. II.3, this Vol.) have been conducted on this aspect, initiated primarily with the work of Beasley (1971, 1977). Cotton ovules cultured in the presence of IAA and GA_3 increase in size and produce fibers; however, abscisic acid inhibits in-vitro growth of fiber (Beasley and Ting 1974). It is suggested that abscisic acid inhibits fiber growth by interfering with malate metabolism (Dhindsa et al. 1976). Boron deficiency also retarded fiber growth (Dugger and Palmer 1985). Cell suspensions derived from ovules, which pro-

vide a unique opportunity to study the development of fibers from single cells, were used by Trolinder et al. (1987). These free cells underwent the stages of fiber development described for in vivo fiber development. The maximum length was observed when 1 mg/l each of NAA and kinetin were present in the culture medium. Later, Triplett et al. (1989) studied the growth and development of cotton fibers in a developmental mutant, and reported that the length attained in culture is affected by the genotype of the explant tissue.

Seagull (1990) used cotton ovule culture to study changes in microtubules occurring during primary to secondary wall formation in the fibers. The effect of cycling warm-cool temperatures showed that cellulose synthesis in the fiber increased above 18 °C, and reached a plateau between 28 and 37 °C (Haigler et al. 1991). Treatment of 6–11-day postanthesis cultures with monensin, a Golgi function inhibitor, inhibited secondary wall deposition but not fiber elongation (Davidonis 1993a,b). Recently, the work of Ramsey et al. (1996) suggests that IAA is the critical phytohormone, and stimulates increased production of fiber-specific PEP carboxylase.

2. *Peroxidases.* Soluble peroxidase activity is secreted into the medium of 5- to 35-day-old cotton cultures (Mellon 1986). All the ovule-secreted proteins, including peroxidases, are de novo synthesized (Mellon and Triplett 1989). The peroxidase isolated from the culture medium was purified about 150-fold (Mellon 1991). Plant anionic peroxidases tend to be cell-wall-associated. Since maximum peroxidase secretion rates in culture correspond closely with seed coat lignification, cotton ovule peroxidase probably functions in cottonseed coat lignification.

3. *Plant-Fungal Interaction.* The cotton ovule culture system has also been utilized as a model system for investigation of interactions between cotton and the toxigenic fungus *Aspergillus flavus*. This fungus infects developing seeds and produces the carcinogen aflatoxin. Ovule cultures inoculated with *A. flavus* conidia accumulate peroxidase up to 20-fold higher than noninoculated controls (Mellon 1988). The response range of cotton ovule cultures to fungal challenge includes the production of low molecular weight stress metabolites. Inoculated ovule tissue produces a complex mixture of stress metablites. The crude extracts derived from *A. flavus*-inoculated cotton ovules inhibit aflatoxin production in axenic cultures of *A. flavus* by as much as 93% (Mellon 1992).

2.8 Haploid Production Through Anther/Pollen Culture

The invitro production of haploids is valuable for obtaining homozygous lines in one generation, and these homozygous lines can be evaluated in the field (see Bajaj 1990a). Because doubled haploids possess additive variance only, early-generation testing of segregating populations is possible. Haploids can also increase the efficiency of existing breeding methods through improved reliability of selection. Natural occurrence of cotton haploids has been occasionally observed, and also induced by conventional methods, i.e., polyembryony and semigamy (Blank and Allison 1963; Lee 1970), but the frequency of production is very low. Thus anther/pollen culture studies have been undertaken (see Bajaj and Gill 1996b; Chap. III.1.2, this Vol.). Early attempts on anther culture resulted only in callus formation. Later, Bajaj and Gill (1989), working with *G. arboreum*, reported the induction of pollen embryos and callus with haploid cells as well as a wide range of variation in chromosome number (Bajaj and Gill 1985). The excised anthers (at the uninucleate stage of the microspore) of *G. arboreum* and *G. hirsutum* in culture underwent profuse proliferation (Bajaj 1982). Highest callus formation was observed on MS medium supplemented with NAA (2 mg/l) and BAP (1 mg/l). Cold treatment (4 °C) to excised anthers promoted nuclear division in pollen, resulting in the formation of multinucleate pollen, and pollen embryos (Bajaj and Gill 1989). Moreover, anther-derived callus frozen in liquid nitrogen revived and resumed growth (Bajaj 1982). This method may be used to check instability in haploid cell lines.

Attempts were also made to culture isolated pollen (see Chap. III.2, this Vol.), but so far haploid plants have not been obtained.

2.9 Somaclonal Variation/Genetic Variability

The extent of genetic variability in the base population determines the success of any crop improvement program. On prolonged storage at ordinary temperatures or on periodical subculturing, the cell cultures undergo genetic erosions, such as endomitosis, polyloidy, chromosome loss, translocations, gene amplification, mutations, etc. (D'Amato 1985; Bajaj 1990b), and thus are a rich source of genetic variability. In cotton cell cultures, various changes in chromosomes have been reported (Bajaj and Gill 1985; Stelly et al. 1989), and plants showing tolerance to high temperature, and cell lines tolerant to salt have been obtained. Moreover, variability in cotton through embryo culture has been induced and hybrid callus obtained (Gill and Bajaj 1987). Recently, through genetic transformation, cotton plants resistant to insects, herbicides, and invitro induction of male sterility have been reported (Reynaerts and De Sonville 1994). The work on various chromosomal changes in callus cultures, and the regeneration of somaclonal plants in cotton is summarized here.

1. *Chromosomal Variation in Callus Cultures.* Callus derived from various explants, such as hypocotyls, anthers, ovules, hybrid embryos, etc. (Bajaj and Gill 1985) of both diploid and tetraploid *Gossypium* species showed a wide range of variation in chromosome number and structure. The number varied from haploid to more than hexaploid, highly polyploid, and aneuploid. Several mitotic abnormalities, such as multinucleate and micronuclei in interphase; chromosome rings and telophase bridges, etc. were observed. It is speculated (Larkin and Scowcroft 1981) "that tissue culture may generate environments for enhancing chromosome breakage and reunion events, and thus a tissue culture cycle of the hybrid material may provide the means

for obtaining the genetic exchange needed between two genomes in the interspecific hybrid. The hybrid callus may enhance the frequency of requisite exchange".

2. *Variation in Somaclonal Plants.* Somaclonal plants regenerated from callus cultures of *G. hirsutum* were extremely varied in phenotype (Stelly et al. 1989). Chromosome number studied in 32 plants ranged from 49 to 53. Many somaclones (34%) formed strikingly abnormal flowers that lacked a corolla and an androecium, and sometimes a gynoecium. Fertility was generally poor; 84% of the somaclonal plants failed to set any bolls (Trolinder and Goodin 1988b).

3. *High Temperature-Resistant Somaclones.* High temperatures and drought stress affect both quantity and quality of the fiber. In this regard, selection of high temperature-resistant somaclonal cells and their subsequent regeneration is of great significance. Trolinder and Shang (1991) regenerated high temperature-resistant plants from selected somaclones of *G. hirsutum* cv. Coker. The cells were exposed to 45 °C for 3 h each day, and after the stress treatment, they were plated onto embryo development medium, and plants were regenerated. Leaf tissue of stress-selected plants formed and maintained callus growth when incubated at 38 °C; whereas tissue excised from nonselected controls rarely formed callus, or quickly became necrotic (for details see Chap. IV.3, this Vol.).

4. *Salt-Tolerant Cell Lines.* Salinity induces nutritional imbalance (Leidi et al. 1991) and affects cotton yield and fiber quality (Razzouk and Whittington 1991). Thus, in vitro induction of genetic variability for salt tolerance in cotton is an important aspect; the work on cotton has only just begun. Trolinder and Allen (1994) observed that transgenic cotton cells overexpressing Mn-superoxide dismutase grew between than control cells when cultured in polyethylene glycol-containing medium. Gossett et al. (1994) studied the effect of salt-tolerant and salt-sensitive cotton cultivars, and linked antioxidant enzyme activities to salt tolerance. The results from their study showed that varietal differences in salt tolerance observed in whole plants were also apparent at the cellular level.

In addition to the above-mentioned aspects of somaclonal variation in cell cultures, variability has been induced through genetic transformation and cotton plants resistant to insects, herbicides, and male sterile lines have been obtained (these aspects are dealt with in Sect. V, this Vol.).

5. *Nematode Resistance in Cotton.* Nematodes cause huge losses in cotton yield and quality. In 1992, in the USA alone, nematodes accounted for yield loss of 528 000 bales valued at millions of US$ (Goodell 1993). With a view to understanding the mode of infection/resistance, studies have been conducted on the penetration, reproduction, and development of *Meloidogyne incognita* in cotton (Creech et al. 1995; Jenkins et al. 1995). However, gnotobiotic studies through cell and tissue culture have their own advantages (Tanda et al. 1980).

The in vitro system enables investigations to be carried out under strictly controlled environments, the individual effect of various factors, such as temperature, moisture, pH, etc. can be studied, most important being that the somaclones selected from these cultures can be regenerated into nematode-resistant plants. Studies conducted on potato and eggplant have resulted in lines showing resistance to nematodes (Wenzel and Uhrig 1981; Gleddie et al. 1985; Fassuliotis 1990). Such studies in cotton may lead in future to the regeneration of nematode-resistant plants.

2.10 Genetic Transformation

Genetic transformation has now become an established method of gene transfer in crops. There are more than 150 species of plants on which such studies have been reported. Transformation enables modification of the genetic makeup of a plant, alters gene expression, and enhances agronomical traits. In cotton, numerous studies have been conducted, resulting in the production of plants resistant to insects, herbicides, etc. Primarily two methods have generally been used, i.e., *Agrobacterium tumefaciens* and particle bombardment.

1. *Agrobacterium-Mediated Transformation. Agrobacterium tumefaciens*, a soil-borne bacterium, is known to incite crown gall tumors when inoculated on plants. A portion of the Ti tumor-inducing plasmid, T-DNA, is transferred and incorporated in the nuclear DNA of plant cells transferred during the infection process. This method has been used by a number of workers to transform cotton (Firoozabady et al. 1987; Umbeck et al. 1987; Perlak et al. 1990; Revenkova et al. 1990).

Various explants, i.e., segments of hypocotyl (Umbeck et al. 1987) and cotyledons (Firoozabady et al. 1987) of *G. hirsutum* were inoculated with a suspension of *Agrobacterium tumefaciens*, and incubated at ambient temperature and light. After 7–10 days, cotyledon pieces initiated transformed kanamycin-resistant micro-calli at wound sites. Of the total of 221 putative transformed calli, 75–98% were octopine-positive (Firoozabady et al. 1987). Such calli differentiated somatic embryos, which eventually germinated and developed into mature plants. The percentage of octopine-positive plants regenerated varied from 75–85%, and kanamycin-resistant plants from 85–100%.

2. *Particle Bombardment.* This method of gene transfer, developed by Klein et al. (1987), demonstrated that tungsten particles coated with DNA can be accelerated through onion cell walls, using a gun-powder-driven device. Particle bombardment was first used for cotton transformation by Finer and McMullen (1990), who bombarded embryogenic cell suspensions of cv. Coker 310 with the hygromycin gene, and later regenerated hygromycin-resistant plants. This method was refined by McCabe and Martinell (1993), who described a protocol for variety-independent transformation of cotton by meristem bombardment. Southern blot analysis demonstrated the presence of

the GUS gene in transformants. Progeny analysis established transmission of transgene in a Mendelian fashion. Recently, Chlan et al. (1995) used this method to introduce the kanamycin gene into meristems of *G. hirsutum*, and plants were recovered under selection pressure.

2.11 Transgenic Cotton

One of the most exciting developments in the biotechnology of cotton is the production of transgenic plants showing resistance to insects and herbicides. Such plants have been field-tested for their performance, and approved for release to the public.

2.11.1 Insect-Resistant Cotton

There are more than 100 insect species that attack the cotton plant (Berger 1969). Some of the most important, causing considerable damage to the crop, are white fly (*Bemisia tabaci*), pink bollworm (*Pectinophora gossypiella*), boll weevil (*Anthonomous grandis*), American bollworm (*Heliothis virescens*), etc. Insect damage to the cotton yield in the US alone has been estimated to cost consumers over 600 million US$ per year (Perlak et al. 1990), and over 200 million US$ are spent annually to protect cotton from insects (Jenkins et al. 1991). To overcome these losses, various strategies have been applied to induce tolerance/resistance to insects (see Chap. V.2, this Vol.).

Cotton contains a number of alleo-chemicals that inhibit the growth of insects. Gossypol is one such compound which adversely affects the growth of tobacco budworm and cotton leafworm (Lukefahr and Martin 1966; Meisner et al. 1977). Tannins and gossypol together greatly affect oviposition rates (Sharma and Agarwal 1984). The red color of some cotton varieties had a negative effect on adult boll weevils (Isley 1928). Flavonoid content may also play a role in conferring insect resistance (Hedin et al. 1992). Thus, combining feeding, visual and structural cues, pigments, and other alleochemicals present in cotton may discourage insect feeding.

Another approach for insect resistance is the accumulation of protease inhibitors and the use of the genes encoding them. These peptides are induced in plants by insect predation, and interfere with the breakdown of proteins ingested by insects, causing starvation (Ryan 1990). Thomas et al. (1995) reported that protease inhibitor genes from *Maduca sexta* were expressed in 35 elite transgenic lines of *G. hirsutum*. Whitefly assays demonstrated a lower degree of reproduction of *Bemisia tabaci*. Thus, the overexpression of protease inhibitors in transgenic cotton may decrease insect damage.

The most successful approach for insect resistance has been through the use of the gram-positive bacterium *Bacillus thuringiensis* (B.t.) which produces proteins that are toxic to insects. These proteins, after being eaten by the insect larvae, release endotoxins in the midgut, causing death. The advantages of B.t. toxins are that they break down quickly and safely in the environment, and lack toxicity in mammals, including man. In earlier studies, the introduction and expression of B.t. toxins provided some degree of protection from insects in transgenic tobacco (Barton et al. 1987), tomato (Fischoff et al. 1987), rice (Fujimoto et al. 1993), and maize (Koziel et al. 1993).

In cotton, the B.t. gene has also been successfully incorporated and expressed (Perlak et al. 1990, 1991), and up to 100-fold expression of B.t. toxins compared to the native gene was achieved in transgenic plants. Extensive studies on cotton expressing B.t. proteins has shown that some of the most damaging insect pests of cotton can be controlled (Jenking et al. 1991; Wilson et al. 1994).

2.11.2 Herbicide-Resistant Transgenic Cotton

Genetic transformation has been successfully used to induce resistance to various herbicides in crops such as potato, (De Block 1988), brassica (De Block et al. 1989), maize (Spencer et al. 1992) etc.

2,4-dichlorophenoxyacetic acid (2,4-D) is the most commonly used herbicide; however, one drawback is its volatility. It tends to drift with the wind, and can cause severe damage to nearby crops. In cotton, 2,4,-D-resistant plants have been engineered (Bayley et al. 1992; Lyon et al. 1993). Bayley et al. (1992) introduced the 2,4-D resistance trait into *G. hirsutum*. The 2,4-D monooxygenase gene *tfd*A from *Alcaligenes eutrophus* plasmid pJP5 was isolated, modified, and expressed in transgenic cotton plants. Analysis of the transgenic progeny showed stable transmission of the chimeric *tfd*A gene and production of active 2,4-D monooxygenase. The cotton plants obtained were tolerant to three times the field level of 2,4-D used for wheat, corn, sorghum, and pasture crops. Whereas all 18 wild-type plants were killed within 5 days, 83 out of 107 progeny plants derived from the transformant were unaffected by the herbicide treatment. Cultivation of 2,4-D-tolerant varieties of cotton would reduce annual losses caused by drift-levels of the herbicide (for details see Chap. V.3, this Vol.). Recently, Rajasekaran et al. (1996) transformed Coker and Acala varieties, and observed that transgenic progeny plants were resistant to imidazolinone herbicides at five times the field application level.

2.11.3 Engineered Male Sterility and Hybrid Cotton

Cotton is a self-pollinated plant, but, depending on the presence of suitable insects, cross-pollination can occur. In cotton, hybrid vigor has been demonstrated (see Roupakias et al. Chap. II.5, this Vol.), and male sterile cytoplasms have been developed (Meyer 1974). Although the advantages of hybrid cotton are evident, it has not been commercialized in many countries due to a number of problems. The lack of adequate insect pollinators and the high cost involved in hand pollination (Srinivasan et al. 1972; Davis 1978) are the major setbacks, though in India a 213000-ha area was under hybrid cotton cultivation in 1992

(Basu and Patil 1993), and it has increased since then. Thus, biotechnology could be of immense importance. For instance, studies have been conducted on the production of several hybrid crops by somatic hybridization (Akamatsu et al. 1988), somaclonal variation (Ling 1996), and gene transfer (Mariani et al. 1990). However, in cotton, work on in vitro induction or engineered male sterility has only been initiated.

Reynaerts and De Sonville (1994) constructed genes which can restore fertility to genetically engineered male sterility in tobacco. The same system was extended by them to cotton. Thus, to introduce male sterility and restoration of fertility genes in *G. hirsutum* lines, *Agrobacterium*-mediated transformation was carried out. Fifty six transgenic callus lines were established, and of these 42 became embryogenic. From most of these lines, 4 or 5 plants were regenerated. Two types of male sterility were observed. All male sterile flowers were smaller than their fertile counterparts (Reynaerts and De Sonville 1994).

2.12 Field Trials/Performance of Transgenic Cotton

Since 1990, extensive field tests conducted on transgenic B.t. cotton have shown that the B.t. gene is effective against tobacco budworm (Jenkins and McCarty 1995) and pink bollworm (Flint et al. 1995). Thomas mentions (see Chap. V.2, this Vol.) that commercialization of insect-protected transgenic crops producing B.t. has been undertaken by several companies, including Monsanto. Field tests showed that some of the most damaging cotton insects can be controlled by this approach (Jenkins et al. 1991; Wilson et al. 1994). Jenkins et al. (1991) observed that insect damage to bolls of B.t.-expressing cotton was between 2 and 12% of all bolls examined, in contrast to 29% damage recorded for control plants. In addition, one line contained a higher percentage of lint than the controls. As a result of extensive field studies Environmental Protection Agency (EPA) approved B.t. cotton registration for Monsanto, the first insectprotection gene to be approved for cotton. The EPA also granted permission to Monsanto to make large-scale plantings of cotton varieties containing the B.t. gene to produce seed for 1996 sales to America's farmers (Anonymous 1996). In 1996, Monsanto's Bollgard cotton containing the B.t. gene comprised 13% of the US cotton (over 1.8 million acres). About 60% of the growers did not spray insecticides for tobacco budworms, bollworms, or pink bollworms. Moreover, Bollgard cotton had an average yield improvement of 7% (Anonymous 1997).

According to Ow and Quisenberry (Chap. V.3, this Vol.), transgenic cotton lines harboring a single copy of the *tfd*A (herbicide-resistant) construct were released to a number of investigators for field trials, and the results obtained substantiated their greenhouse-based conclusions. Some of the lines tolerated up to three times the recommended 2,4-D level for cereal crops. Cotton breeders who received the seeds are currently using a backcrossing method to move the transgene into varieties adapted to the different production climates of the US cotton belt.

The successful field performance of transgenic plants showing resistance to insects and herbicides has opened up new avenues for the release of new improved cotton cultivars in the next few years.

2.13 Risk Assessment Studies on Transgenic Plants

Although transgenic crops have been obtained showing resistance to herbicides, pests, and environmental stress, there is, however, the potential for deterimental effects on the soil ecosystem from residual plant material following harvesting. Thus, Donegan et al. (1995) evaluated this concern by placing leaves of three different lines of cotton genetically engineered to produce the *Bacillus thuringiensis* var. *kurstaki* (B.t.k.) endotoxin in soil and monitoring numbers and species of indigenous soil bacteria and fungi. Two of the three transgenic cotton lines caused a transient increase in total bacterial and fungal population levels that were significantly higher. In contrast, neither the third transgenic line nor the purified B.t.k. toxins had any significant effects on the total numbers of bacteria and fungi. The plant line specificity of the response and the lack of effects from the purified B.t.k. toxins suggest that the observed effects of the two transgenic lines on soil microorganisms may not have resulted from the plants' production of B.t.k. toxin. Donegan et al. (1995) therefore suggested that genetic manipulation or tissue culturing of the plants may have produced a change in plant characteristics, aside from B.t.k. toxin production, that can influence growth and species composition of soil microorganisms.

Risk assessment studies have also been conducted on potato (Donegan et al. 1995). Risk assessment of cross-fertilization between transgenic crops and wild relatives have also been studied (Kareiva et al. 1994).

2.14 Genetic Engineering of Cotton Fiber

Some of the 'specific modifications of fiber useful for various applications include superior strength, length, dye binding, dimensional stability, and thermal adaptability (John 1994a,b). There are two general strategies for fiber modifications. One approach is to increase or decrease levels of fiber proteins or enzymes (John and Stewart 1992). A second, and perhaps more feasible, is to select potential genes from sources other than cotton, because cotton genes responsible for fiber strength and length are not known. Therefore, attempts are being made to isolate and characterize genes that are preferentially expressed in fiber.

Four mRNAs, E6, H6, B6, and FbL2A that are preferentially expressed in fiber have been characterized (John and Crow 1992; John and Keller 1995; John 1996). There are at least two closely related E6 genes in the upland cotton varieties. E6 is a cytoplasmic protein that is expressed during early fiber development. In transgenic cotton, through antisense technology, E6 expression was reduced by 98%. Thus, it appears that the E6 protein does not play a critical role in fiber development (John 1996). An mRNA (H6), encoding a protein-rich (35 mol%) protein was identified from fiber (John and Keller 1995). This may function in the development of secondary walls. Another mRNA, FbL2A, is expressed during late fiber development (Rinehart et al. 1996). Recently, Pear et al. (1996) identified two cotton cDNA clones that contain conserved amino acid regions that are similar to those of the catalytic subunit of bacterial cellulose synthase.

John and Keller (1996) genetically engineered cotton to produce an aliphatic polyester compound, polyhydroxyalkanoate (PHA), in fiber. The transgenic fiber extract contains a compound with a retention time identical to that of bacterial PHB derivative. The presence of PHB in fibers resulted in measurable changes in their thermal properties. A strategy to modify fiber using metabolic pathway engineering to produce aliphatic polyester compounds is under development (see Chap. V.5, this Vol.).

2.15 Molecular Markers – RFLP, RAPD, Chromosome Mapping

In addition to isozymes (see Penel et al. 1993), newly developed techniques, such as RFLP (restriction fragment length polymorphism; Tanksley et al. 1989), RAPD (randomly amplified polymorphic DNA; Williams et al. 1990), and AFLP (amplification fragment length polymorphism; Vos et al. 1996) have attracted much attention as molecular markers. They are more precise, and a microquantity of DNA can be screened. They have been used for a variety of purposes.

2.15.1 RFLP

Brubaker and Wendel (1994) studied the origin of domesticated cotton (*G. hirsutum*) using RFLPs, and discussed its use to estimate levels and patterns of genetic diversity and the congruence of RFLP- and allozyme-based diversity estimates. Levels of RFLP variation in *G. hirsutum* are low relative to other plant taxa, and, in contrast to the few comparable studies, levels of allozyme variation are higher than levels of RFLP variation. Despite assaying 205 loci, only 6 of the 23 upland cultivars were found to have unique multilocus genotypes. Because RFLPs offer a theoretically unlimited number of screenable loci relative to the finite number of allozyme or morphological markers, Brubaker and Wendel (1994) hoped that the data would provide greater resolution of, and hence new insights into the earliest stages of cotton domestication. In a later study, Reinisch et al. (1994) worked on a detailed RFLP map of *G. hirsutum* × *G. barbadense* chromosome organization and evolution in a disomic polyploid genome.

Paterson (1993) reported a detailed RFLP map of the cotton chromosomes. The map is based on an F_2 population of a hybrid between *G. hirsutum* and *G. barbadense*, and contains about 700 markers with a genomic length of approximately 5000 centimorgans and 39 linkage groups.

2.15.2 RAPD

This technique uses a single short oligonucleotide primer (10 bp) of arbitary DNA sequence and polymerase chain reaction (PCR)-mediated amplification of random fragments from genomic DNA. The main advantage of RAPD includes the ease and rapidity of analyses: it does not require preparation of genomic or cDNA clones, restriction enzyme digestion of DNA, Southern blotting, or hybridization with radioactively labeled probes, as data are directly scored from agarose gels. RAPD promises to be a cost-effective diagnostic technology for plant breeding. Moreover, because RAPD analyses require much smaller amounts of DNA and less pure DNA than RFLP analyses, this technique is particularly attractive to geneticists working with species recalcitrant to nucleic acid extraction, such as cotton (Tatieni et al. 1996). Murgeai et al. (1995) used this technique to investigate the potential of molecular markers to monitor the introgression of diploid *Gossypium* genomes into upland cotton. It is in this context that the RAPD marker was developed by them to facilitate the transfer of *G. sturtianum* genes into *G. hirsutum* genome by hybridization and backcrossing. This technique was further improved by the use of activated charcoal during DNA extraction (Vroh et al. 1996). It was shown that RAPD markers specific to wild diploid species can be generated quickly and efficiently. The RAPD technique provides a highly effective means to fingerprint cotton species and interspecific hybrids (see Chap. II.4, this Vol.).

2.16 Secondary Metabolites

In addition to the various compounds, such as lipids, fatty acids, proteins, carbohydrates, vitamins, etc., cotton contains two very important secondary metabolites, i.e., gossypol and tannins, which are discussed here.

2.16.1 Gossypol

Gossypol, a sesquiterpenoid, is of importance, (1) as a toxic contaminant present in industrial products derived from cottonseed, (2) as a deterrent to insect pests in cotton (Lukefahr and Houghtaling 1969), and (3) because of its pharmaceutical effects in various drugs. During the past few years, it has attracted much attention, especially due to its antifertility (Hong et al. 1989), antiparasitic (Eid et al. 1988), antitumor (Jaroszewski et al. 1990; Gilbert et al. 1995), and anti-HIV (Royer et al. 1995) properties.

Gossypol is located in the glands present in the embryo, leaves, flower buds, etc. The seed may contain as much as 10% gossypol (Fisher et al. 1988). Cottonseed meal is a rich fodder supplement; however, the presence of gossypol is toxic for non-ruminant animals. Chemical removal is costly and also reduces protein quality. Thus, methods have to be developed to reduce/eliminate gossypol from seed, and biotechnology may play an important role.

Gossypol formation has been observed in cell suspensions of *G. hirsutum* (Heinstein and El-Shagi 1981), and sesquiterpene aldehyde was stimulated by conidia of *Verticillium dahliae* (Heinstein 1985).

In order to prevent successful invasion by phytopathogenic fungi, plants have developed elaborate defence mechanisms which include the formation of potentially toxic secondary metabolites called phytoalexins (Albersheim and Valent 1978). The induction of phytoalexin formation is thought to be initiated by the interaction of the fungal conidia with the plant cell. Apostol et al. (1987) reported that the addition of an elicitor preparation from *Verticillium dahliae* to cotton cell suspension induced the formation of sesquiterpene aldehydes. Using fluorescence assays, they screened several plant metabolites for activity as suppressors or enhancers of elicitation, and reported that citrate is a potent inhibitor of elicitation. The physiological significance of the inhibition by citrate of phytoalexin formation in cell suspension was supported by the observation that a similar inhibition of sesquiterpene aldehyde formation occurs in cotton plantlets elicited by cold shocks or *V. dahliae* stress (Apostol et al. 1987). These experiments not only demonstrate that citrate can act as a specific inhibitor of elicitation, but they further confirm the validity of monitoring elicitation and its modulation with fluorescent probes.

As already mentioned, protein products and oil from cottonseed have restricted use because of the presence of gossypol in the lysigenous glands in the embryo. Higher gossypol content in leaves and flower buds is correlated with improved resistance to bollworm, *Heliothis* (Benedict et al. 1985). A principal deterrent to using high gossypol germplasm for insect control is that gossypol levels in seed are concomitantly increased, limiting cottonseed utilization. Combined agricultural and seed-processing needs would be satisfactorily met by developing *G. hirsutum* genotypes that form glanded aerial parts and glandless seed. A cotton plant with this attribute could also be extremely useful for developing insect-resistant cotton with higher gossypol (Altman et al. 1987). Plants of several wild Australian diploid species, including *G. sturtianum*, are glanded, but their seed are glandless and do not contain gossypol. Taking these observations into consideration, Altman et al. (1987) studied introgression of the glanded-plant and glandless-seed trait from *G. sturtianum* into *G. hirsutum*, using ovule culture. One BC_2 plant had no glands while in culture, but was fully glanded after the first true leaf emerged. From this "glandless seed" plant, 125 BC_4 plants were derived, of which 5 were completely glandless. Germplasm derived from this material could be important for developing cotton with higher levels of insect resistance and/or improved nutritional use (Altman et al. 1987).

2.16.2 Tannins

Tannins are polyphenolic compounds (Legé et al. 1992, 1995) that are considered to be the most important secondary metabolites involved in plant-defence mechanism (Swain 1979). They have been utilized in host-resistant breeding programs in cotton (White et al. 1982), and are reported to exhibit resistance characteristics to many species of arthropods (Smith et al. 1992). The mode of action of tannins is primarily that of a feeding deterrent to these pests. The bolls containing higher levels of tannins are less preferred than those with lower levels of tannins (Chakravorty and Basu 1981). Diets containing condensed tannins severely inhibited larval growth of cotton bollworm (Chan et al. 1978), and many other arthropods (for references see Chap. V.2, this Vol.). Resistance to various fungal pathogens (Bell et al. 1992) and nematodes (Veech and McClure 1977) has also been reported.

In transgenic cotton (Navon et al. 1993), however, it was observed that the resistance imparted by condensed tannins deterred feeding by larvae, which ultimately reduced the amount of B.t. toxin ingested by the insect. As a result, larvae feeding on high-tannin, B.t. toxin-containing diets had lower mortality rates than insects feeding on diets containing condensed tannins or B.t. toxin. Since tannins impart resistance to diseases mainly through critical enzyme inhibition, biotechnology studies on these lines would be rewarding in unraveling the mechanism of action of tannins.

3 Summary and Prospects

Cotton is a very important oilseed and fiber crop, being grown in a number of countries. There are 50 species of *Gossypium*, of which only four are cultivated; the rest, though wild, are a rich source of genetic variability for disease and pest resistance, fiber quality, gossypol-free germplasm, etc. Although numerous excellent varieties of cotton have been produced through conventional methods of breeding, they seem insufficient to exploit the existing germplasm due to various incomptatibility barriers. The lack of genetic variability hampers the breeding objectives. Thus, methods need to be developed to induce and incorporate new genetic variability in cotton-breeding programs.

In view of the recent advances in plant biotechnology, and the achievements already made in many crops, including cotton, it is highly desirable that these new technologies be incorporated into existing plant-breeding strategies to generate variability. In addition to numerous academic excercises in basic research, in vitro culture studies, and molecular genetics in cotton, some of the most outstanding developments of far-reaching implication are: (1) interspecific hybridization between cultivated and wild species through embryo rescue, (2) regeneration of plants from somatic embryos and protoplasts for genetic manipulation studies, (3) somaclonal variant plants showing tolerance to high temperature, and (4) the last but not the least, although perhaps the most important contribution of biotechnology to cotton is the production of transgenic plants. Transgenic cotton showing resistance to insects/herbicides has been patented, field-tested, and is being released to the farmers.

The ongoing studies on the improvement of fiber quality through genetic engineering are commendable, and genes are being isolated and characterized for fiber improvement. Another striking area in genetic manipulation would be to obtain gossypol-free cotton by gene regulation. Attention must also be focused towards the induction of haploids and male sterile lines, which can give great impetus to the production of hybrid cotton; biotechnology would be of great assistance to achieve this goal. Relatively new techniques, such as RFLP, RAPD, AFLP, and genome mapping will facilitate achieving the desired goals in the genetic engineering of cotton.

References and Further Reading

(Also included are references not cited in the text but which deal with the tissue culture/ biotechnology of cotton)

Akamatsu T, Todhita M, Ohawa Y, Shiga T (1988) Introduction of cytoplasmic male sterility (ogura cytoplasm) into cabbage (*Brassica oleracea* L.) by protoplast fusion. Jpn J Plant Breed 38:14–15

Albersheim P, Valent B (1978) Host-pathogen interactions in plants. Plants, when exposed to oligosaccharides of fungal origin, defend themselves by accumulating antibiotics. J Cell Biol 78:627–643

Altman DW, Stelly DM, Kohel RJ (1987) Introgression of the glanded-plant and glandless-seed trait from *Gossypium sturtianum* Willis into cultivated upland cotton using ovule culture. Crop Sci 27:880–884

Altman DW, Fryxell PA, Koch SD, Howell CR (1990) *Gossypium* germplasm conservation augmented by tissue culture techniques for field collection. Econ Bot 44:106–113

Anonymous (1996) Online Available. http://www.monsanto.com/crop-protection/html-files/news.html

Anonymous (1997) Monsanto in the news – additional reading. Bollgard Cotton Update, Jan 6, 1997

Apostol I, Low PS, Heinstein P, Stipanovic RD, Altman DW (1987) Inhibition of elicitor-induced phytoalexin formation in cotton and soybean cells by citrate. Plant Physiol 84:1276–1280

Bajaj YPS (1979a) Test-tube fertilization and development of maize (*Zea mays* L.) plants. Indian J Exp Biol 17:475–478

Bajaj YPS (1979b) Establishment of germplasm banks through freeze-storage of plant tissue culture and their implications in agriculture. In: Sharp WR, Larsen PO, Paddock EF, Raghavan V (eds) Plant cell and tissue culture: principles and applications. Ohio State Univ Press, Columbus, pp 745–774

Bajaj YPS (1982) Survival of anther- and ovule-derived cotton callus frozen in liquid nitrogen. Curr Sci 51:139–140

Bajaj YPS (1986) In vitro preservation of genetic resources. In: Proc Int Symp Nuclear techniques and in vitro culture for plant improvement. IAEA, Vienna, pp 43–57

Bajaj YPS (1990a) In vitro production of haploids and their use in cell genetics and plant breeding. In: Bajaj YPS (ed) Haploids in crop improvement I. Biotechnology in agriculture and forestry, vol 12. Springer, Berlin Heidelberg New York, pp 3–44

Bajaj YPS (1990b) Somaclonal variation – origin, induction, cryopreservation, and implications in plant breeding. In: Bajaj YPS (ed) Somaclonal variation in crop improvement I. Biotechnology in agriculture and forestry, vol 11. Springer, Berlin Heidelberg New York, pp 3–48

Bajaj YPS (1994) Somatic hybridization – a rich source of genetic variability. In: Bajaj YPS (ed) Somatic hybridization in crop improvement I. Biotechnology in agriculture and forestry, vol 27. Springer, Berlin Heidelberg New York, pp 3–32

Bajaj YPS (1995) Somatic embryogenesis and its applications for crop improvement. In: Bajaj YPS (ed) Somatic embryogenesis and synthetic seed I. Biotechnology in agriculture and forestry, vol 30. Springer, Berlin Heidelberg New York, pp 105–125

Bajaj YPS (1996) Somaclonal variation in crop improvement II. Biotechnology in agriculture and forestry, vol 36. Springer, Berlin Heidelberg New York
Bajaj YPS, Gill MS (1985) In vitro induction of genetic variability in cotton (*Gossypium* spp.). Theor Appl Genet 70:363–368
Bajaj YPS, Gill MS (1986a) Micropropagation and germplasm preservation of cotton (*Gossypium* spp.) through shoot tip and meristem culture. Indian J Exp Biol 24:581–583
Bajaj YPS, Gill MS (1986b) Biotechnology of cotton improvement. In: Crocomo OJ, Sharp WR, Evans DA, Bravo JE, Tavers FCA, Paddock EF (eds) Biotechnology of plants and microorganisms. Ohio State Univ Press, Columbus, pp 118–151
Bajaj YPS, Gill MS (1989) Pollen embryogenesis and chromosomal variation in anther culture of a diploid cotton (*Gossypium arboreum* L.). SABRAO J 21:57–63
Bajaj YPS, Gill MS (1992) Micropropagation of cotton (*Gossypium* species). In: Bajaj YPS (ed) Biotechnology in agriculture and forestry, vol 19. High-tech and micropropagation III. Springer, Berlin Heidelberg New York, pp 483–504
Bajaj YPS, Gill MS (1996a) Hybridization between *Gossypium herbacenm* and *Gossypium stocksii* through embryo rescue. Cell Res (Beijing) 6:85–90
Bajaj YPS, Gill MS (1996b) In vitro induction of haploidy in cotton. In: Jain SM, Sopory SK, Veilleux RE (eds) In vitro haploid production in higher plants, vol 5. Kluwer, Dordrecht, pp 165–174
Bajaj YPS, Gupta RK (1987) Plants from salt-tolerant cell lines of napier grass *Pennisetum purpureum* Schum. Indian J Exp Biol 25:58–60
Bajaj YPS, Gill MS, Mohapatra D (1986) Somaclonal and gametoclonal variation in wheat, cotton and brassica. In Semal J (ed) Somaclonal variations and crop improvement. Martinus Nijhoff, Dordrecht, pp 160–169
Barrow JF, Katterman D, Williams D (1978) Haploid and diploid callus from cotton anthers. Crop Sci 18:619–622
Barton KA, Whitely HR, Yang KS (1987) *Bacillus thuringiensis* endotoxins expressed in transgenic *Nicotiana tabacum* provides resistance to lepidopteran insects. Plant Physiol 85:1103–1109
Basu AK, Patil NB (1993) Producing quality cotton from commercial cotton hybrids grown under rainfed conditions in India. Technical Seminar, 52nd Plenary meeting Int Cotton Advisory Committee, New Delhi
Bayley C, Trolinder N, Ray C, Morgan M, Quisenberry JE, Ow DW (1992) Engineering 2,4-D resistance into cotton. Theor Appl Genet 83:645–649
Beasley CA (1971) In vitro culture of fertilized cotton ovules. Bio Science 21:906–907
Beasley CA (1977) Ovule culture: fundamental and pragmatic research for the cotton industry. In: Reinert J, Bajaj YPS (eds) Applied and fundamental aspects of plant cell, tissue, and organ culture. Springer, Berlin Heidelberg New York, pp 160–178
Beasley CA, Egli E (1977) Fiber production in vitro from a conditional fiberless mutant of cotton. Dev Biol 57:234–237
Beasley CA, Ting IP (1974) Effects of plant growth substances on in vitro fiber development from unfertilized cotton ovules. Am J Bot 61:188–194
Beasley JO (1940) The origin of American tetraploid *Gossypium* species. Am Nat 74:285–286
Beasley JO (1942) Meiotic chromosome behaviour in species, species hybrids, haploids, and induced polyploids of *Gossypium*. Genetics 27:25–54
Bell AA, El-Zik KM, Thaxton PM (1992) Chemistry, biological significance and genetic control of proanthocyanidins in cotton (*Gossypium* spp.). In: Hemingway RW, Laks PE (eds) Plant phenols. Synthesis, properties, significance. Plenum Press, New York, pp 571–595
Benedict JH, Altman DW, Walmsley MH, Schmidt KM (1985) Plant-insect interactions of high terpenoid cottons and bollworm. In: Nelson TC (ed) Proc Beltwide Cotton Prod Res Conf, New Orleans, pp 392–394
Beradi LC, Cherry JP (1980) Textured properties of cottonseed proteins. J Food Sci 45:377–382
Berger J (1969) The world's major fiber crops, their cultivation and manuring. Conzett and Huber, Zurich
Birnbaum EH, Dugger WM, Beasley CA (1977) Interaction of boron with components of nucleic acid metabolism in cotton ovules cultured in vitro. Plant Physiol 59:1034–1038

Blank ML, Allison DC (1963) Frequency of polyembryony in certain strains of *Gossypium hirsutum* L. Crop Sci 3:97–98
Brubaker CL, Wendel JF (1994) Reevaluating the origin of domesticated cotton (*Gossypium hirsutum*; Malvaceae) using nuclear restriction fragment length polymorphisms (RFLPs). Am J Bot 81:1309–1326
Buchala AJ, Roulin S, Meier H (1989) Polysaccharides in the culture medium of cotton (*Gossypium hirsutum* L.) ovules cultured in vitro. Plant Cell Rep 8:25–28
Chakravorty SC, Basu AK (1981) Changes in tannin during growth of bolls in three cotton cultivars known for their differential susceptibilities to pink bollworm (*Pectinophora gossypiella*) incidence. Sci Cult 47:373–375
Chan BG, Waiss AC, Lukefahr M (1978) Condensed tannins, an antibiotic chemical from *Gossypium hirsutum*. Insect Physiol 24:113–118
Chen Zhixian, Li SJ, Yue JX, Liu SX, She JM, Wu JY, Wang HB (1989) Plantlet regeneration from protoplasts isolated from embryogenic suspension culture of cotton (*Gossypium hirsutum* L.). Acta Bot Sin 31:966–969
Cherry JP, Katterman FRH, Endrizzi JE (1970) Comparative studies of seed proteins of species of *Gossypium* by gel-electrophoresis. Evolution 24:431–477
Chlan CA, Lin J, Cary JW, Cleveland TE (1995) A procedure for biolistic transformation and regeneration of transgenic cotton from meristematic tissue. Plant Mol Biol Rep 13:31–37
Creech RG, Jenkins JN, Tang B, Lawrence GW, McCarty JC (1995) Cotton resistance to root-knot nematode. I. Penetration and reproduction. Crop Sci 35:365–368
D'Amato F (1985) Cytogenetics of plant cell and tissue cultures and their regenerates. CRC Criticical Rev Plant Sci 3:73–112
Davidonis GH (1993a) Cotton fiber growth and development in vitro: effects of tunicamycin and monensin. Plant Sci 88:229–236
Davidonis GH (1993b) A comparison of cotton ovule and cotton cell suspension cultures: response to gibberellic acid and 2-chloroethylphosphonic acid. J Plant Physiol 141:505–507
Davidonis GH, Hamilton RH (1983) Plant regeneration from callus tissue of *Gossypium hirsutum* L. Plant Sci Lett 32:89–93
Davis DD (1978) Hybrid cotton: specific problems and potentials. Adv Agron 30:129–157
Davis DG, Dusbabek KE, Hoerauf RA (1974) In vitro culture of callus tissues and cell susnensions from okra (*Hibiscus esculentus* L.) and cotton (*Gossypium hirsutum* L.). In Vitro 9:395–398
De Block M (1988) Genotypic-independent leaf disc transformation of potato (*Solanum tuberosum*) using *Agrobacterium tumefaciens*. Theor Appl Genet 76:767–774
De Block M, De Brouwer D, Tenning P (1989) Transformation of *Brassica napus* and *Brassica oleracea* using *Agrobacterium tumefaciens* and the expression of the bar and neo genes in transgenic plants. Plant Physiol 91:694–701
Denchev PD, Kuklin AI, Scragg AH (1992) Somatic embryo production in bioreactors. J Biotechnol 26:99–109
Dhindsa RS, Beasley CA, Ting IP (1976) Effects of abscisic acid on in vitro growth of cotton fiber. Planta 130:197–201
Donegan KK, Palm CJ, Fieland VJ, Porteous LA, Ganion LM, Schaller DL, Bucao LQ, Seidler RJ (1995) Changes in levels, species and DNA fingerprints of soil microorganisms associated with cotton expressing the *Bacillus thuringiensis* var. *kurstaki* endotoxin. Appl Soil Ecol 2:111–124
Dugger WM, Palmer RL (1985) Effect of boron on the incorporation of glucose by cotton fibers grown in vitro. J Plant Nutr 8:311–325
Edwards GA, Mirza MA (1979) Genomes of the Australian wild species of cotton II. The designation of a new G-genome for *Gossypium bickii*. Can J Genet Cytol 21:367–372
Eid AAH, DeLange E, Waterkeyn L (1973) In vitro culture of fertilized cotton ovules. I. The growth of cotton embryos. Cellule 69:361–371
Eid JE, Ueno H, Wang CC, Donelson JE (1988) Gossypol-induced death of African trypanosomas. Exp Parasit 66:140–142
El-Nockrashy AS, Simmons JG, Frampton VL (1969) A chemical survey of seeds of the genus *Gossypium*. Phytochemistry 8:1949–1958

Endrizzi JE, Turcotte E, Kohel RJ (1984) In: Kohel RJ, Lewsis CF (eds) Cotton. Agronomy No 24. Am Soc Agron, Madison, pp 81–129

FAO (1993) Production yearbook. FAO, Rome

Fassuliotis G (1990) Somaclonal variation for nematode resistance. In: Bajaj YPS (ed) Biotechnology in agriculture and forestry, vol 11. Somaclonal variation in crop improvement I. Springer, Berlin Heidelberg New York, pp 258–268

Finer J (1988) Plant regeneration from somatic embryogenic suspension cultures of cotton (*Gossypium hirsutum* L.). Plant Cell Rep 7:399–402

Finer J, McMullen M (1990) Transformation of cotton (*Gossypium hirsutum* L.) via particle bombardment. Plant Cell Rep 8:586–589

Firoozabady E (1989) Transformation in cotton. In: Bajaj YPS (ed) Plant protoplasts and genetic engineering II. Biotechnology in agriculture and forestry, vol 9. Springer, Berlin Heidelberg New York, pp 140–154

Firoozabady E, DeBoer DL (1993) Plant regeneration via somatic embryogenesis in many cultivars of cotton (*Gossypium hirsutum* L.). In Vitro Cell Dev Biol 29:166–173

Firoozabady E, DeBoer D, Merlo D, Halk E, Amerson L, Rashka KE, Murray EE (1987) Transformation of cotton (*Gossypium hirsutum* L.) by *Agrobacterium tumefaciens* and regeneration of transgenic plants. Plant Mol Biol 10:105–116

Fischhoff DA, Bowdish KS, Perlak FJ, Marrone PG, McCormick SM, Niedermeyer JG, Dean DA, Kusano-Kretzmer K, Mayer EJ, Rochester DE, Rogers SG, Fraley RT (1987) Insect tolerant transgenic tomato plants. Bio/Technology 5:807–813

Fisher GS, Frank AW, Cherry JP (1988) Total gossypol content of glandless cottonseed. J Agric Food Chem 36:42–44

Francey Y, Jaquet JP, Cairoli S, Buchala AJ, Meier H (1989) The biosynthesis of p-glucans in cotton (*Gossypium hirsutum* L.) fibres of ovules cultured in vitro. J Plant Physiol 134:485–491

Fryxell PA (1992) A revised taxonomic interpretation of *Gossypium* L. (Malvaceae). Rheedea 2:108–165

Fujimoto H, Itoh K, Yamamoto M, Kyozuka J, Shimamoto K (1993) Insect resistant rice generated by introduction of a modified δ-endotoxin gene of *Bacillus thuringiensis*. Bio/Technology 11:1151–1155

Gawel NJ, Robacker CD (1995) Somatic embryogenesis in cotton (*Gossypium* spp.). In: Bajaj YPS (ed) Somatic embryogenesis and synthetic seed II. Biotechnology in agriculture and forestry, vol 31. Springer, Berlin Heidelberg New York, pp 256–266

Gawel NJ, Rao AP, Robacker CD (1986) Somatic embryogenesis from leaf and petiole callus cultures of *Gossypium hirsutum* L. Plant Cell Rep 5:457–459

Gerstel DU (1953) Chromosome translocations in interspecific hybrids of the genus *Gossypium*. Evolution 7:234–244

Gilbert NE, O'Reilly JE, Chang CJG, Lin YC, Brueggemeier RW (1995) Antiproliferative activity of gossypol and gossypolone on human breast cancer cells. Life Sci 57:61–67

Gill MS, Bajaj YPS (1984a) In vitro production of interspecific hybrids in cotton. Curr Sci 53:102–104

Gill MS, Bajaj YPS (1984b) Interspecific hybridization in the genus *Gossypium* through embryo culture. Euphytica 33:305–311

Gill MS, Bajaj YPS (1986) In vitro production of interspecific hybrids in *Gossypium* and their utilization in backcrossing. Z Pflanzenzuecht 96:174–176

Gill MS, Bajaj YPS (1987) Hybridization between diploid (*Gossypium arboreum*) and tetraploid (*Gossypium hirsutum*) cotton through ovule culture. Euphytica 36:625–630

Gill MS, Bajaj YPS (1991) In vitro hybridization between *Gossypium arboreum* and *G. anomalum* and their backcross progeny. Plant Tissue Cult 1:101–108

Gill MS, Bajaj YPS (1996) Hybridization between *Gossypium herbaceum* and *Gossypium stocksii* through emberyo rescue. Cell Res (Beijing) 6:85–90

Gleddie S, Fassuliotis G, Keller WA, Setterfield G (1985) Somatic hybridization as a potential method of transferring nematode and mite resistance into eggplant. Z Pflanzenzuchtg 94:348–351

Goodell PB (1993) Cotton nematodes. In: Cotton nematodes – our hidden enemies. Beltwide Cotton Nematode Survey and Education Committee. The Cotton Foundation, Memphis, TN, pp 1–3

Gossett DR, Millhollon EP, Lucas MC, Banks SW, Marney MM (1994) The effects of NaCl on antioxidant enzyme activities in callus tissue of salt-tolerant and salt-sensitive cotton cultivars (*Gossypium hirsutum* L.). Plant Cell Rep 13:498–503

Gould J, Banister S, Hasegawa O, Fahima M, Smith R (1991) Regeneration of *Gossypium hirsutum* and *G. barbadense* from shoot apex tissue for transformation. Plant Cell Rep 10:12–16

Gould JH, Palmer RL, Dugger WM (1986) Isolation and culture of cotton ovule epidermal protoplasts (prefiber cells) and analysis of the regenerated wall. Plant Cell Tissue Organ Cult 6:47–59

Haigler CH, Rao NR, Roberts EM, Huang JY, Upchurch DR, Trolinder NL (1991) Cultured ovules as models for cotton fiber development under low temperatures. Plant Physiol 95:88–96

Harland SC (1940) New polyploids in cotton by the use of colchicine. Trop Agric (Trin) 17:53–54

Hedin PA, Jenkins JN, Parrott WL (1992) Evaluation of flavonoids in *Gossypium arboreum* (L.) cottons as potential source of resistance to tobacco budworm. J Chem Ecol 18:105–114

Heinstein PF (1985) Stimulation of sesquiterpene aldehyde formation in *Gossypium arboreum* cell suspension cultures by conidia of *Verticillium dahliae*. J Nat Prod 48:907–915

Heinstein PF, El-Shagi H (1981) Formation of gossypol by *Gossypium hirsutum* L. cell suspension cultures. J Nat Prod 44:1–6

Hong CY, Huang JJ, Wo P (1989) The inhibitory effect of gossypol on human sperm motility: relationship with time, temperature and concentration. Human Toxicol 8:49–51

Isley D (1928) The relation of leaf color and leaf size to boll weevil infestation. J Econ Entomol 21:553–559

Jaroszewski JW, Kaplan O, Cohen JS (1990) Action of gossypol and rhodamine 123 on wild-type and multidrug-resistant MCF-7 human breast cancer cells: ^{31}P NMR and toxicity studies. Cancer Res 50:6936–6934

Jenkins JN, Parrot WL, McCarty JC Jr, Barton KA, Umbeck PF (1991) Field test of transgenic cottons containing a *Bacillus thuringiensis* gene. Mo Agric For Exp Stn Tech Bull 174

Jenkins JN, Creech RG, Tang B, Lawrence GW, McCarty JC (1995) Plant resistance in cotton to *Meloidogyne incognita*. II. Post-penetration development. Crop Sci 35:369–373

John ME (1994a) Re-engineering cotton fiber. Chem Ind (Lond): 676–679

John ME (1994b) Cotton futures. ChemTech: 27–30

John ME (1996) Structural characterization of genes corresponding to cotton fiber mRNA, E6: reduced E6 protein in transgenic plants by antisense gene. Plant Mol Biol 30:297–306

John ME, Crow LJ (1992) Gene expression in cotton (*Gossypium hirsutum* L.) fiber: cloning of the mRNAs. Proc Natl Acad Sci USA 89:5769–5773

John ME, Keller G (1995) Characterization of mRNA for a proline-rich protein of cotton fiber. Plant Physiol 108:669–676

John ME, Keller G (1996) Metabolic pathway engineering in cotton: Biosynthesis of polyhydroxybutyrate in fiber cells. Proc Natl Acad Sci USA 93:12768–12773

John M, Stewart J McD (1992) Genes for jeans: biotechnological advances in cotton. Tibtech 10:165–170

Joshi PC, Johri BM (1972) In vitro growth of ovules of *Gossypium hirsutum*. Phytomorphology 22:195–209

Kareiva P, Morris W, Jacobi CM (1994) Studying and managing the risk of cross-fertilization between transgenic crops and wild relatives. Mol Ecol 3:15–21

Katterman F, Williams MD, Clay WF (1977) The influence of a strong reducing agent upon the initiation of callus from the germinating seedlings of *Gossypium barbadense* L. Phyiol Plant 40:198–200

Klein TM, Wolf ED, Wu R, Sanford JC (1987) High-velocity micro-projectiles for delivering nuclear acids into living cells. Nature 327:70–73

Klocke JA, Chan BG (1982) Effects of cotton condensed tannin on feeding and digestion in the cotton pest *Heliothis zea*. J Insect Physiol 28:911–915

Koziel MG, Beland GL, Bowman C, Carozzi NB, Crenshaw R, Crossland L, Dawson J, Desai N, Hill M, Kadwell S, Launis K, Lewis K, Maddox D, McPherson K, Meghji E, Merlin MR, Rhodes R, Warren GW, Wright M, Evola SV (1993) Field performance of elite transgenic maize plants expressing an insecticidal protein derived from *Bacillus thuringiensis*. Bio/Technology 11:194–200

Kranz E, Loerz H (1993) In vitro fertilization with isolated, single gametes results in zygotic embryogenesis and fertile maize plants. Plant Cell 5:739–746

Kuo CC, Zurakowski K, Carlson K, Kilner P, Wang A (1989) Plant regeneration from root-derived callus of *Gossypium hirsutum* L. In: Brown JM (ed) Proc Beltwide Cotton Production Rcs Conf. Natl Cotton Council of America, Memphis, pp 638–640

Larkin PJ, Scowcroft WR (1981) Somaclonal variation – a novel source of variability from cell culture for plant improvement. Theor Appl Genet 60:197–214

Lawhon JT, Carter CM, Mattil KF (1977) Evaluation of the food use potential of sixteen varieties of cottonseed. J Am Oil Chem Soc 54:75–80

Lee JA (1970) On the origin of haploid/diploid twining in cotton. Crop Sci 10:453–454

Lee JA (1984) Cotton as a world crop. In: Kohel RJ, Lewis CF (eds) Cotton. Am Soc Agron, Madison, pp 1–25

Legé KE, Smith CW, Cothren JT (1992) Genotypic and cultural effects of condensed tannin concentration of cotton leaves. Crop Sci 32:1024–1028

Legé KE, Cothren JT, Smith CW (1995) Phenolic acid and condensed tannin concentrations of six cotton genotypes. Environ Exp Bot 35:241–249

Leidi EO, Nogales R, Lips SH (1991) Effect of salinity on cotton plants grown under nitrate or ammonium nutrition at different calcium levels. Field Crops Res 26:35–44

Li R, Stelly D, Trolinder N (1989) Cytogenetic abnornamilities in cotton (*Gossypium hirsutum* L.) cell cultures. Genome 32:1128–1134

Ling DH (1996) In vitro production of male sterile rice plants. In: Bajaj YPS (ed) Somaclonal variation in crop improvement II. Biotechnology in agriculture and forestry, vol 36. Springer, Berlin Heidelberg New York, pp 20–45

Liu C, Shun J, Liu J (1992) In vitro interspecific fertilization, embryo development and formation of hybrid seedlings between *Gossypium hirsutum* and *G. arboreum*. Euphytica 60:79–88

Liu G, Wu J, Bao W, Hsi Y (1986) Somatic embryogenesis of cotton in vitro and its cytological observations. J Agric Sci China 2 (Suppl):36–41

Lu ZX, Xia ZA (1991) Study on tissue and protoplast culture of wild cotton (*Gossypium davidsonii*). Acta Bot Sin 33 (2):98–103

Lukefahr MJ, Houghtaling JE (1969) Resistance of cotton strain with high gossypol content to *Heliothis* spp. J Econ Entomol 62:588–591

Lukefahr MJ, Martin DF (1966) Cotton-plant pigments as a source of resistance to the bollworm and tobacco budworm. J Econ Entomology 59:176–179

Lyon BR, Llewellyn DJ, Huppatz JL, Dennis ES, Peacock WJ (1989) Expression of a bacterial gene in transgenic tobacco plants confers resistance to the herbicide 2,4-dichlorophenoxy acetic acid. Plant Mol Biol 13:533–540

Lyon BR, Cousins Y, Llewellyn DJ, Dennis ES (1993) Cotton plants transformed with a bacterial degradation gene are protected from accidental spray drift damage by the herbicide 2,4-dichlorophenoxyacetic acid. Transgen Res 2:162–169

Mariani C, Beuckeleer MD, Truettner J, Leemans J, Goldberg BR (1990) Induction of male sterility in plants by a chimeric ribonuclease gene. Nature 347:737–741

Matlin SA (1994) Prospects in pharmacological male contraception. Drugs 48:851–863

McCabe DE, Martinell BJ (1993) Transformation of elite cotton cultivars via particle bombardment of meristems. Biotechnology 11:596–598

Mehboob Ali S, Krishnayya, VV, Satyanarayayana Murthy K (1964) A note on an interspecific hybrid in cotton (*Gossypium arboreum* × race *indicum* × *G. stocksii*). Indian Cotton Grow Rev 18:107–108

Meisner J, Zur M, Kabonci E, Ascher KRS (1977) Influence of gossypol content of leaves of different cotton strains on the development of *Spodoptera littoralis* larvae. J Econ Entomol 70:714–716

Mellon JE (1986) Some characteristics of peroxidase secreted by cotton ovule cultures. Plant Cell Rep 5:338–341

Mellon JE (1988) Use of ovule culture as a model for cotton-*Aspergillus flavus* interactions. In: Brown JM (ed) Proc Beltwide Cotton Prod Res conf. National Cotton Council of America, Memphis, pp 35–37

Mellon JE (1991) Purification and characterization of isoperoxidases elicited by *Aspergillus flavus* in cotton ovule cultures. Plant Physiol 95:14–20
Mellon JE (1992) Inhibition of aflatoxin production in *Aspergillus flavus* by cotton ovule extracts. J Am Oil Chem Soc 69:945–947
Mellon JE, Triplett BA (1989) De novo synthesis of peroxidase in cotton ovule culture medium. Physiol Plant 77:302–307
Meng GD, Zhu JC, Chen ZW, Wong LT, Zhang GY, Hu YZ, Ding JH, Wang XH, Qian SZ, Wang C, Manchin D, Pinol A, Waites GMH (1988) Recovery of sperm production following the cessation of gossypol treatment: a two-centre study in China. Int J Androl 11:1–11
Mergeai G, Vroh Bi Irie, Du Jardin P, Baudoin JP (1995) Development of an interspecific breeding programme in cotton assisted by RAPD markers. In: Cotton biotechnology. REUR Technical Series, FAO, Rome
Meyer VG (1974) Interspecific cotton breeding. Econ Bot 28:56–60
Murashige T, Skoog F (1962) A revised medium for rapid growth and bio-assays with tobacco tissue cultures. Physiol Plant 15:473–497
Navon A, Hare JD, Federici BA (1993) Interactions among *Heliothis virescens* larvae, cotton condensed tannin and the CryIA (c) δ-endotoxin of *Bacillus thuringiensis*. J Chem Ecol 19:2485–2499
Parmar AL, Shroff VN, Deshpande NV (1978) Studies on vegetative propagation in cotton. Crop Improv 5:73–76
Paterson A (1993) Molecular markers in cotton improvement. Proc Beltwide Conf, National Cotton Counc, New Orleans
Pear JR, Kawagoe Y, Sehreckengost WE, Delmer DP, Stalker DM (1996) Higher plants contain homologs of the bacterial cel A genes encoding the catalytic subunit of cellulose synthase. Proc Natl Acad Sci USA 93:12637–12642
Peeters MC, Willems K, Swennen R (1994) Protoplast-to-plant regeneration in cotton (*Gossypium hirsutum* L. cv. Coker 312) using feeder layers. Plant Cell Rep 13:208–211
Penel C, Gaspar Th, Greppin H (eds) (1993) Plant peroxidases 1980–1990. Topics and detailed literature on molecular, biochemical, and physiological aspects. University of Geneva Press, Geneva
Percival AE, Kohel RH (1990) Distribution, collection, and evaluation of *Gossypium*. Adv Agron 44:225–256
Perlak FJ, Deaton R, Armstrong T, Fuchs R, Sims S, Greenplate J, Fischoff D (1990) Insect-resistant cotton plants. Biotechnology 8:939–943
Perlak FJ, Fuchs RL, Dean DA, McPherson SL, Fischhoff DA (1991) Modifications of the coding sequence enhances plant expression of insect control proteins. Proc Natl Acad Sci USA 88:3324–3328
Phillips LL (1963) The cytogenetics of *Gossypium* and the origin of New World cottons. Evolution 17:460–469
Phillips LL, Strickland MA (1966) The cytology of a hybrid between *Gossypium hirsutum* and *G. longicalyx*. Can J Genet Cytol 8:91–95
Prentice AN (1972) Cotton with special reference to Africa. Longmann, London, pp 68–76
Price HJ, Smith RH (1979) Somatic embryogenesis in suspension culture of *G. klotzschianum* Anderss. Planta 145:305–307
Price HJ, Smith RH (1994) Cotton. In: Ammirato PV, Evans DA, Sharp WR, Yamada Y (eds) Handbook of plant cell culture, vol 3. Crop species. Macmillan, New York, pp 487–510
Rafaat M, Rossignol L, Demarly Y (1984) Interspecific hybrid of *Gossypium hirsutum* L. × *G. barbadense* L. via in vitro fertilization and ovule culture. Z Planzenzuecht 93:137–146
Rajasekaran K, Grula JW, Hudspeth RL, Pofelis S, Anderson DM (1996) Herbicide-resistant Acala and Coker cottons transformed with a native gene encoding mutant forms of acetohydroxyacid synthase. Mol Breed 2:307–319
Ramsey P, Menke S, Zeiher C (1996) Stimulation of PEP carboxylase by auxin and gibberellin in cultured cotton ovules. Plant Physiol 111(S):114
Randel RD, Case CC, Wyse SJ (1992) Effects of gossypol and cottonseed products on reproduction of mammals. J Anim Sci 70:1628–1638

Razzouk S, Whittington WJ (1991) Effects of salinity on cotton yield and quality. Field Crops Res 26:305–314

Ryan CA (1990) Protease inhibitors in plants: genes for improving defences against insects and pathogens. Annu Rev Phytopathol 28:425–449

Reinisch AJ, Dong JM, Brubaker CL, Stelly DM, Wendel JF, Paterson AH (1994) A detailed RFLP map of cotton, *G. hirsutum* × *Gossypium barbadense*: chromosome organization and evolution in a disomic polyploid genome. Genetics 138:829–847

Revenkova EV, Kraev AS, Skryabin KG (1990) Transformation of cotton (*Gossypium hirsutum* L.) with a supervirulent strain of *Agrobacterium tumefaciens* A281. Mol Biol 24:1017–1023

Reynaerts A, De Sonville A (1994) Potential application of engineered male sterility in production of cotton hybrids. Proc Meet Working Group on Cotton Biotechnology. REUR Tech Series 32, FAO, Rome, pp 37–42

Rinehart JA, Petersen MW, John ME (1996) Tissue-specific and developmental regulation of cotton gene Fb2L2: Plant Physiol 112:1331–1341

Roberts EM, Rao NR, Huang JY, Trolinder NL, Haigler CH (1992) Effects of cycling temperatures on fiber metabolism in cultured cotton ovules. Plant Physiol 100:979–986

Rohrbach MS, Kreofsky TJ, Vuk-pavlovic Z, Lauque D (1992) Cotton condensed tannin: A potent modulator of alveolar macrophage host-defence function. In: Hemingway RW, Laks PE (eds) Plant polyphenols. Synthesis, properties, significance. Plenum Press, New York

Royer ER, Deck LM, Jagt TJV, Martinez FJ, Mills RG, Young SA, Jagt DLV (1995) Synthesis and anti-HIV activity of 1,1-dideoxygossypol and related compounds. J Med Chem 38:2427–2432

Saka K, Katterman RF, Thomas JC (1987) Cell regeneration and sustained division of protoplasts from cotton (*Gossypium hirsutum* L.). Plant Cell Rep 6:470–472

Schenk RU, Hildebrandt AC (1972) Medium and techniques for induction and growth of monocotyledonous and dicotyledonous plant cell cultures. Can J Bot 50:199–204

Seagull RW (1990) The effects of microtubule and microfilament disrupting agents on cytoskeletal arrays and wall deposition in developing cotton fibers. Protoplasma 159:44–59

Seagull RW (1992) A quantitative electron microscopic study of changes in microtubule arrays and wall microfibril orientation during in vitro cotton fiber development. J Cell Sci 101:561–577

Shamina ZB, Turaev AM, Musaev DA (1986) Capacity for induction of callus formation in cotton anther culture. Cytol Genet 20:40–44

Shang XM, Huang JY, Haigler CH, Trolinder NL (1991) Buffer capacity of cotton cells and effects of extracellular pH on growth and somatic embryogenesis in cotton cell suspensions. In Vitro Cell Dev Biol 27B:147–152

Sharma HC, Agarwal RA (1984) Factors imparting resistance to stem damage by *Earias vittella* F. (Lepidoptera: Noctuidae) in some cotton phenotypes. Protect Ecol 6:35

She JM, Wu JY, Wang HB, Zhou HY, Chen ZX, Li SJ, Yue JX (1989) Somatic embryogenesis and plant regeneration from protoplast culture of cotton. Genet Manipul Plants 5(2):63–68

She JM, Wu JY, Zhou HY (1993) High frequency of plating efficiency and plant regeneration from protoplasts of cotton (*Gossypium hirsutum* L.). Acta Genet Sin 20(6):536–540

She JM, Wu JY, Zhou HY (1995) Regeneration of plants from protoplasts of *Gossypium* species (Cotton). In: Bajaj YPS (ed) Plant protoplasts and genetic engineering VI. Biotechnology in agriculture and forestry. vol 34. Springer, Berlin Heidelberg New York, pp 63–69

Shoemaker RC, Couche LJ, Galbraith DW (1986) Characterization of somatic embryogenesis and plant regeneration in cotton (*Gossypium hirsutum* L.). Plant Cell Rep 3:178–181

Skovsted A (1937) Cytological studies in cotton. IV. Chromosome conjugation in interspecific hybrids. J Genet 34:97–134

Smith CW, McCarty JC, Altamarino TP, Lege KE, Schuster MF, Phillips JR, Lopez JD (1992) Condensed tannins in cotton and bollworm (Lepidoptera: Noctuidae) resistance. J Econ Entomol 85:2211–2217

Smith RH, Price HJ, Thaxton JE (1977) Defined conditions for the initiation and growth of cotton callus in vitro I. *Gossypium arboreum*. In Vitro 13:329–334

Spencer MT, O'Brien JV, Start WG, Adams TR, Gordon-Kamm WJ, Lemaux PG (1992) Segregation of transgenes in maize. Plant Mol Biol 18:201–210

Srinavasan K, Santhanam V, Rajasekaran S (1972) Development of hybrid cotton utilizing male-sterile line. Cotton Dev 2:37–39

Stelly DM, Altman DW, Kohel RJ, Rangan TS, Commiskey E (1989) Cytogenetic abnormalities of cotton somaclones from callus cultures. Genome 32:762–770

Stewart JM, Hsu CL (1977) In-ovulo embryo culture and seedling development of cotton (*Gossypium hirsutum* L.). Planta 137:113–117

Streber WR, Willmitzer L (1989) Transgenic tobacco plants expressing a bacterial detoxifying enzyme are resistant to 2,4-D. Biotechnology 7:811–816

Swain T (1979) Phenolics in the environment. In: Swain T, Harborne JB, Van Sumere CF (eds) Recent advances in phytochemistry, vol 12. Plenum Press, New York, pp 624–637

Tal M (1990) Somaclonal variation for salt tolerance. In: Bajaj YPS (ed) Somaclonal variation in crop improvement I. Biotechnology in agriculture and forestry, vol 11. Springer, Berlin Heidelberg New York, pp 236–257

Tan XL, Qian YG (1988) Effect of explant sources and cultural condition on plant regeneration in *Gossypium gossypioides* (Ulbrich) standley. Acta Genet Sin 15(2):81–85

Tanda AS, Atwal AS, Bajaj YPS (1980) Reproduction and maintenance of root-knot nematode (*Meloidogyne incognita*) in excised roots and callus cultures of okra (*Abelmoschus esculentus*). Indian J Exp Biol 18:1340–1342

Tanksley SD, Young ND, Pat AH, Bonierbale MW (1989) RFLP mapping new tools for an old science. Biotechnology 7:257–264

Tatieni V, Cantrell RG, Davis DD (1996) Genetic diversity in elite cotton germplasm determined by morphological characteristics and RAPDs. Crop Sci 36:186–192

Thomas JC, Adams DG, Keppenne VD, Wasmann CC, Brown JK, Kanost MR, Bohnert HJ (1995) Protease inhibitors of *Manduca sexta* expressed in transgenic cotton. Plant Cell Rep 14:758–762

Triplett BA, Busch WH, Goynes WR Jr (1989) Ovule and suspension culture of a cotton fiber development mutant. In Vitro Cell Dev Biol 25:197–200

Trolinder NL, Allen RD (1994) Molecular approaches to stress tolerance in cotton. Int Cotton Res Conf, Brisbane, Australia, p 57

Trolinder NL, Chen ZX (1989) Genotype specificity of the somatic embryogenesis respones in cotton. Plant Cell Rep 8:133–136

Trolinder NL, Goodin JR (1988a) Somatic embryogenesis in cotton (*Gossypium*) I. Effect of source of explant and hormone regime. Plant Cell Tissue Organ Cult 12:31–42

Trolinder NL, Goodin JR (1988b) Somatic embryogenesis in cotton (*Gossypium*). II. Requirements for embryo development and plant regeneration. Plant Cell Tissue Organ Cult 12:43–53

Trolinder NL, Shang Xaiomin (1991) In vitro selection and regeneration of cotton resistant to high-temperature stress. Plant Cell Rep 10:448–452

Trolinder NL, Berlin JD, Goodin JR (1987) Differentiation of cotton fibers from single cells in suspension culture. In Vitro Cell Dev Biol 23:789–794

Turaev AM, Shamina ZB (1993) Culture of cotton microspores in agarose. Russ Plant Physiol 40:294–297

Umbeck P, Johnson G, Barton K, Swain W (1987) Genetically transformed cotton (*Gossypium hirsutum* L.) plants. Biotechnology 5:263–266

Umbeck PF, Stewart J McD (1985) Substitution of cotton cytoplasm from wild diploid species for cotton germplasm improvement. Crop Sci 25:1015–1019

Vaeck M, Reynaerts A, Hofte H, Jansens S, de Beukler M, Dean C, Zabeau M, Van Montagu M, Leemans J (1987) Transgenic plants protected from insect attack. Nature 328:33–37

Veech JA, McClure MA (1977) Terpenoid aldehydes in cotton roots susceptible and resistant to the root-knot nematode, *Meloidogyne incognita*. J Nematol 9:225–229

Vos P, Hogers R, Bleekers M, Reijans M, Lee T, Hornes M, Frijters A, Pot J, Peloman J, Kulper M, Zabcau M (1996) AFLP: a new technique for DNA fingerprinting. Nucleic Acids Res 23:4407–4414

Vroh Bi I, Harvengt L, Chandelier A, Mergeai G, du Jardin P (1996) Improved RAPD amplification of recalcitrant plant DNA by the use of activated charcoal during DNA extraction. Plant Breed 115:205–206

Wainwright IM, Palmer RL, Dugger WM (1980) Pyrimidine pathway in boron-deficient cotton fiber. Plant Physiol 65:893–896

Wang N, Zhou L, Guan M, Lei H (1987) Effects of (−)- and (+)-gossypol on fertility in male rats. J Ethnopharmacol 20:21–24

Wenzel G, Uhrig H (1981) Breeding for nematode and virus resistance in potato via anther culture. Theor Appl Genet 59:333–340

Weaver JB Jr (1958) Embryological studies following interspecific crosses in *Gossypium*. II. *G. arboreum* × *G. hirsutum*. Am J Bot 45:10–16

White PR (1934) Potentially unlimited growth of excised tomato root tips in a liquid medium. Plant Physiol 9:585–600

White WH, Jenkins JN, Parrot WL, McCarty JC, Collum DH, Hedin PA (1982) Strain and within-season variability of various allelochemics within a diverse group of cottons. Crop Sci 22:1235–1238

Williams JGK, Kubelik AR, Livak KJ, Rafalski JA, Tingcy SV (1990) DNA polymorphisms amplified by arbitrary primers are useful as genetic markers. Nucleic Acids Res 18:7213–7218

Wilson AD, Flint HM, Deaton WR, Buehler RE (1994) Yield, yield components, and fiber properties of insect-resistant cotton lines containing a *Bacillus thuringiensis* toxin gene. Crop Sci 34:38–41

Wu JY, She JM (1990) Shoot apex and axillary bud culture of *Gossypium* spp. Jiangsu J Agric Sci 6 (2):22–26

Wu JY, She JM, Zhou HY (1990) Effects of the growth of regenerated plantlets from somatic cell cultures on transplantation in *G. hirsutum* L. Jiangsu J Agric Sci 6(Suppl):33–38

Wu JY, She JM, Zhu WM, Cai XN, Wu HM (1994) Bud cluster formation and plantlet regeneration from in vitro leaves of Simian 3 (*Gossypium hirsutum* L.). Jiangsu Agric Sci 2:18–19

Xie W, Trolinder NL, Haigler CH (1993) Cool temperature effects on cotton fiber initiation and elongation clarified using in vitro cultures. Crop Sci 33:1258–1264

Yoder JI, Goldsbrough AP (1994) Transformation systems for generating marker-free transgenic plants. Bio/Technology 12:263–267

Zenkteler M (1990) In vitro fertilization and wide hybridization in higher plants. Crit Rev Plant Sci 9:267–279

Zhang JM, Sun JZ, Liu JL, Zhang XL (1994) Studies on plant regeneration from somatic cells and transferring technique of plantlets in upland cotton. Acta Agron Sin 20(2):210–216

Zhang XL, Sun JH, Liu JL (1991) Somatic embryogenesis and plant regeneration in upland cotton. Acta Gen Sinc 18 (5):461–467

Zhou G-Y, Weng J, Zeng Y, Huang J, Qian S, Liu G (1983) Introduction of exogenous DNA into cotton embryos. Methods Enzymol 101:433–481

Zimmerman TW, Robacker CD (1988) Media and gelling agent effect on cotton callus initiation from excised seed hypocotyls. Plant Cell Tissue Organ Cult 15:269–274

I.2 Establishment of Callus Culture, Somatic Embryogenesis, and the Regeneration of Cotton Plants

J.Y. Wu[1], J.M. She[1], X.N. Cai[1], and Y.P.S. Bajaj[2]

1 Introduction

For the induction of genetic variability and genetic manipulation studies, it is essential that the cells in culture develop into complete plants. In cotton, callus capable of differentiation has been established in a number of wild and cultivated species of *Gossypium*. There are two modes of differentiation, i.e., organogenesis through bud formation and somatic embryogenesis, the latter being more prevalent. Recently, numerous publications have appeared on the induction of somatic embryogenesis in cotton, and the subject reviewed (Gawel and Robacker 1995). Plants regenerated from callus derived from segments of hypocotyl and leaf, and isolated protoplasts transferred to pots/field matured, and some of them bloomed and set bolls (Shoemaker et al. 1986; Stelly et al. 1989; Zhang et al. 1994). In this chapter, our work on the establishment of callus, induction of somatic embryogenesis, and plant regeneration in various species of *Gossypium* is discussed.

2 Establishment of Callus

Callus cultures have been induced from almost every part (i.e., ovary, ovule, embryo, anthers, segments of young seedlings, root, leaf, and isolated protoplasts) of cotton plants. The results are summarized in Table 1, and discussed below.

2.1 Callus from Excised Ovaries, Ovules, and Embryos (Figs. 1, 2)

Young flower buds denuded of petals and sepals (Fig. 1A) cultured on MS (Murashige and Skoog 1962) containg 5 mg/l of 2,4-D started to proliferate, and callus appeared within 2 weeks (Fig. 1B). This callus could be easily

[1] Institute of Agrobiological Genetics and Physiology, Jiangsu Academy of Agricultural Sciences, Nanjing, Jiangsu Province 210014, China
[2] Biotechnology in Agriculture and Forestry, A-137 New Friends Colony, New Delhi 110065, India

Biotechnology in Agriculture and Forestry, Vol. 42
Cotton (ed. by Y.P.S. Bajaj)

Table 1. Some references on the establishment of callus cultures of various species of *Gossypium*

Explant cultured	*Gossypium* spp.	Reference
1. Anthers	*G. arboreum*	Bajaj (1982); Bajaj and Gill (1989, Fig. 3)
2. Ovary	*G. hirsutum*	Bajaj, unpubl., Fig. 1
3. Ovule	*G. hirsutum*	Bajaj (1982); Bajaj and Gill (1985, Fig. 2)
	G. arboreum	
4. Embryo	*G. arboreum*	Gill and Bajaj (1991)
5. Hypocotyl	*G. arboreum*	Smith et al. (1977); Bajaj and Gill (1985);
	G. hirsutum	Shoemaker et al. (1986); Trolinder and Goodin (1988a); Zimmerman and Robacker (1988)
	G. klotzschianum	Price and Smith (1979)
6. Mesocotyl	*G. hirsutum*	Schenk and Hildebrandt (1972)
7. Cotyledon	*G. barbadense*	Katterman et al. (1977); Firoozabady et al. (1987)
	G. davidsonii	Lu and Xia (1991)
8. Leaves	*G. hirsutum*	Davis et al. (1974); Wu et al. (1994)
9. Root	*G. hirsutum*	Kuo et al. (1989)
10. Petiole	*G. hirsutum*	Gawel et al. (1986)
11. Stem	*G. hirsutum*	Saka et al. (1987)
	G. gossypiodes	Tan and Qian (1988)
12. Protoplasts	*G. hirsutum*	She et al. (1989)
		Peeters et al. (1994), see also She et al. (1995), Fig. 4, for more references

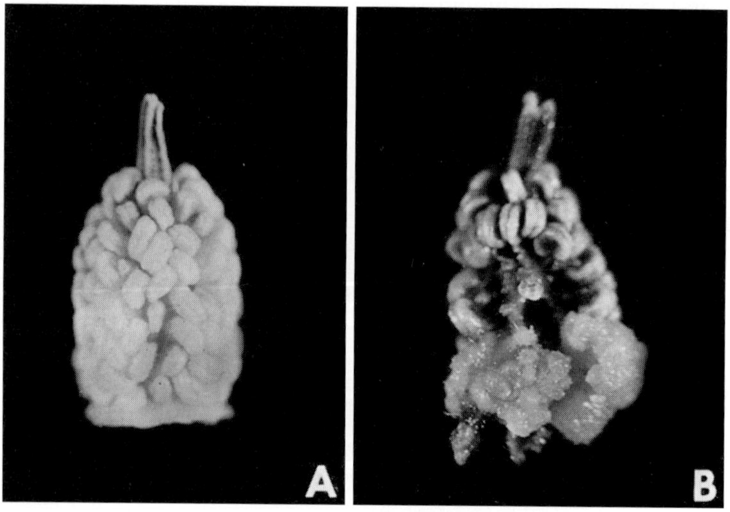

Fig. 1 A,B. Induction of callus on excised ovary (without wall) of *G. hirsutum* cultured on MS + 2,4-D (2 mg/l). **A** At time of culture. **B** Two weeks later, showing callusing. (Bajaj, unpubl)

subcultured, and was maintained on MS + 1 mg/l of 2,4-D. Likewise, young ovules (3 days after pollination) of various crosses (Bajaj and Gill 1985) cultured on MS + IAA (1.5 mg/l) + kin (0.5 mg/l) + casein hybrolysate (250 mg/l) attained double their size within 6 days of culture, and callus forma-

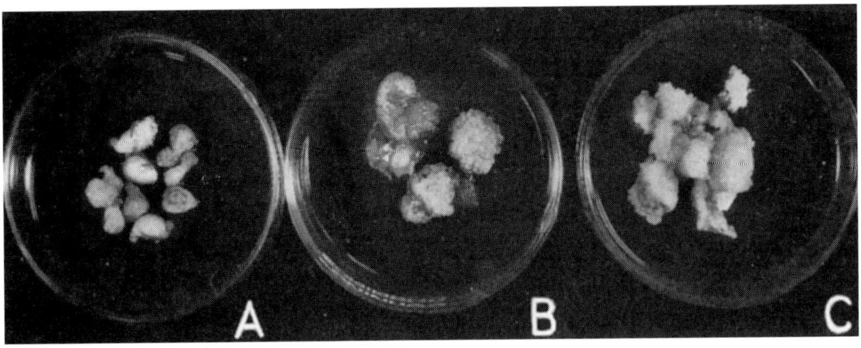

Fig. 2A–C. Ovules of *G. hirsutum* × *G. arboreum* (cultured 3 days after pollination) after 14, 21, and 28 days of culture. (Bajaj and Gill 1985)

tion started within 1 week. There was profuse proliferation in some cases, and the callus was soft and friable (Fig. 2A–C). The callus formed from hybrid ovules with *G. arboreum* cv. G27 as the female parent was pinkish, whereas the ovules with *G. herbaceum* as the female parent formed creamy white callus. In both species, the ovule-derived callus grew faster than that from anthers and the hypocotyl. *Gossypium herbaceum* callus was whitish, whereas that of *G. arboreum* was reddish – the color of the donor variety, G27. (Bajaj and Gill 1985).

The embryos of a cross, *G. hirsutum* × *G. arboreum*, cultured on Smith et al. medium (1977), proliferated to form callus. This was then subcultured periodically on MS + NAA (2 mg/l) + BAP (0.5 mg/l). The callus was creamy white, soft, and friable. Cytological examination of 1-year-old callus showed a wide range of chromosome number (Bajaj and Gill 1985).

2.2 Callus from Excised Anthers (Fig. 3)

The anthers of *G. hirsutum* and *G. arboreum* excised from the flower buds were cultured on MS or SH medium supplemented with various concentrations and combinations of 2,4-D, NAA, BAP, kinetin, and coconut milk (Bajaj and Gill 1989). Within 2 weeks of culture they started to proliferate (Fig. 3; Bajaj 1982), and a mass of subculturable callus was formed within 4 weeks. The callus of *G. arboreum* formed on the medium containing NAA and BAP was compact and had a pinkish tinge; however, on 2,4-D medium it was friable and off-white. Substitution of sucrose by glucose (3%) in the medium resulted in fairly fast growth.

2.3 Callus from Seedling Explants

Segments from in vitro-grown seedlings, i.e., hypocotyl, cotyledons, and mesocotyl, have been used for the induction of callus; however, in general,

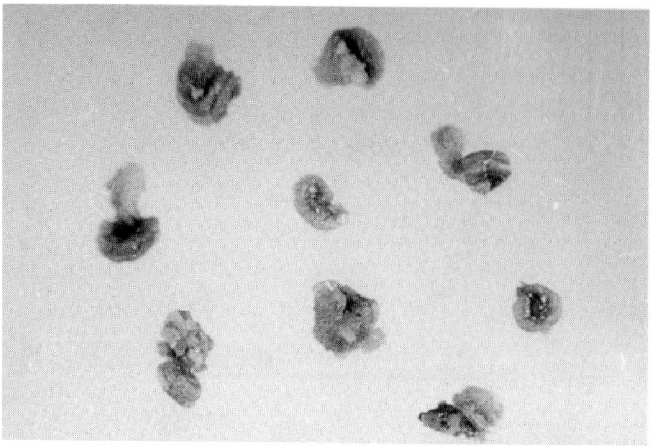

Fig. 3. Callus formation from excised anthers of *G. arboreum* cultured at the uninucleate stage of pollen; age 2 weeks. (Bajaj 1982)

hypocotyls proved to be superior to other segments. After 10–15 days of culture, grayish white callus appeared at the cut ends of the hypocotyl segments in all of the five Chinese varieties of *G. hirsutum*, i.e., Simian 3, Lumian 9554, Chuan 109, Sumian 3118, and Glandless 2031 (Cai et al. 1997). Sometimes, roots appeared along with callus from the cut ends of the hypocotyl. In most of the studies conducted by various workers on regeneration/somatic embryogenesis hypocotyl-derived callus has been used. The *G. herbaceum* callus showed various levels of ploidy (Bajaj and Gill 1985).

2.4 Callus from Protoplasts (Fig. 4)

The protoplasts were released by enzymatic treatment from cell suspension cultures (hypocotyl-derived callus) of *G. hirsutum* cvs. Sumian 3118, Lumian 9554, and Jinmian 4 (Fig. 4A). The first division occurred within 3–4 days of culture, and the second or third division was observed in about 1 week (Fig. 4B). Cell colonies were formed on the 12th day (Fig. 4C). After 20 days of culture, the protoplast-derived calli could be observed with the naked eye (Fig. 4D). Calli of yellow, compact pellets were selected and cultured on MS-1 medium supplemented with 0.1 mg/l each of zeatin, IAA, and 2,4-D (MS-1 = MS salts, B5 vitamins; MS-2 = MS-1, 2 × KNO_3; MS-3 = MS-1, 2 × KNO_3, half NH_4NO_3 – She et al. 1993).

2.5 Callus from Leaf Segments (Fig. 5A)

Young leaves of *G. davidsonii*, *G. raimondii*, *G. herbaceum*, *G. arboreum*, *G. bickii*, and *G. anomalum* were surface sterilized in a saturated bleach solution for 30 min, and washed four times with sterile water. Leaves were cut into pieces of 0.5 cm^2 and cultured. During callus induction and subcultures, leaves

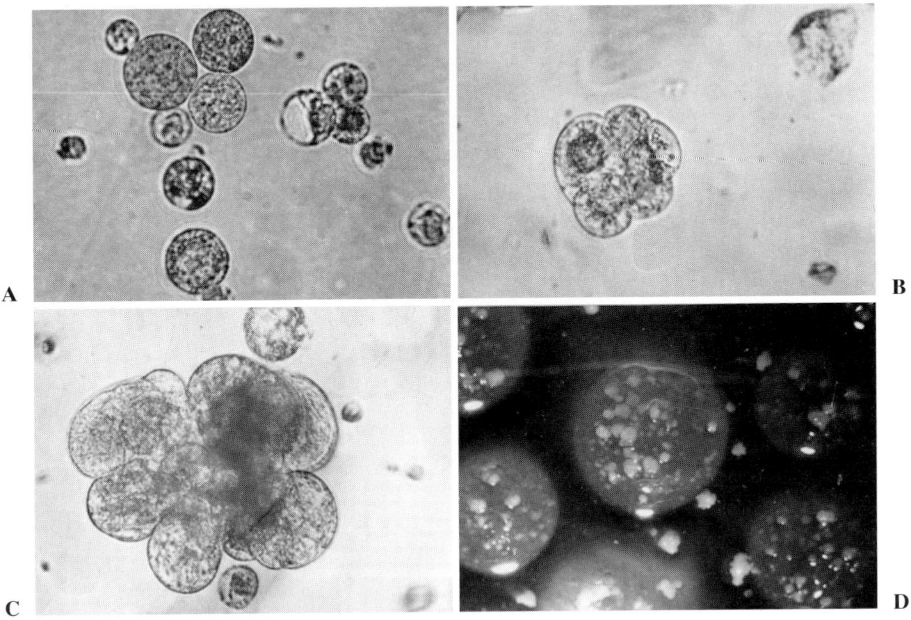

Fig. 4A–D. Callus formation from isolated protoplasts of *G. hirsutum* cultured on K_3 + 2,4-D (0.5 mg/l) + kinetin (1 mg/l). **A** Freshly isolated protoplasts. **B** Third division after 1 week of culture. **C** Cell cluster, after 12 days of culture. **D** Calli after 3 weeks of culture. (She et al. 1989)

and calli were maintained under a 16:8-h light:dark photoperiod with a light intensity of 1000 lx at 28 ± 1 °C.

After 1 week, the leaf segments of *G. davidsonii* cultured on solid MS medium supplemented with 1 mg/l zeatin, 0.2 mg/l IAA, 500 mg/l CH, and 30 g/l sucrose turned from greenish blue to yellowish in color. Yellow embryogenic calli appeared at the edges and vein of leaf segments within 2 weeks. Within 5–6 weeks of culture, embryogenic calli covered both surfaces of the leaf segments (Fig. 5A). On the same medium, embryogenic calli emerged in *G. raimondii*, *G. arboreum* L., and organogenic calli were induced in *G. bickii*.

3 Differentiation of Callus, Somatic Embryogenesis, and Plant Regeneration (Fig. 5)

There are two modes of differentiation of callus in cotton, i.e., through organogenesis by shoot bud formation, and through somatic embryogenesis. Since the first reports of plant regeneration through differentiation of callus (Davidonis and Hamilton 1983), and through somatic embryogenesis (Price and Smith 1979), numerous studies have been conducted. Cotton plants are

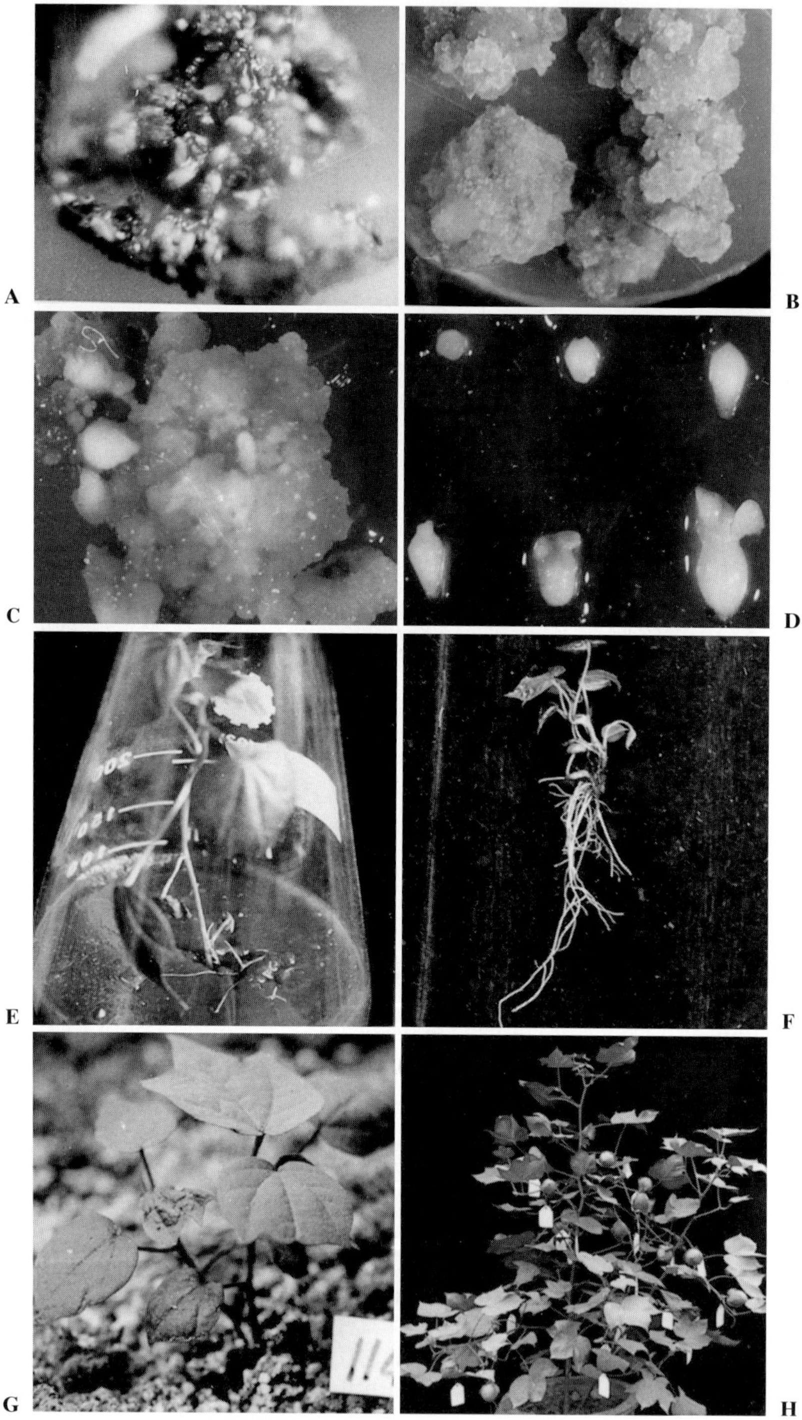

Fig. 5. **A** Embryogenic callus and somatic embryos formed from leaf segments of *G. davidsonii* on MS + zeatin (1 mg/l) + IAA (0.2 mg/l) after 6 weeks of culture. **B–H** *Gossypium hirsutum*. **B** Grayish white calli. **C** Embryogenic calli and embryoids. **D** Globular, heart-shaped, and torpedo embryos. **E** Regenerated plantlet. **F** Strong plantlet in the strengthening medium (Table 4). **G** Transplanted plant (Wu et al. 1990). **H** Fertile plant

also obtained from cultured explants by direct formation of adventitious buds (Tan and Qian 1988) and multiple shoot formation (Bajaj and Gill 1992).

Induction of somatic embryogenesis has been a significant achievement in plant tissue culture. Its applications for the mass production of plants, genetic transformation, and synthetic seed formation have been highlighted by the author (see Bajaj 1995). In cotton, somatic embryogenesis was first reported in a wild species, *G. klotzchianum*, in 1979 by Price and Smith, followed by numerous reports on various cultivars of *G. hirsutum* (Firoozabady and DeBoer 1993), reviewed by Gawel and Robacker (1995) and discussed in this chapter. Genotype plays an important role in determining the induction of somatic embryogenesis in cotton. Trolinder and Goodin (1987, 1988a,b) defined the culture parameters for somatic embryogenesis in the Coker varieties, and suggested the importance of genotype. Trolinder and Chen (1989) subsequently clearly showed that cotton somatic embryogenesis is genotype-dependent. Genetic analysis of cotton somatic embryogenesis sustained that the trait was heritable (Gawel and Robacker 1990). Shang et al. (1991) observed that a pH of 4, 7, or 8 in the medium stimulated somatic embryogenesis. These somatic embryos matured in culture, germinated on sterile vermiculite (Umbeck et al. 1987; Firoozabady et al. 1987), and some of the plants, on transfer to the field, bloomed and set seed (Wu et al. 1990).

Plants have also been regenerated from calli of *G. gossypoides* (Tan and Qian 1988) and *G. davidsonii* (Lu and Xia 1991).

3.1 Plants from Leaf-Derived Callus

In leaf-derived callus, the relation of embryoid induction and subculture generation appeared to be a linear regression, $Y = 26.2 + 3 \times$ ($r = 0.7846$, $p < 0.01$). The callus of *G. davidsonii* subcultured on a medium containing 2,4-D (0.01 mg/l) + IAA (0.1 mg/l) + zeatin (0.1 mg/l) maintained somatic embryogenic potency for 37 months; 44.5% of the leaf segments produced embryoids directly from the yellow pellet embryogenic callus in *G. davidsonii*. Embryoids from callus of *G. davidsonii* and *G. raimondii* developed to the torpedo stage on callus induction medium (Fig. 5D). Embryos germinated and regenerated on MS supplemented with IBA (0.5 mg/l) + kinetin (0.1 mg/l) + GA_3 (0.1 mg/l). The rate of embryo germination of *G. davidsonii* was 47.3%, and 12% resulted in plant regeneration. Gawel et al. (1986) also regenerated plants via somatic embryos from leaf-derived callus of *G. hirsutum*.

3.2 Plants from Hypocotyl-Derived Callus

The grayish-white calli of *G. hirsutum* L. cvs. when transferred from MS-1 medium to the hormone-free MS-2 medium (MS-1 medium, $2 \times KNO_3$), formed yellow pellet proembryoids after 4–6 weeks of culture (Fig. 5C). SH medium (Schenk and Hildebrandt 1972) containing IAA (0.5 mg/l) and kinetin (0.5 mg/l) proved better for embryogenic callus induction than MS-1 medium.

Table 2. Effect of IAA + kinetin on the induction rate of embryogenic callus in *G. hirsutum* L. cv. Simian 3

Hormone combination (mg/l)		No. of calli	No. of calli with embryogenesis potency	Induction rate of embryogenic calli (%)
IAA	kinetin			
0.5	0.5	72	16	22.0
1.0	0.5	56	4	7.1
2.0	0.5	56	2	3.6
4.0	0.5	56	0	0

Table 3. Effect of activated charcoal (AC) on embryoid germination in *G. hirsutum* L. cv. Glandless 2031

Treatment	Total no. of embryoids	No. of germinating embryoids	Rate of embryoid germination (%)
MS	30	2	6.7
MS + AC	56	11	19.6
MS-2	30	0	0
MS-2 + AC	30	5	16.7

While the concentration of kinetin (0.5 mg/l) remained the same, the embryogenesis potency of Simian 3 calli reduced gradually as IAA increased from 0.5 mg/l to 4 mg/l on SH medium (Table 2).

Embryoids of *G. hirsutum* L. cv. Sumian 3118 developed from the globular to the torpedo stage on hormone-free MS-2 medium. The mature embryos germinated on hormone-free White's medium (White 1934) containing 5 g/l activated charcoal (Fig. 5E). The rate of embryo germination was 55.5%, and 20% in plantlet regeneration. Activated charcoal had an outstanding effect on increasing embryo germination in *G. hirsutum* L. cv. Glandless 2031 (Table 3). In most of the studies conducted on the induction of somatic embryos and plant regeneration, hypocotyl-derived callus was used (Shoemaker et al. 1986; Trolinder and Goodin 1988b; Stelly et al. 1989).

3.3 Plants from Protoplast-Derived Callus

Protoplast-derived calli (Fig. 4D) of yellow, compact pellets were selected and cultured on MS-1 medium supplemented with 0.1 mg/l 2,4-D, 0.1 mg/l IAA and 0.1 mg/l zeatin. A number of embryoids presented the white heart or torpedo shape on MS-3 medium (MS-1, 2 × KNO_3, 1/2 NH_4NO_3) containing 0.1–0.5 NAA. The mature embryos developed hypocotyls on hormone-free MS-1 medium, then cotyledons and roots and complete plantlets were formed in *G. hirsutum* L. cvs. Lumian 9554, Sumian 3118, Jinmian 4, and Coker 312 (Chen et al. 1989; She et al. 1989, 1993). Recently, plants were also regenerated from protoplasts of Coker 312 by culturing them on feeder layers (Peeters et al. 1994; see also Chap. I.3, this Vol.).

The rate of plantlet regeneration from mature embryos of cv. Sumian 3118 on solid MS-1 medium was 20%, and 1.2% in liquid MS-1 medium. Through the strengthening culture medium, the regenerated plantlets had stronger growth potential and prosperous roots. In different seasons, the survival frequency of transplantation was 70–80%.

4 Transfer of Plants to the Field

Generally, the survival rate of regenerated plantlets after transplantation was rather low (1.5–5.5%). However, when the regenerated plantlets were first transferred to the strengthening medium (Table 4), strong plants with dark green leaves and prosperous roots were formed (Fig. 5F). The number of leaves, stem nodes, and plant height of these plants was 3–13.5 times that of the control. Of 56 (Sumian 3118) strengthened plants, 44 survived (Fig. 5G) under natural field conditions. The survival rate was 78.8%, and 72% of these plants bloomed and set bolls (Fig. 5H; Wu et al. 1990). However, Stelly et al. (1989) reported that fertility in callus-derived plants of *G. hirsutum* was generally poor, and 84% failed to set bolls.

5 Summary

Callus cultures have been established from almost every part (i.e., ovary, ovule, embryo, anthers, segments of young seedlings, excised roots, leaf, and isolated protoplasts) of various species of *Gossypium* plants on media fortified with hormones. Callus underwent two modes of differentiation, i.e., organogenesis through shoot bud formation, and somatic embryogenesis, the latter being more common. Extensive studies have been conducted on the induction of somatic embryos, and factors affecting differentiation of callus have been studied. The buds and somatic embryos eventually developed into complete plants. Plants were regenerated from embryogenic calli of *G. davidsonii*, *G. raimondii*, and *G. hirsutum* cvs. Simian 3, Lumian 9554, Chuan

Table 4. Composition of strengthening culture medium

Constituent	mg/l
KNO_3	81
$Ca(NO_3)_2 \cdot 4H_2O$	98
KH_2PO_4	12
$MgSO_4 \cdot 7H_2O$	96
$FeSO_4 \cdot 7H_2O$	2.8
Na-EDTA	3.7
Activated charcoal	5000 (0.5%)
Glucose	10000 (1%)

109, Sumian 3118, and Glandless 2031 through somatic embryogenesis. On transfer to the pots/field, the plants showed a survival rate of about 78%, of which 72% set flowers and bolls.

References

Bajaj YPS (1982) Survival of anther-, and ovule-derived cotton callus frozen in liquid nitrogen. Curr Sci 51:139–140

Bajaj YPS (1995) Somatic embryogenesis and its applications for crop improvement. In: Bajaj YPS (ed) Somatic embryogenesis and synthetic seed I. Biotechnology in agriculture and forestry, vol 30. Springer, Berlin Heidelberg New York, pp 105–125

Bajaj YPS, Gill MS (1985) In vitro induction of genetic variability in cotton (*Gossypium* spp.). Theor Appl Genet 70:363–368

Bajaj YPS, Gill MS (1989) Pollen embryogenesis and chromosomal variation in anther culture of a diploid cotton (*Gossypium arboreum* L.). SABRAO J 21:57–63

Bajaj YPS, Gill MS (1992) Micropropagation of cotton (*Gossypium* species). In: Bajaj YPS (ed) Biotechnology in agriculture and forestry, vol 19. High-tech and micropropagation III. Springer, Berlin Heidelberg New York, pp 483–504

Cai XN, Wu JY, She JM (1997) Induction of embryogenic callus and plant regeneration from upland cotton. Jiangsu Agric Sci (in press)

Chen Zhixian, Li SJ, Yue JX, Liu SX, She JM, Wu JY, Wang HB (1989) Plantlet regeneration from protoplasts isolated from embryogenic suspension culture of cotton (*Gossypium hirsutum* L.). Acta Bot Sin 31:966–969

Davidonis GH, Hamilton RH (1983) Plant regeneration from callus tissue of *Gossypium hirsutum* L. Plant Sci Lett 32:89–93

Davis DG, Dusbabek KE, Hoerauf RA (1974) In vitro culture of callus tissues and cell suspensions from okra (*Hibiscus esculentus* L.) and cotton (*Gossypium hirsutum* L.). In Vitro 9:395–398

Firoozababy E, DeBoer DL (1993) Plant regeneration via somatic embryogenesis in many cultivars of cotton (*Gossypium hirsutum* L.). In Vitro Cell Dev Biol 29:166–173

Firoozababy E, DeBoer D, Merlo D, Halk E, Amerson L, Rashka K, Murray E (1987) Transformation of cotton (*Gossypium hirsutum* L.) by *Agrobacterium tumefaciens* and regeneration of transgenic plants. Plant Mol Biol 10:105–116

Gawel NJ, Robacker CD (1990) Genetic control of somatic embryogenesis in cotton petiole callus cultures. Euphytica 49:249–254

Gawel NJ, Robacker CD (1995) Somatic embryogenesis in cotton (*Gossypium* spp.). In: Bajaj YPS (ed) Biotechnology in agriculture and forestry, vol 31. Somatic embryogenesis and synthetic seed II. Springer, Berlin Heidelberg New York, pp 256–266

Gawel NJ, Rao AP, Robacker CD (1986) Somatic embryogenesis from leaf and petiole callus cultures of *Gossypium hirsutum* L. Plant Cell Rep 5:457–459

Gill MS, Bajaj YPS (1987) Hybridization between diploid (*Gossypium arboreum*) and tetraploid (*Gossypium hirsutum*) cotton through ovule culture. Euphytica 36:625–630

Gill MS, Bajaj YPS (1991) In vitro hybridization between *Gossypium arboreum* and *G. anomalum* and their backcross progeny. Plant Tissue Cult (Dhaka) I(2):101–108

Katterman F, Williams MD, Clay WF (1977) The influence of a strong reducing agent upon the initiation of callus from the germinating seedlings of *Gossypium barbadense* L. Physiol Plant 40:198–200

Kuo CC, Zurakowski K, Carlson K, Kilner P, Wang A (1989) Plant regeneration from root-derived callus of *Gossypium hirsutum* L. In: Brown JM (ed) Proc Beltwide Cotton Production Res Conf, Natl Cotton Council of America, pp 638–640

Lu ZX, Xia ZA (1991) Study on tissue and protoplast culture of wild cotton (*Gossypium davidsonii*). Acta Bot Sin 33(2):98–103

Marashige T, Skoog F (1962) A revised medium for rapid growth and bio-assays with tobacco tissue cultures. Physiol Plant 15:463–497

Peeters MC, Willems K, Swennen R (1994) Protoplast-to-plant regeneration in cotton (*Gossypium hirsutum* L. cv. Coker 312) using feeder layers. Plant Cell Rep 13:208–211

Price HJ, Smith RH (1979) Somatic embryogenesis in suspension culture of *Gossypium klotzschianum* Anderse. Planta 145:305–307

Saka K, Katterman RF, Thomas JC (1987) Cell regeneration and sustained division of protoplasts from cotton (*Gossypium hirsutum* L.). Plant Cell Rep 6:470–472

Schenk RU, Hildebrandt AC (1972) Medium and techniques for induction and growth of monocotyledonous and dicotyledonous plant cell cultures. Can J Bot 50:199–204

Shang XM, Huang JY, Haigler CH, Trolinder NL (1991) Buffer capacity of cotton cells and effects of extracellular pH on growth and somatic embryogenesis in cotton cell suspensions. In Vitro Cell Dev Biol 27P:147–152

She JM, Wu JY, Wang HB, Zhou HY, Chen ZX, Li SJ, Yue JX (1989) Somatic embryogenesis and plant regeneration from protoplast culture of cotton. Genet Manipul Plants 5(2):63–68

She JM, Wu JY, Zhou HY (1993) High frequency of plating efficiency and plant regeneration from protoplasts of cotton (*Gossypium hirsutum* L.). Acta Gen Sin 20(6):536–540

She JM, Wu JY, Zhou HY (1995) Regeneration of plants from protoplasts of *Gossypium* species (Cotton). In: Bajaj YPS (ed) Plant protoplasts and genetic engineering VI. Biotechnology in agriculture and forestry, vol 34. Springer, Berlin Heidelberg New York, pp 63–69

Shoemaker RC, Couche LJ, Galbraith DW (1986) Characterization of somatic embryogenesis and plant regeneration in cotton (*Gossypium hirsutum* L.). Plant Cell Rep 3:178–181

Smith RH, Price HJ, Thaxton JB (1977) Defined conditions for the initiation and growth of cotton callus in vitro I. *Gossypium arboreum*. In Vitro 13:329–334

Stelly DM, Altman DW, Kohel RJ, Rangan TS, Commiskey E (1989) Cytogenetic abnormalities of cotton somaclones from callus cultures. Genome 32:762–770

Tan XL, Qian YG (1988) Effect of explant sources and cultural condition on plant regeneration in *Gossypium gossypioides* (Ulbrich) Standley. Acta Gen Sin 15(2):81–85

Trolinder NL, Chen ZX (1989) Genotype specificity of the somatic embryogenesis respones in cotton. Plant Cell Rep 8:133–136

Trolinder NL, Goodin JR (1987) Somatic embryogenesis and plant regeneration in cotton (*Gossypium hirsutum* L.). Plant Cell Rep 6:231–234

Trolinder NL, Goodin JR (1988a) Somatic embryogenesis in cotton (*Gossypium*) 1. Effect of source of explant and hormone regime. Plant Cell Tissue Organ Cult 12:31–42

Trolinder NL, Goodin JR (1988b) Somatic embryogenesis in cotton (*Gossypium*). II. Requirements for embryo development and plant regeneration. Plant Cell Tissue Organ Cult 12:43–53

Umbeck P, Johnson G, Barton K, Swain W (1987) Genetically transformed cotton (*Gossypium hirsutum* L.) plants. Bio/Technology 5:263–266

White PR (1934) Potentially unlimited growth of excised tomato root tips in a liquid medium. Plant Physiol 9:585–600

Wu JY, She JM, Zhou HY (1990) Effects of the growth of regenerated plantlets from somatic cell cultures on transplantation in *Gossypium hirsutum* L. Jiangsu J Agric Sci 6(Suppl):33–38

Wu JY, She JM, Zhu WM, Cai XN, Wu HM (1994) Bud cluster formation and plantlet regeneration from in vitro leaves of Simian 3 (*Gossypium hirsutum* L.). Jiangsu Agric Sci 2:18–19

Zhang JM, Sun JZ, Liu JL, Zhang XL (1994) Studies on plant regeneration from somatic cells and transferring technique of plantlets in upland cotton. Acta Agron Sin 20(2):210–216

Zimmerman TW, Robacker CD (1988) Media and gelling agent effect on cotton callus initiation from excised seed hypocotyls. Plant Cell Tissue Organ Cult 15:269–274

I.3 Regeneration of Plants from Cotton Protoplasts

M.-C. PEETERS and R. SWENNEN[1]

1 Introduction

Protoplasts are recognized to be ideal for genetic transformation as direct DNA transfer mediated via polyethylene glycol, electroporation or microinjection can result in transgenic plants (see Bajaj 1996). *Agrobacterium*-mediated transformation and particle bombardment of target tissue followed by regeneration through somatic embryogenesis, and particle bombardment of shoot tips are the techniques used to transform cotton (Firoozabady et al. 1987; Umbeck et al. 1987; Finer and McMullen 1990; Perlak et al. 1990; Bayley et al. 1992; McCabe and Martinell 1993, also see Chaps. V.1–3, this Vol.). Both techniques, however, have some limitations. Indeed, the regeneration through somatic embryogenesis is genotype-dependent (Trolinder and Chen 1989). Therefore only *Gossypium hirsutum* L. cv. Coker could be transformed, resulting in time-consuming and laborious backcrossing to introduce the novel characteristic of transgenic Coker into commercial cultivars. Alternatively, meristem transformation is genotype-independent, but transformation frequency is very low (0.001%; McCabe and Martinell 1993). Also, because frequency of chimera formation is high, meristem transformation is not appealing. Transformed protoplasts offer the prospect of producing nonchimeric plants, as they regenerate from a single cell. The single-cell origin of regenerated plants is also useful in the development of mutants with valuable characteristics resulting from somaclonal variation or from induced mutagenesis during tissue culture.

A rich genetic potential of useful characteristics such as fiber quality, disease and pest resistance, and drought and salt tolerance still remains untapped in wild cotton species and relatives. However, further genetic improvement of cotton through interspecific hybridization is cumbersome because of incompatibility barriers. By in vitro culture of fertilized ovules, interspecific hybrids have been obtained from an otherwise incompatible cross between a diploid (*G. arboreum*) and a tetraploid (*G. hirsutum*) cultivated cotton (Gill and Bajaj 1987), but, these hybrids were sterile. Fertile in vitro hybrids between *G. arboreum* and *G. anomalum* using ovule culture were obtained; however, this was only for crosses which also result in sterile hybrids when

[1] Laboratory of Tropical Crop Improvement, Catholic University of Leuven, Kardinaal Mercierlaan 92, B-3001 Heverlee, Belgium

Biotechnology in Agriculture and Forestry, Vol. 42
Cotton (ed. by Y.P.S. Bajaj)

Table 1. Different protoplast isolation protocols in cotton

Species/cultivar	Donor tissue	Medium	Enzymes	Digestion condition	Digestion time (h)	Isolation method	Yield ($\times 10^5$ pp/g)	Viability (%)	Reference
G. hirsutum	Hypocotyl callus	11% mannitol inorganic salts pH 5.6	4% Meicelase P 0.5% Driselase 0.4% Macerozyme	Darkness 31 °C	3	Centrifugation (100 *g*, 10 min) sieving, sucrose flotation	4.6	–	Bhojwani et al. (1977)
G. klotzschianum	Hypocotyl callus	0.7 M mannitol pH 4.7	1% Cellulysin 0.5% Macerase	Darkness 27 ± 2 °C 40 rpm	14	Sieving, centrifugation (100 *g*, 5 min)	15	–	Finer and Smith (1982)
G. hirsutum Stoneville 2B	Anther callus	0.71 M mannitol pH 5.6	2% Cellulase 0.5% Pectinase	Darkness 29 °C	2–3	Sucrose flotation, centrifugation (100 *g*, 2 min)	20	–	Thomas and Katterman (1984)
G. hirsutum 3 cvs. *G. barbadense* 2 cvs.	Preincubated cotyledons	MS-modified 2% glucose 7% mannitol pH 5.7	1% Cellulysin 0.5% Macerase	Vacuum infiltration, darkness 30 °C 50 rpm	17	Sieving, centrifugation (100 *g*, 5 min), sucrose flotation	25	70	Firoozabady and DeBoer (1986)
G. hirsutum Deltapine	Stem-callus	0.7 M mannitol pH 5	0.5% Cellulase 0.6% Macerase	Darkness 25 °C 40 rpm	6	Sieving, sucrose flotation, centrifugation (121 *g*, 5 min)	–	70	Saka et al. (1987)
G. hirsutum 8 cvs.	Embryogenic suspension	CPW9	1–1.5% Cellulase 0.5–0.7% Semi-cellulase 0.5–0.7% Macerozyme	Darkness 28 °C 50 rpm	12–16	Sieving, centrifugation (800 rpm, 5–10 min), sucrose flotation	100–150	90	Chen et al. (1989) She et al. (1993)
G. hirsutum Coker 312	Embryogenic hypocotyl suspension	MS-modified 7% mannitol 2% glucose pH 5	0.5% Cellulase 0.25% Macerozyme	Darkness 28 °C 50 rpm	18	Sieving, centrifugation (146 *g*, 5 min), sucrose flotation	40	90	Peeters et al. (1994)

Table 2. Different protoplast culture protocols in cotton

Species/cultivar	Donor tissue	Culture technique	Medium	Plant growth regulators mg/l	Plating density $\times 10^5$ pp/ml	Plating efficiency (%)	Growing conditions	Type of regenerant	Reference
G. hirsutum	Hypocotyl callus	Liquid	B5 0.5% glucose 1% sucrose 11% mannitol pH 5.6	0.5 2,4-D 0.2 NAA 0.2 BAP	1	1.2	Darkness 27–29 °C	Macrocolonies (25–30 cells)	Bhojwani et al. (1977)
G. klotzschianum	Hypocotyl callus	Plating in thin layer	MS-modified 1.5% glucose 0.6 M mannitol 0.3% agar pH 5.7	1 2iP 10 NAA	0.1–1	30	4 days darkness, then 50 µE/m²/s during 16 h per day 29 °C	Macrocolonies (40–50 cells)	Finer and Smith (1982)
G. hirsutum Stoneville 2B	Anther callus	Liquid	B5 0.71 M glucose pH 5.6	0.5 NAA 0.5 2iP	–	–	0.5 µ/E/m²/s 29 °C	Callus	Thomas and Katterman (1984)
G. hirsutum 3 cvs.	Preplasmolyzed cotyledon	Liquid	MS-modified 7% mannitol	0.2 BAP 0.2 kinetin	0.25	0–13	Darkness 30 °C	Microcolonies (2–3 cells)	Firoozabady and DeBoer (1986)
G. barbadense 2 cvs.			2% glucose 0.3% galactose pH 5.7	0.5 2,4-D 1 NAA 1 zeatin		20–25		Microcolonies (5–8 cells)	
G. hirsutum Deltapine 16	Stem callus	Liquid or plating or droplet	K3-modified 0.35 M sucrose	0.2 BAP 0.96 IAA	5	2–4	4 days darkness, then 50 µE/m²/s during 16 h per day 30 °C	Callus	Saka et al. (1987)
G. hirsutum Coker 312, Coker 201 J4 and Lu9554, JiHe312, Lul	Embryogenic cell suspension	Liquid thin layer or embedded in agarose	K3-modified with Km8P organics 9% mannitol 3% glucose	0.1 2,4-D 0.5 2iP or 0.1 KIN	5	2.5	Darkness 28 ± 2 °C	Plant Microcolonies	Chen et al. (1989) She et al. (1993)
G. hirsutum Coker 312	Embryogenic cell suspension	Feeder layer	MS-modified 7% mannitol 2% glucose pH 5	0.2 BAP 0.2 kinetin 0.5 2,4-D 1 NAA or without	0.5	12	Darkness 28 ± 2 °C	Fertile plant	Peeters et al. (1994 and unpubl.)
G. hirsutum Coker 312	Cotyledons	Feeder layer	MS-modified 9% mannitol 2% glucose pH 5.8	0.2 BAP 0.2 kinetin 0.5 2,4-D 1 NAA or without	0.5	<0.1	Darkness 28 ± 2 °C	Microcolonies (5–10 cells)	Peeters et al. (unpubl.)

conventional breeding techniques such as backcrossing with *G. arboreum* are used (Gill and Bajaj 1991). Fusion of protoplast through in vitro hybridization offers prospects for genetic improvement. These hybrids can then be used in classical cotton-breeding programs.

Numerous studies have been conducted on the isolation and culture of cotton protoplasts, and the subject has been recently reviewed (She et al. 1995). Various studies have been summarized in Tables 1 and 2.

2 Isolation of Protoplasts

Much research has gone into medium composition and physical isolation conditions to improve yield and viability (Table 1). Highest yield and viability are obtained for protoplasts isolated from embryogenic cell suspensions of *G. hirsutum* (Chen et al. 1989; She et al. 1993; Peeters et al. 1994). However, large differences exist in protoplast yield and viability between the different *G. hirsutum* cv. Coker 312 hypocotyl-derived cell lines (M.-C. Peeters et al., unpubl.). Yield ranges between 4×10^6 and 10^4 protoplasts/g cells. In general, dark-colored cell suspensions are less productive than pale yellow suspensions. A positive correlation is also observed between yield and viability. High-yielding cell lines maintain their high protoplast yield but regeneration capacity is lost after 2–3 years of subculturing. High yield with 70% viability was obtained for cotyledon-derived protoplasts (Firoozabady and DeBoer 1986). Yields up to 5×10^6 protoplasts/g cotyledon fresh weight and viability up to 94% were obtained (M.-C. Peeters et al., unpubl.). Genotype, quality, and seedling age influence protoplast yield. However, great variability is observed between isolation experiments under identical conditions, indicating that yield is influenced by unknown factors. Yield and viability of callus-derived protoplasts was lowest in all reported protocols (Bhojwani et al. 1977; Finer and Smith 1982; Saka et al. 1987), except for anther callus as protoplast donor (Thomas and Katterman 1984).

Cellulase (0.5–2%) and Macerozyme (0.4–0.7%) are commonly used as digestion enzyme (Table 1). In *G. barbadense* cv. GSA 78, cotyledon protoplast yield increased with the concentration of Cellulysin up to 1% and Macerase up to 0.5%. At higher concentrations, enzyme solutions were toxic, and protoplast viability was significantly reduced (Firoozabady and DeBoer 1986). Pectinase in combination with cellulase was used to release protoplasts from anther callus (Thomas and Katterman 1984). Addition of pectinase to the digestion medium, however, had no enhancing influence on the yield of protoplasts released from cotyledons or cell suspensions of *G. hirsutum* cv. Coker 312 (unpubl.).

The isolation of cotton anther callus protoplasts is greatly enhanced when the amino acids arginine, serine, or glycine, or the divalent cations Ca^{2+} or Mg^{2+} are included in the enzyme mixture (Thomas and Katterman 1984).

These compounds prevent the spontaneous lysis of the protoplasts caused by the RNase found in the cellulase enzyme mixture used.

A combination of 7–9% mannitol and 2–3% glucose is used as osmoticum for digestion of cell suspension and cotyledon protoplasts (Firoozabady and DeBoer 1986; Chen et al. 1989, 1993; Peeters et al. 1994). In the absence of a C-source, a higher concentration (0.7 M) of mannitol is used (Finer and Smith 1982; Thomas and Katterman 1984; Saka et al. 1987).

Digestion time varied between 3 and 18 h (Table 1). An increase in incubation time from 12 to 18 h had no significant influence on yield of protoplasts released from cotyledons and cell suspensions of Coker 312, but reduced viability in suspension-derived protoplasts (unpubl.). In all reports, digestion was done in darkness at 28 ± 2 °C under slight agitation (40–50 rpm). Even low light intensity (3–5 $\mu E/m^2/s$) significantly depressed the yield of protoplasts isolated from anther callus (Thomas and Katterman 1984).

After digestion, protoplasts are isolated by filtration and centrifugation. An additional purification, which is beneficial for further culture, is often done by flotation on sucrose.

3 Culture of Protoplasts

Several factors like genotype, donor tissue, medium composition, culture technique, and environmental culture condition influence protoplast culture (Table 2). An association between donor tissue and protoplast response could be demonstrated. Lowest response, i.e., microcolonies consisting of a few cells, was obtained from protoplasts released from cotyledons (Firoozabady and DeBoer 1986). Macrocolonies of more than 25 cells (Bhojwani et al. 1977; Finer and Smith 1982; Saka et al. 1987) or callus (Thomas and Katterman 1984) were obtained from protoplasts released from callus. Plant regeneration was obtained only from protoplasts isolated from embryogenic cell suspensions (Chen et al. 1989, 1993; Peeters et al. 1994). Plating efficiency up to 12% was obtained for protoplasts isolated from embryogenic cell suspensions of *G. hirsutum* L. cv. Coker 312 when cultured on a feeder layer (Peeters et al. 1994). Plating efficiency also depends upon the cell line used as protoplast donor, and decreases with its age (M.-C. Peeters et al., unpubl.).

Firoozabady (1986) studied the relationship between cell-cycle stage and growth and development of isolated cotyledon protoplasts in *G. hirsutum* cv. Gisa 75. Cell-cycle parameters of protoplasts are very similar to those of donor tissue. In cotyledons, as well as in cotyledon-derived protoplasts, the proportion of cells in the G_1 phase increased with seedling age. In addition, a positive correlation between G_2 or S and G_2 and cell-wall regeneration and initial cell division was demonstrated, suggesting that young cotyledons from rapidly growing seedlings are the best donor tissue material for protoplast culture.

Several culture techniques like liquid medium, embedding in agarose, plating on semisolid medium, or thin layer have been used. Chen et al. (1989,

1993) obtained plant regeneration using culture in liquid or in 1% agarose. Feeder layers are commonly used to improve the protoplast culture efficiency in recalcitrant species or to reduce the protoplast density required for culture. This technique was successfully applied by Peeters et al. (1994) to regenerate plants from protoplasts in Coker 312. Embryogenic cell suspensions of Coker 312 were used as feeder layer.

The modified Murashige and Skoog (1962) medium with 2–3% glucose and 7–11% mannitol is the commonly used culture medium (Table 2). All media contain cytokinin and auxin as plant growth regulator. By using the feeder-layer technique in cotton protoplast culture, however, exogenous plant growth regulators have no stimulating effect on the plating efficiency, i.e., the percentage of original plated protoplasts that formed microcolonies (Peeters et al. 1994). The nature of the growth-promoting factors released by the feeder cells is still unknown.

Initially, all protoplast cultures are kept in darkness at 28 ± 2 °C. Transfer to light after 4 days is sometimes done (Finer and Smith 1982; Saka et al. 1987). Plant regeneration was obtained for cultures which were kept in darkness for a longer time (Chen et al. 1993; Peeters et al. 1994).

Polyploidization was observed in some protoplast-derived cell lines of Coker (M.-C. Peeters et al. unpubl.). This polyploidization probably started during the initial stages of protoplast culture as a spontaneous fusion of nuclei because a high frequency of multinucleate protoplasts is observed in freshly isolated protoplast suspensions.

4 Regeneration of Plants

To the authors' knowledge, plant regeneration in cotton is reported only for protoplasts released from embryogenic cell suspensions (Chen et al. 1989; She et al. 1993; Peeters et al. 1994). Because plant regeneration through somatic embryogenesis was demonstrated to be highly genotype-dependent and superior in Coker 312 and 201 compared to 38 other tested cultivars, protoplast-to-plant regeneration was first obtained in these cultivars. Chen et al. (1993) reported plant regeneration for two additional cultivars, J4 Jiangsu 3118 and Lu9554.

The regeneration capacity of the cell suspension used as donor for protoplast isolation is positively correlated with the regeneration capacity of the protoplasts. Direct somatic embryogenesis after 2–3 months of culture was occasionally obtained using young, highly regenerable Coker 312 cell suspensions. With aging of the donor suspensions, regeneration capacity of the protoplast-derived suspension decreases. Protoplasts released from cell suspensions which lost their regeneration capacity due to the long culture time developed callus, but no plants (M.-C. Peeters et al., unpubl.).

The production of abnormal somatic embryos is very common in cotton (Firoozabady and DeBoer 1993; Zheng Sijun 1994). This was also observed in

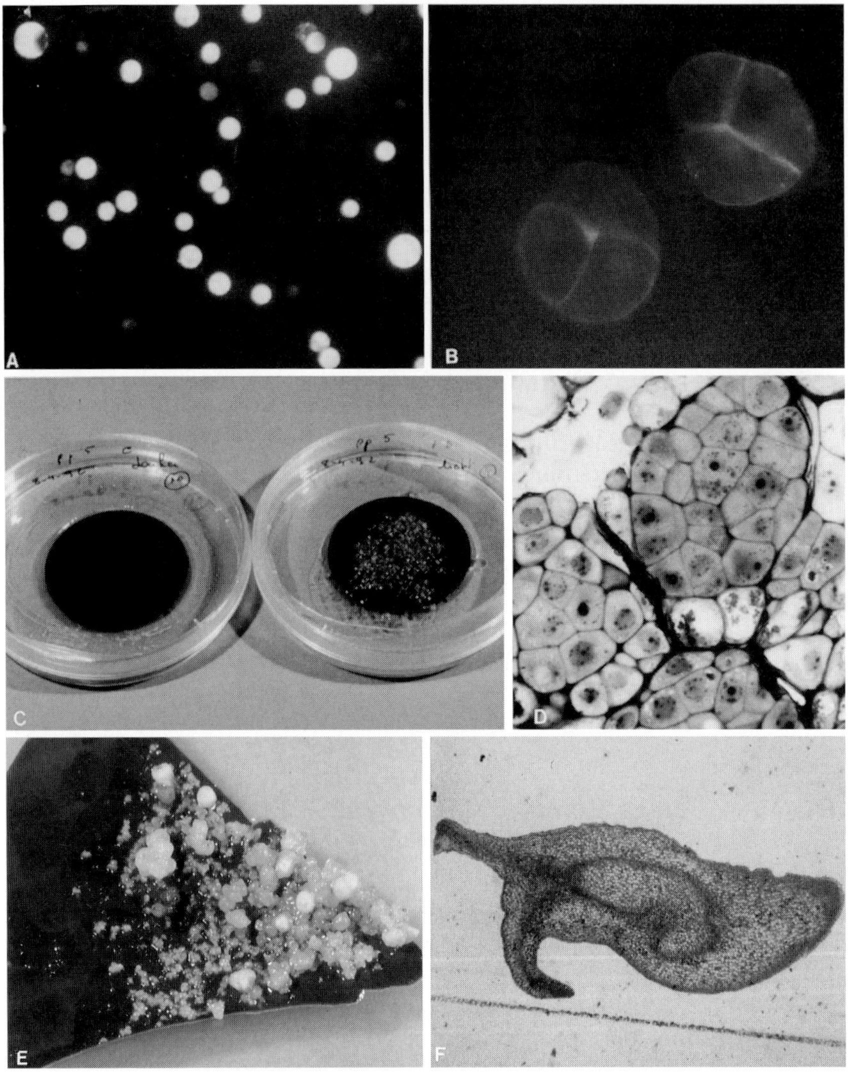

Fig. 1A–I. Protoplast-to-plant regeneration in Coker 312 using the feeder-layer technique. **A** Protoplasts released from embryogenic cell suspensions after 18 h treatment with digestion medium containing MS salts, B5 vitamins, 2% glucose, 7% mannitol, 400 mg/l myoinositol, 1% cellulase and 0.5% Macerozyme, and stained with fluorescein diacetate. **B** First divisions in protoplast-derived cells after 1 week of culture using the feeder-layer technique on semisolid culture medium, which is the same as the digestion medium without enzymes and supplemented with 2% Phytagel, and stained with Calcofluor white. **C** Development of microcallus after 3 weeks of culture using the feeder-layer technique (*right*) and without feeder layer (*left*). **D** Histological section of protoplast-derived callus (cf. **C**). **E** Development of protoplast-derived somatic embryos after 2 months of culture. **F** Histological section of protoplast-derived somatic embryo. **G** Protoplast-derived rooted plantlet 2 months after transfer to tube with germination S medium containing 0.5% agar, 0.15% Gelrite, and 0.5% sucrose. **H** Protoplast-derived plant 2 months after transfer to hydroculture in greenhouse. **I** Protoplast-derived bushy off-type plant

Fig. 1A–I. *Continued*

cotton protoplast culture. Off-types at vegetative (bushy type, lancet leaves, twin bolls) and regenerative level (sterile flowers) were obtained. The frequency of off-types increases with age of the suspension line used as protoplast donor (M.-C. Peeters et al., unpubl.).

Culture technique does not seem to play a decisive role, because plant regeneration was obtained when protoplasts were cultured in liquid medium or embedded in 1% agarose (Chen et al. 1993). Peeters et al. (1994), however, used a feeder layer. The different stages of regeneration using a feeder-layer technique are presented in Fig. 1. Without the feeder layer, which consists of embryogenic cell suspensions of Coker 312, plating effciency was even zero.

Both research groups used a protocol based on the one reported by Trolinder and Goodin (1987) for the further differentiation of the protoplast-derived callus into somatic embryos and plants. Chen et al. (1993) transferred 20 protoplast-derived plantlets to the greenhouse, but only five survived.

5 Summary

Protoplasts have numerous applications in crop improvement, such as direct transformation using electroporation, microinjection or polyethylene glycol, somatic hybridization after protoplast fusion, and mutant selection after stress induction. Protoplast isolation protocols have been developed for different cotton explants. Highest yield is obtained for protoplasts isolated from embryogenic cell suspensions of *G. hirsutum*, although significant differences are also observed between different cell lines. Age of the cell line had no influence on yield and viability. High yield was also obtained for cotyledon-derived protoplasts of different cotton cultivars, but great differences are observed between different isolation experiments under identical conditions.

Plant regeneration is, until now, only feasible for protoplasts isolated from embryogenic cell suspensions. Chen et al. (1993) used a thin-layer and agarose-embedding technique, whereas Peeters et al. (1994) used a feeder-layer technique.

6 Protocol

The protocol presented here is based on the paper of Peeters et al. (1994).

6.1 Cell Suspension Cultures

Cell suspensions of *G. hirsutum* L. cv. Coker 312 are developed from hypocotyl-derived callus as described by Trolinder and Goodin (1987), and recently reviewed by Gawel and Robacker (1995). The suspensions are subcultured bimonthly into a basic medium (BM) containing MS salts (Murashige and Skoog 1962), B5 vitamins, 3% glucose, and 100 mg/l myoinositol. The pH is adjusted to 5.8 prior to autoclaving. The cell suspensions are maintained at 28 °C on an orbital shaker at 70 rpm with a 16 h photoperiod (25 μE/m²/s).

6.2 Protoplast Isolation

Cell suspensions in the exponential growth phase, i.e., 8–14 days after the last subculture, are used for protoplast isolation. The suspension is sieved (100-mesh) and the filtrate is sieved again (60-mesh). The residue is washed and placed in filter-sterilized digestion medium at a concentration of 10 ml/g cells. This digestion medium contains MS salts, B5 vitamins, 400 mg/l myoinositol, 2% glucose, 1.47 g/l $CaCl_2$, 7% mannitol, 0.25–0.5% Macerozyme R10 (Serva, Heidelberg) and 0.5–1% cellulase Onozuka R10 (Serva). Its pH is adjusted to 5. Incubation takes place for 18 h in darkness at 28 °C on an orbital shaker (50 rpm). After incubation, the digestion medium is centrifuged (5 min at 146 *g*). The pellet is gently suspended and rinsed by centrifugation with washing medium, i.e., digestion medium without enzymes. The protoplasts are suspended in 10 ml of a 22% sucrose solution. This solution is overlayed with 1 ml of washing medium and centrifuged (5 min at 146 *g*). The protoplasts are gently collected with a boll pipet from the washing medium-sucrose interface and washed two times by centrifugation with washing medium. The yield of viable protoplasts is determined by counting the number of protoplasts stained with fluorescein diacetate in a Neubauer hemacytometer.

6.3 Protoplast Culture

The protoplasts are suspended at a density of 5×10^4 viable protoplasts/ml in protoplast culture medium. This medium is the same as the protoplast digestion medium without enzymes but with 146 mg/l glutamin. One ml of this protoplast suspension is plated on a black gridded membrane filter (Millipore AABG, 0.8 μm pore size) over a feeder layer. The feeder layer is prepared as follows: 1 ml culture medium containing 0.30 g cell of an embryogenic cell suspension of Coker is placed over 20 ml culture medium solidified with 0.2% Phytagel (Sigma) in 9-cm Petri plates. The plates are wrapped with Nescofilm and incubated at 28 °C in the dark for 1 month.

One month after protoplast isolation, the filter with the plated protoplasts is transferred for callus proliferation on 20 ml semisolid BM containing 0.2% Phytagel in 9-cm Petri plates. The plates are maintained at 28 ± 3 °C with a 16-h/day photoperiod (50–75 μE/m²/s) for 1 month.

6.4 Plant Regeneration

Cell suspensions in BM are initiated from the protoplast-derived callus. Further regeneration is as described by Trolinder and Goodin (1987). The embryogenic cells are transferred for embryo development at low plating density on semisolid maturation medium (MM), which is BM with only half the amount of MS salts and supplemented with 1.9 g/l KNO_3 (Trolinder and Goodin 1987). Somatic embryos are partially dehydrated (Bayley et al. 1992) by placing them during 2 weeks in darkness on an S medium (Stewart and Hsu 1977) containing 2% agar and 3% glucose. Obtained embryos are placed for germination in 25 × 150 mm culture tubes containing S medium (0.5% agar, 0.15% Gelrite, and 0.5% sucrose). The environmental conditions for embryo development and germination are the same as for protoplast-derived callus. Plantlets with a well-developed rooting system and at least two foliage leaves are transferred to hydroculture in the greenhouse.

Acknowledgement. The authors are grateful to Suzy Voets and Wim Dillemans for technical assistance, and to the Nationaal Fonds voor Wetenschappelijk Onderzoek, the Onderzoeksraad K.U. Leuven, and the Vlaams Ministerie van Wetenschapsbeleid for financial support.

References

Bajaj YPS (1996) Plant protoplasts and genetic engineering VII. Biotechnology in agriculture and forestry, vol 38. Springer, Berlin Heidelberg New York

Bayley Y, Trolinder N, Razy C, Morgan M, Quisenberry J, Ow D (1992) Engineering 2,4-D resistance into cotton. Theor Appl Gen 83:645–649

Bhojwani S, Power J, Cocking E (1977) Isolation, culture and division of cotton callus protoplasts. Plant Sci Lett 8:85–89

Chen ZX, Li SJ, Yue JX, Jiao GI, Liu SX, She JM, Wu JY, Wang HB (1989) Plant regeneration from protoplasts isolated from an embryogenic suspension culture of cotton (*Gossypium hirsutum* L.). Acta Bot Sin 31(12):966–969

Chen ZX, She JM, Wu JY, Li SJ (1993) Plant regeneration through somatic embryogenesis from suspension cultures derived protoplasts of six cotton cultivars (*Gossypium hirsutum* L.). Biotechnology in agriculture series. Kluwer, Dordrecht, pp 283–286

Finer J, Smith R (1982) Isolation and culture of protoplasts from cotton (*Gossypium klotzschianum* Anderss.) callus cultures. Plant Sci Lett 26:147–151

Finer J, McMullen M (1990) Transformation of cotton (*Gossypium hirsutum* L.) via particle bombardment. Plant Cell Rep 8:586–589

Firoozabady E (1986) The effects of cell cycle parameters on cell wall regeneration and cell division of cotton protoplasts (*Gossypium hirsutum* L.). J Exp Bot 37:1211–1217

Firoozabady E, DeBoer D (1986) Isolation, culture and cell division in cotyledon protoplasts of cotton (*Gossypium hirsutum* and *G. barbadense*). Plant Cell Rep 5:127–131

Firoozabady E, DeBoer D (1993) Plant regeneration via somatic embryogenesis in many cultivars of cotton (*Gossypium hirsutum* L.). In Vitro Cell Dev Biol 29P:166–173

Firoozabady E, DeBoer E, Merlo D, Halk L, Amerson L, Rashka E, Murray E (1987) Transformation of cotton (*Gossypium hirsutum* L.) by *Agrobacterium tumefaciens* and regeneration of transgenic plants. Plant Mol Biol 10:105–116

Gawel NJ, Robacker CD (1995) Somatic embryogenesis in cotton (*Gossypium* spp.). In: Bajaj YPS (ed) Biotechnology in agriculture and forestry, vol 31. Somatic embryogenesis and synthetic seed II. Springer, Berlin Heidelberg New York, pp 256–266

Gill MS, Bajaj YPS (1987) Hybridisation between diploid (*Gossypium arboreum*) and tetraploid (*Gossypium hirsutum*) cotton through ovule culture. Euphytica 36:625–630

Gill MS, Bajaj YPS (1991) In vitro hybridisation between *Gossypium arboreum* and *G. anomalum* and their backcross progeny. Plant Tissue Cult 12:101–108

McCabe D, Martinell B (1993) Transformation of elite cotton cultivars via particle bombardment of meristems. Bio/Technology 1:596–598

Murashige T, Skoog F (1962) A revised medium for rapid growth and bio-assays with tobacco tissue cultures. Physiol Plant 15:473–497

Peeters M-C, Willems K, Swennen R (1994) Protoplast-to-plant regeneration in cotton (*Gossypium hirsutum* L. cv. Coker 312) using feeder layers. Plant Cell Rep 13:208–211

Perlak F, Deaton R, Armstrong T, Greenplate J, Fischoff D (1990) Insect-resistant cotton plants. Bio/Technology 8:939–943

Saka K, Katterman F, Thomas J (1987) Cell regeneration and sustained division of protoplasts from cotton (*Gossypium hirsutum* L.). Plant Cell Rep 6:470–472

She JM, Wu JY, Zhou HY (1993) High frequency of plating efficiency and plant regeneration from protoplasts of cotton (*Gossypium hirsutum* L.). Acta Gen Sin 20(6):199–204

She JM, Wu JY, Zhou HY (1995) Regeneration of plants from protoplasts of *Gossypium* species (cotton). In: Bajaj YPS (ed) Plant protoplasts and genetic engineering VI. Biotechnology in agriculture and forestry, vol 34. Springer, Berlin Heidelberg New York, pp 63–69

Stewart JMcD, Hsu CL (1977) In-ovulo embryo culture and seedling development of cotton (*Gossypium hirsutum* L.). Planta 137:113–117

Thomas J, Katterman F (1984) The control of spontaneous lysis of protoplasts from *Gossypium hirsutum* anther callus. Plant Sci Lett 36:149–154

Trolinder N, Chen ZX (1989) Genotype specificity of the somatic embryogenesis response in cotton. Plant Cell Rep 8:133–136

Trolinder N, Goodin J (1987) Somatic embryogenesis and plant regeneration in cotton (*Gossypium hirsutum* L.). Plant Cell Rep 6:231–234

Umbeck P, Johnson G, Barton K, Swain W (1987) Genetically transformed cotton (*Gossypium hirsutum* L.) plants. Bio/Technology 5:263–266

Zheng Sijun (1994) Production of abnormal somatic embryos and their transferring into normal seedlings in cotton tissue culture. In: Adams R, Miller J, Golenberg E, Adams J (eds) Conservation of plant genes II: utilization of ancient and modern DNA. Missouri Botanical Garden, St Louis, pp 205–210

I.4 Micropropagation and the Conservation of Germplasm of Cotton

Y.P.S. BAJAJ[1]

1 Introduction

The cotton plant is generally cultivated as an annual crop and is propagated through seed. The varieties suffer rapid deterioration due to natural crossing and through mechanical mixtures during ginning. Therefore, the need for suitable means of vegetative propagation has always been felt. The plants obtained through cloning are true to type and can be used for producing nucleus seed for maintaining the purity of a variety. The conventional method of vegetative propagation in cotton by cuttings (Parmar et al. 1978) shows a slow rate of multiplication. In vitro methods of micropropagation have been used in a number of crop plants, trees, and ornamentals (see Bajaj 1992) to obtain higher rates of multiplication, and these techniques can also be utilized in cotton. Micropropagation (Bajaj and Gill 1992) can be employed for collection and maintenance of germplasm, rapid multiplication of seed of new varieties, production of clones for evaluation against diseases etc., and for propagating male sterile lines.

The conservation of germplasm of cotton is another important aspect. Since the seed suffer deterioration during storage, alternative methods are being experimented. Moreover, with the recent upsurge in the biotechnology of cotton to induce genetic variability through in vitro culture and genetic transformation, methods need to be developed for storage of desirable cell lines/germplasm. In this regard, tissue cultures for short- and medium-term storage, and cryopreservation for long-term conservation are the alternatives (Bajaj 1986).

2 Micropropagation

2.1 Micropropagation Through Shoot-Tip Culture (Fig. 1, Tables 1, 2)

Shoot-tip/meristem culture is one of the simplest methods of in vitro propagation and the plants obtained are true to type. In cotton, the in vitro culture of

[1] Biotechnology in Agriculture and Forestry, A-137 New Friends Colony, New Delhi 110065, India

Biotechnology in Agriculture and Forestry, Vol. 42
Cotton (ed. by Y.P.S. Bajaj)

excised meristems of *G. hirsutum* was first reported by Chappel and Mauney (1967). Although new leaf primordia were initiated, the root system failed to develop.

The author's work (Bajaj and Gill 1986) on shoot tip and meristem cultures of *G. arboreum*, *G. hirsutum*, and three interspecific hybrids, *G. arboreum* × *G. anomalum*, *G. arboreum* × *G. stocksii*, and *G. herbacium* × *G. stocksii*, is briefed here. The shoot tips (approx 1.5 cm) taken from the field-grown plants were surface sterilized in 0.1% mercuric chloride for 5 min, and then washed repeatedly in sterilized distilled water. The shoot tips (1 cm) and meristems (1 mm) were excised aseptically and cultured on MS medium (Murashige and Skoog 1962) supplemented with various combinations of IAA, NAA, kinetin, or BA. The meristems cultured on MS + IAA (0.5 mg/l) + kinetin (1–2 mg/l) grew slowly and formed callus (Fig. 1A). In the case of shoot tips, the unfolding of the leaves occurred after 4–5 days in both the species (Fig. 1C,D). However, after 10–14 days of culture, these leaves either turned brown or remained green, but all withered. New axillary outgrowths were formed after 20–25 days in *G. arboreum* and 26–30 days in *G. hirsutum*. The response of shoot tips cultured on different media is shown in Tables 1 and 2. When IAA was used in combination with kinetin, both the species responded best at the 2 mg/l kinetin level. With an increase in the concentration of kinetin, growth of shoots slowed down and multiple shoots were formed (Fig. 1B). The in vitro-grown shoots (Fig. 1C,D) formed roots in 4 weeks (Fig. 1E) when transferred to the medium containing 2 mg/l IAA. The plants thus obtained continued further growth upon transfer to the soil (Fig. 1F). The shoot tips of interspecific hybrids responded best on MS + kinetin (1 mg/l) and complete plants were regenerated (Table 2), the success rate being 41.6 to 72.7% in different hybrids. *G. arboreum* × *G. anomalum* hybrids gave the best response, and this hybrid also showed the maximum vigor in the field.

Altman et al. (1990) devised an in vitro method for collecting the germplasm of various wild species of *Gossypium*, especially those where viable seeds could not be obtained. The cuttings obtained from various species were surface sterilized in the field itself and cultured on the medium without any further precautions to assure asepsis. The cuttings were transported in the field in the culture tubes for up to 3 weeks. The cultured shoots grew vigorously, and adventitious roots were initiated at the basal ends. Upon return to the laboratory, the cuttings were rinsed with a 4% bleach solution, treated with Rootone F on their basal ends, and placed in a moistened, sterile soil mix of peat: sand: vermiculite (1:2:1) supplemented with dolomite lime and Osmocote slow-release nutrients. The overall response was 10% shoot formation from six *Gossypium* species. The rooting of cuttings was also possible.

Shoot apex and axillary bud culture has also been used to obtain plants of various Chinese cultivars (Wu and She 1990), and transgenic plants of *G. hirsutum* and *G. barbadense* (Gould et al. 1991).

Table 1. Shoot tip culture of *Gossypium arboreum* and *G. hirsutum* on various media. (Bajaj and Gill 1986)

Medium (mg/l)	*G. arboreum*				*G. hirsutum*			
	No. of shoots cultured	No. of growing shoots	Growth (%)	Remarks	No. of shoots cultured	No. of shoots growing	Growth (%)	Remarks
1. MS + IAA (0.5) + kin (1)	11	3	27.2	Good growth and callus	16	7	43.7	Good growth
2. MS + IAA (0.5) + kin (2)	13	5	38.4	Very good growth	13	6	46.2	Very good growth
3. MS + IAA (0.5) + kin (4)	11	4	36.4	Good growth	11	4	36.4	Poor growth
4. MS + IAA (0.5) + kin (6)	13	2	15.4	Multiple shoots	11	3	27.2	Poor growth

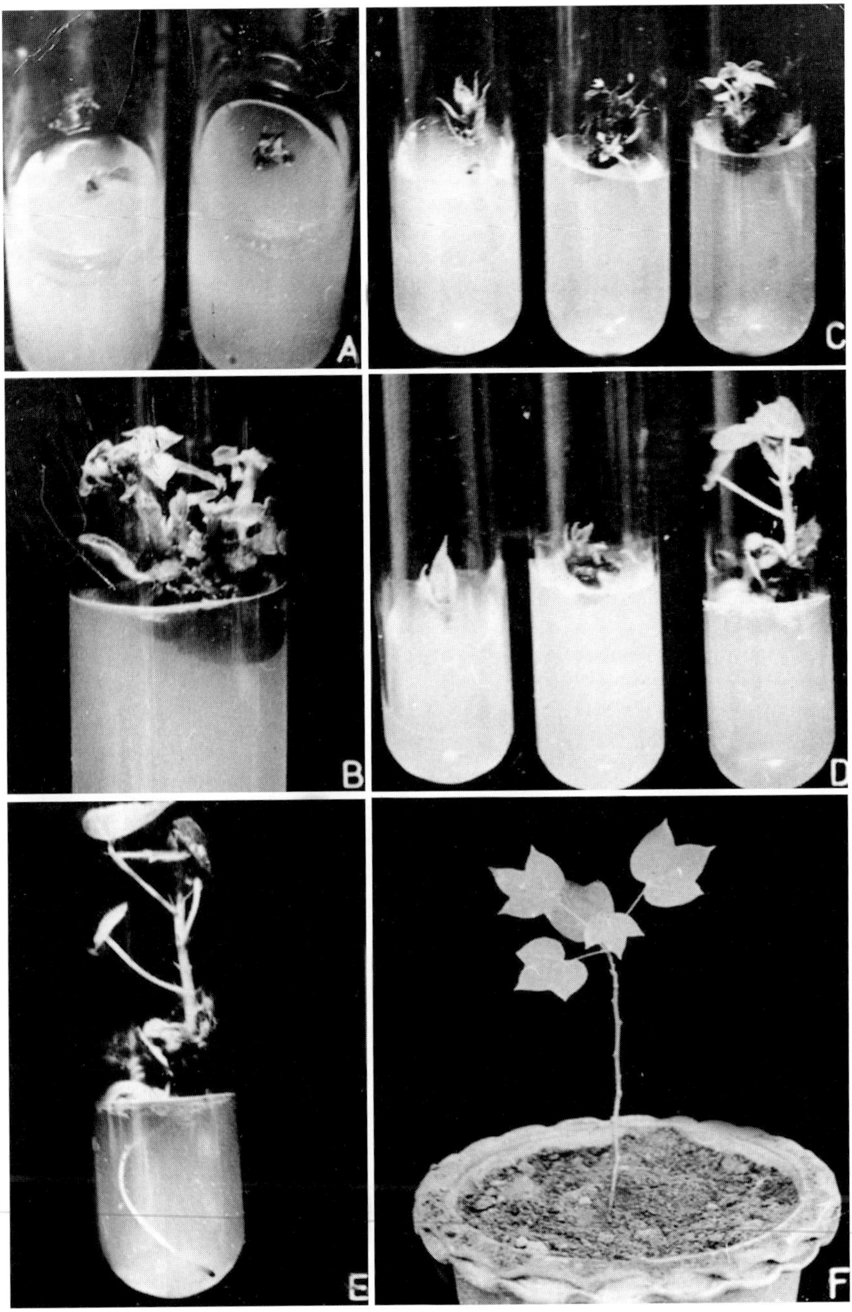

Fig. 1A–F. In vitro culture of shoot tips and meristems of cotton. **A** Two- and 4-week-old cultures of meristems (1 mm) of *Gossypium arboreum* on MS + IAA (0.5 mg/l) + kinetin (1 mg/l). **B** Multiple shoot formation on MS + NAA (0.5 mg/l) + BA (2 mg/l). **C, D** Shoot tip (1 cm) culture of *G. arboreum* and *G. hirsutum* on MS + IAA (0.5 mg/l) + kinetin (2 mg/l), *left* at the time of culture; *right* after 4 weeks. **E** An in vitro-obtained plantlet of *G. hirsutum* in which rooting has been induced on a medium supplemented with IAA (2 mg/l). **F** Same, 3 months after transfer to a pot. (Bajaj and Gill 1986)

Table 2. Shoot-tip culture of various interspecific hybrids of diploid cotton on MS + kinetin (1 mg/l). (Bajaj and Gill 1986)

Cross	No. of explants cultured	No. of shoots growing	Response (%)	Remarks
G. arboreum × *G. anomalum*	11	8	72.7	Very good growth, vigorous shoot formation
G. arboreum × *G. stocksii*	12	5	41.6	Good growth
G. herbaceum × *G. stocksii*	11	6	54.5	Good growth

2.2 Micropropagation Through Embryo and Ovule Culture (Figs. 2, 3, Tables 3, 4)

The technique of embryo/ovule culture can be employed for propagation of species/genotypes from which viable seed cannot be obtained. In our studies (Gill and Bajaj 1984, 1986, 1987, 1991, 1996), plants of *G. hirsutum*, *G. arboreum*, *G. herbaceum*, *G. stocksii*, *G. anomalum*, and various interspecific hybrids were obtained from 3 DAP ovules and 15 DAP embryos cultured on MS supplemented with different concentrations and combinations of IAA, kinetin, coconut milk (CM), casein hydrolysate (CH), sucrose, and agar (0.7%).

The response of three DAP ovules of three species in culture is shown in Table 3. Some of the self-pollinated ovules germinated to produce seedlings even on medium devoid of hormones, CH, and CM. However, germination was considerably enhanced when the medium was supplemented with either CH or CM alone or in combination with IAA and kinetin. On almost all the media, the ovules started to proliferate within 1 week of culture. The ovules from the diploid species produced profuse, friable callus (Fig. 2A). Along with the mass of callus, the ovules kept growing and started germinating after about 50 and 60 days. At the time of germination, the size of the *G. arboreum* or *G. herbaceum* ovules was smaller than those developed in situ, however *G. hirsutum* ovules attained a size almost equal to the in vivo-developed ovules. Seedlings of *G. hirsutum* were also more vigorous. The best response was observed on MS + IAA (1.5 mg/l) + kinetin (0.5 mg/l) + CH (250 mg/l) for *G. arboreum* and *G. herbaceum*, whereas for *G. hirsutum* the best medium was MS + IAA (1 mg/l) + kinetin (0.2 mg/l) + CH (250 mg/l).

The 15 DAP embryos responded better than 12 DAP embryos. In the latter case, the embryos generally turned brown and died, only occasionally forming weak seedlings. In *G. hirsutum* 6% sucrose proved to be better than 3% sucrose for younger embryos. The 15 DAP embryos produced callus or plants on most of the media. The best response with respect to plant formation

Table 3. Percentage of ovules giving seedlings and the callus growth response of excised ovules (3 days after pollination) of various species of *Gossypium* cultured on various media (figures in parentheses indicate the number of ovules cultured). (Gill and Bajaj 1987)

Medium (mg/l)	*G. arboreum* Germination (%)	Callus[a] development	*G. hirsutum* Germination (%)	Callus[a] development	*G. herbaceum*[a] Germination (%)	Callus[a] development
MS basal	17.6 (125)	+	8.3 (96)	–	7.9	+
MS + CH 250	29.2 (96)	+	22.2 (99)	–	18.8	+
MS + CW 7%	21.2 (80)	+	25.4 (110)	–	18.0	+
MS + CH 250 + IAA 0.5 + kin 0.2	22.2 (108)	+	23.8 (122)	–	16.6	+
MS + CH 250 + IAA 0.5 + kin 0.5	23.7 (93)	+	25.0 (120)	–	12.5	+ +
MS + CH 250 + IAA 1 + kin 0.2	28.1 (121)	+	44.4 (133)	+ +	22.5	+ + +
MS + CH 250 + IAA 1 + kin 0.5	20.3 (103)	+ + +	24.1 (108)	+ +	19.4	+ + +
MS + CH 250 + IAA 1.5 + kin 0.5	37.5 (112)	+ + +	18.8 (117)	+ +	25.0	+ + +

[a] – No callus; + little callus; + + good callus; + + + profuse callus.

Fig. 2A–D. In vitro culture of excised ovules of *G. arboreum*, and the transfer of plants to field. **A** A mass of callus obtained from an ovule-cultured 3 DAP on MS + IAA (2 mg/l) + kinetin (1 mg/l), 4-week-old culture. **B, C** Plantlets obtained from excised ovules initially cultured 3 DAP on MS + IAA (1.5 mg/l) + kinetin (0.5 mg/l) + CH (250 mg/l), and then transferred to half-strength MS + 1% sucrose. **D** Embryo-derived plants of *G. arboreum* 5 months after transfer to the field; note the normal growth and flowering. (Bajaj and Gill 1992)

was observed on MS + IAA (1.5 mg/l) + kinetin (0.5 mg/l) + CH (250 mg/l) + sucrose 3% in *G. arboreum* and *G. herbaceum*. The 12 and 15 DAP embryos of *G. hirsutum* responded best on MS + IAA (1.5 mg/l) + CW (7%) and sucrose 6%. *G. arboreum* embryos produced multiple shoots (Fig. 3A,B) on

Fig. 3A,B. Induction of multiple shoots. **A** *Gossypium arboreum* embryo cultured on MS + IAA (1 mg/l) + BAP (2 mg/l) showing the initiation of buds and multiple shoots. **B** A hybrid embryo of *G. arboreum* × *G. anomalum* 15 days after culture on MS + CH (500 mg/l) + IAA (1.5 mg/l) + kinetin (1 mg/l) showing multiple shoots. (Bajaj and Gill 1992)

MS + IAA (1 mg/l) + BA (2 mg/l). These shoots could be separated and subcultured on medium to induce rooting. It was also observed that when 15 DAP embryos of *G. arboreum* and *G. herbaceum* were kept in complete dark for the first 15 days of culture, plant formation was considerably increased. When light was provided during the first 15 days, the embryos turned green and produced weak and abnormal seedlings (Gill and Bajaj 1984).

The response of hybrid ovules and embryos from various interspecific crosses (Table 4) cultured on different media was similar to their respective maternal parent, though the development was somewhat slower than the self-pollinated ovules. On transfer to the pots/field these plants continued growth, and set flowers and bolls.

2.3 Micropropagation Through Somatic Embryos

Somatic embryogenesis is one of the fastest methods for large-scale propagation (Attree and Fowke 1991). In cotton, somatic embryogenesis was first

Table 4. In vitro growth response of the hybrid embryos cultured (15 days after pollination) on MS + IAA (1.5 mg/l) + kinetin (0.5 mg/l) + casein hydrolysate (250 mg/l) and 3% sucrose. (Gill and Bajaj 1984)

Hybrid	No. of embryos cultured	No. of plants formed	Plant formation (%)
1. *G. arboreum* × *G. stocksii*	31	22	71
2. *G. arboreum* × *G. anomalum*	74	51	69
3. *G. herbaceum* × *G. stocksii*	39	19	48

observed in suspension cultures of *G. klotzschianum* (Price and Smith 1979). Since then, numerous reports have appeared on *G. hirsutum* (see Gawel and Robacker 1995; Chap. I.2, this Vol.). In *G. hirsutum* the percentage of callus forming proembryos increased to about 30% by prolonged culture without NAA and kinetin (Davidonis and Hamilton 1983). Proembryoid development was further enhanced by transfer to media lacking NH_4NO_3 but containing double the standard KNO_3 and GA. Shoemaker et al. (1986) evaluated 17 cultivars of *G. hirsutum* and identified 2 cultivars, Coker 201 and Coker 315, as embryogenic, and their mature embryos germinated upon transfer to auxin-free medium with reduced sugar level. Trolinder and Goodin (1987) obtained somatic embryos in all the eight strains of *G. hirsutum* studied. The mature embryos germinated on sterile vermiculite saturated with medium. Likewise, Umbeck et al. (1987) used vermiculite saturated with growth regulators to induce germination. Gelrite medium also induced germination (Firoozabady et al. 1987; Finer 1988). The percent of germination varied, the degree of fertility in the regenerated plants ranging from 15 to 100% (Trolinder and Goodin 1987; Finer 1988). Somatic embryos obtained from protoplast-derived callus also formed plants (She et al. 1995).

Embryos of *G. davidsonii* germinated and formed plants; the rate of germination was 47.3%, and 12% resulted in plants. Likewise, embryos of *G. hirsutum* cv. Sumian 3118 and Glandless 203 germinated on hormone-free White's medium supplemented with charcoal. Out of 56 Sumian 3118 plants, 44 survived under natural field conditions. Some of these plants bloomed and set bolls (Wu et al.; Chap. I.2, this Vol.).

3 In Vitro Storage/Conservation of Germplasm

The germplasm of cotton is customarily preserved through seed. The USDA/ ARS Southern Crops Research Laboratory, College Station, Texas, maintains the National Collection of *Gossypium* germplasm (Percival 1987), and, with about 5100 accessions (Percival and Kohel 1990), is one of the largest collections in the world. However, with the passage of time, the stored seed are likely to suffer deterioration, and "many of the species and individual accessions cannot be properly maintained within the limitations of present technology" (Altman et al. 1990). Thus, alternative methods have to be developed. In this regard, storage of tissue cultures and cryopreservation of cell cultures has been experimented.

3.1 In Vitro Storage of Cultures

In vitro-cultured material has demonstrated abundant potential for short- and medium-term storage in a number of plant species (Bajaj 1986, 1991). In cotton, shoot-tip cultures could be maintained at room temperature, without change of medium, for more than 6 months (Bajaj and Gill 1986). This period may be prolonged by growing them on minimal medium at lower temperatures. Altman et al. (1990) developed a simple tissue-culture method which enables the collection/storage of living vegetative cotton material from the field. Since the availability of viable seed cannot always be assured during botanical expeditions, this method has proved useful.

3.2 Cryopreservation of Germplasm

Plant cell cultures on periodic subculturing undergo genetic erosions, resulting in numerous different cell lines; thus, it is difficult to maintain either the clone or the genetic stability of the desired cell lines. Most of these cell lines may not be of any value and are discarded; however, some, which may not be needed today but are likely to be useful in the future, are also lost. In this regard, cryopreservation has proved to be of immense value. Complete plants in a number of plant species have been obtained from cell cultures, embryos, meristem tips, etc. and freeze-preserved in liquid nitrogen (see Bajaj 1995).

Recent advances in the biotechnology of cotton, especially somaclonal variation in cell cultures (Bajaj and Gill 1985; Stelly et al. 1989) and genetic transformation (see Chaps. V. 1–5, this Vol.) have resulted in cultures which need to be preserved. In cotton, only one study has been reported on cryopreservation (Bajaj 1982). The proliferated anthers and ovules of *G. hirsutum* and *G. arboreum* after treatment with a mixture of cryoprotectant (5% each of dimethylsulfoxide, glycerol, and sucrose) were subjected to freezing in liquid nitrogen. The retrieved cultures turned brown and showed no sign of growth for the first 3–4 weeks. The callus then underwent localized activity, and, at places, especially at the periphery, resumed growth and formed islets of meristems. A mass of callus was formed in 6–7 weeks. It was further observed that the cultures with little callus revived better than those with masses of callus. In the latter, the callus became spongy and died. The cell suspension obtained from the retrieved callus resumed active growth, and contained highly cytoplasmic aggregates of cells. The anther-derived callus was more sensitive than that obtained from the ovules, as evident by 34 and 42% survival, respectively (Bajaj 1982).

It is envisaged that cryopreservation will be increasingly employed for the conservation of rare, elite, and desirable cultures, and also for the establishment of germplasm banks and the international exchange of material (Bajaj 1979).

4 Summary

Various aspects of micropropagation and conservation of germplasm of cotton are discussed. Cotton is conventionally propagated through seed and cuttings, and the germplasm is stored through seed. However, seed suffer genetic deterioration due to natural cross-pollination and during ginning. Thus, in vitro micropropagation was adopted, and the plants were regenerated from excised meristems, somatic embryos, zygotic embryos, and ovules. Shoot-tip-cultured plants are true to type. Shoot-tip culture has also been used for collection of germplasm of wild species of *Gossypium* in which seeds could not be obtained.

In vitro cultures stored, without change of media, for various lengths of time, resumed luxuriant growth on transfer to fresh media. Also callus cultures cryopreserved at −196 °C survived, and resumed growth.

References

Altman DW, Fryxell PA, Koch SD, Howell CR (1990) *Gossypium* germplasm conservation augmented by tissue culture techniques for field collection. Econ Bot 44:106–113

Attree SM, Fowke LC (1991) Micropropagation through somatic embryogenesis in conifers. In: Bajaj YPS (ed) Biotechnology in agriculture and forestry, vol 17. High-Tech and micropropagation I. Springer, Berlin Heidelberg New York, pp 53–70

Bajaj YPS (1979) Establishment of germplasm banks through freeze-storage of plant tissue culture and their implications in agriculture. In: Sharp WR, Larsen PO, Paddock EF, Raghavan V (eds) Plant cell and tissue culture – principles and applications. Ohio State Univ Press, Columbus, Ohio, pp 745–774

Bajaj YPS (1982) Survival of anther and ovule-derived cotton callus frozen in liquid nitrogen. Curr Ser 51:139–140

Bajaj YPS (1986) In vitro preservation of genetic resources: techniques and problems. Proc Int Symp Nuclear techniques and in vitro culture for plant improvement. IAEA/FAO, Vienna, pp 43–57

Bajaj YPS (1991) Storage and cryopreservation of in vitro cultures. In: Bajaj YPS (ed) Biotechnology in agriculture and forestry, vol 17: High-tech and micropropagation I. Springer, Berlin Heidelberg New York, pp 361–381

Bajaj YPS (1992) High-tech and micropropagation II–IV. Biotechnology in agriculture and forestry, vol 18–20. Springer, Berlin Heidelberg New York

Bajaj YPS (1995) Cryopreservation of plant germplasm I. Biotechnology in agriculture and forestry, vol 32. Springer, Berlin Heidelberg New York

Bajaj YPS, Gil MS (1985) In vitro induction of genetic variability in cotton (*Gossypium* spp.) Theor Appl Genet 70:363–368

Bajaj YPS, Gill MS (1986) Micropropagation and germplasm preservation of cotton (*Gossypium* spp.) through shoot tip and meristem culture. Indian J Exp Biol 24:581–583

Bajaj YPS, Gill MS (1992) Micropropagation of cotton (*Gossypium* spp.). In: Bajaj YPS (ed) Biotechnology in agriculture and forestry, vol 19. High-tech and micropropagation III. Springer, Berlin Heidelberg New York, pp 483–504

Bajaj YPS, Gill MS, Mohapatra D (1986) Somaclonal and gametoclonal variation in wheat, cotton and brassica. In: Semal J (ed) Somaclonal variations and crop improvement. Nijhoff, Dordrecht, pp 160–169

Chappel EJ, Mauney JR (1967) Culture of apical meristem of *Gossypium hirsutum* in vitro. Phyton 24:93–100

Davidonis GH, Hamilton RH (1983) Plant regeneration from callus tissues of *Gossypium hirsutum* L. Plant Sci Lett 32:89–93

Finer J (1988) Plant regeneration from somatic embryogenic suspension cultures of cotton (*Gossypium hirsutum* L.). Plant Cell Rep 7:399–402

Firoozabady E, DeBoer D, Merlo D, Halk E, Amerson L, Rashka K, Murray E (1987) Transformation of cotton (*Gossypium hirsutum* L.) by *Agrobacterium tumefaciens* and regeneration of transgenic plants. Plant Mol Biol 10:105–116

Gawel NJ, Robacker CD (1995) Somatic embryogenesis in cotton (*Gossypium* spp.). In: Bajaj YPS (ed) Biotechnology in agriculture and forestry, vol 31. Somatic embryogenesis and synthetic seed II. Springer, Berlin Heidelberg New York, pp 256–266

Gill MS, Bajaj YPS (1984) Interspecific hybridization in the genus *Gossypium* through embryo culture. Euphytica 33:305–311

Gill MS, Bajaj YPS (1986) In vitro production of interspecific hybrids in *Gossypium* and their utilization in back crossing. Z Pflanzenzücht 96:174–176

Gill MS, Bajaj YPS (1987) Hybridization between diploid (*Gossypium arboreum*) and tetraploid (*Gossypium hirsutum*) cotton through ovule culture. Euphytica 36:625–630

Gill MS, Bajaj YPS (1991) In vitro hybridization between *Gossypium arboreum* and *G. anomalum* and thier backcross progeny. Plant Tissue Cult (Dhaka) 1(2):101–108

Gill MS, Bajaj YPS (1996) Hybridization between *Gossypium herbaceum* and *Gossypium stocksii* through embryo rescue. Cell Res (Beijing) 6:85–90

Gould J, Banister S, Hasegawa O, Fahima M, Smith RH (1991) Regeneration of *Gossypium hirsutum* and *G. barbadense* from shoot apex tissues for transformation. Plant Cell Rep 10: 12–16

Murashige T, Skoog F (1962) A revised medium for rapid growth and bio-assays with tobacco tissue cultures. Physiol Plant 15:473–497

Parmar AL, Shroff VN, Deshpande NV (1978) Studies on vegetative propagation in cotton. Crop Improv 5:73–76

Percival AE (1987) The national collection of *Gossypium* germplasm. South Coop Ser Bull 321. Texas Agric Exp Stn, College Station, Texas

Percival AE, Kohel RH (1990) Distribution, collection, and evaluation of *Gossypium*. Adv Agron 44:225–256

Price HJ, Smith RH (1979) Somatic embryogenesis in suspension cultures of *Gossypium klotzschianum* Andress. Planta 145:305–307

She JM, Wu JY, Zhou HY (1995) Regeneration of plants from protoplasts of *Gossypium* species (cotton). In: Bajaj YPS (ed) Biotechnology in agriculture and forestry, vol 34. Plant protoplasts and genetic engineering VI. Springer, Berlin Heidelberg New York, pp 63–69

Shoemaker RC, Couche LJ, Galbraith DW (1986) Characterization of somatic embryogenesis and plant regeneration in cotton (*Gossypium hirsutum* L.). Plant Cell Rep 3:178–181

Stelly DM, Altman DW, Kohel RJ, Rangan TS, Commiskey E (1989) Cytogenetic abnotmalities of cotton somaclones from callus cultures. Genome 32:762–770

Trolinder NL, Goodin JR (1987) Somatic embryogenesis and plant regeneration in cotton. (*Gossypium hirsutum* L.). Plant Cell Rep 6:231–234

Umbeck P, Johnson G, Barton K, Swain W (1987) Genetically transformed cotton (*Gossypium hirsutum* L.) plants. Biotechnology 5:263–266

Wu JY, She JM (1990) Shoot apex and axillary bud culture of *Gossypium* spp. Jiangsu J Agric Sci 6(2):22–26

Section II
Interspecific Hybridization, Embryo/Ovule Culture; In Vitro Fertilization; RAPD; Heterosis

II.1 Interspecific Hybridization in Cotton Through Embryo Rescue and Culture

Y.P.S. BAJAJ[1] and M.S. GILL[2]

1 Introduction

There are about 50 species of *Gossypium* (Fryxell 1992); however, only four are cultivated (*G. arboreum*, *G. herbaceum*, *G. hirsutum*, and *G. barbadense*), the rest are wild ones. Some wild species with useful characters are given in Table 1. The genus *Gossypium* is divided into seven genome groups that are diploid (2n = 2x = 26), and an eighth amphidiploid (2n = 4x = 52) group that combines A and D genomes. These groups show cytogentic separation as well as distinctive geographic distribution (Percival and Kohel 1990):

A genome – Two cultivated species from The Far East, The Middle East, and Africa
B genome – Six wild species from Africa and the Cape Verde Islands
C genome – Ten wild species from Australia
D genome – Thirteen wild species from Mexico, Peru, and the Galapagos Islands
E genome – Four wild species from the Arabian Peninsula and Northeast Africa
F genome – One wild species from East Central Africa
G genome – One wild species from Australia
AD genome – Six (two cultivated and four wild) species from Mexico, South America, and The Hawaiian Islands, the Galapagos Islands, and Brazil (the cultivated two having recently attained worldwide distribution through cultivation) (Fryxell 1984).

There are more than 100 species of insects which cause damage to cotton (Berger 1969). Two thirds of the loss due to insects is by boll weevil, cotton bollworm, and tobacco bollworm. In addition, several bacteria and fungi contribute to reduced yield. Some common diseases are *Fusarium* wilt, *Verticillium* wilt, bacterial blight, *Ascochyta* blight, boll rot, cotton rust, etc. Some of the wild species (Table 1) are a rich source of genetic variability. Although short-fibered or lintless, they have a number of desirable traits such

[1] Biotechnology in Agriculture and Forestry, A-137 New Friends Colony, New Delhi 110065, India
[2] Punjab Agricultural University, Ludhiana (Punjab), India

Biotechnology in Agriculture and Forestry, Vol. 42
Cotton (ed. by Y.P.S. Bajaj)

Table 1. Some wild species of cotton possessing useful characters

Species	Genome	Useful characters
1. *G. anomalum* Wawrex Wawra & Peyr	B_1	Lint quality, resistance to black arm and jassid
2. *G. harkensii* Brendeg.	D_2^{-2}	Resistance to drought, source of cytoplasmic male sterility, resistance to spider mites
3. *G. raimondii* Ulbr.	D_5	Resistance to black arm and jassids, high density of seed hair, and drought tolerance
4. *G. armourianum* Kearn.	D_2^{-1}	Resistance to black arm and boll worm, increased number of ovules per loculus
5. *G. thurberi* Tod.	D_1	Resistance to boll worms and potentiality for imparting high fiber strength
6. *G. aridum* (Rose & Standl.) Skov.	D_4	Drought tolerance
7. *G. somalense* (Gurke) J.B. Hutch.	E_1	Resistance to boll worm
8. *G. stocksii* Mast. in Hook.	E_2	Resistance to drought
9. *G. tomentosum* Nutt. ex Seem.	AD	Resistance to drought and jassids, lint quality, and strength
10. *G. bickii* Prokh.	$2G_1$	Resistance to insects, gossypol-free seeds

as fiber quality and resistance to insects, drought, and diseases (Hutchinson et al. 1947; Prentice 1972).

Another point worth mentioning is that the protein products and oil from cottonseed are restricted in use because of the presence of gossypol in the lysigenous glands. However, plants of several wild diploid species, such as *G. sturtianum*, are glanded, but their seed are functionally glandless and do not contain gossypol. Thus, crossing this species with *G. hirsutum* would be rewarding (Altman et al. 1987). Hybrids between diploid and tetraploid *Gossypium* species are useful for introducing improved agronomic and quality traits into commercial cotton (see Meyer 1974).

Numerous attempts have been made to transfer the desirable characters from wild to cultivated cotton by conventional breeding methods. However, either there has been a low frequency of crossing followed by the eventual hybrid breakdown (Gerstel and Phillips 1958), or there is complete failure due to incompatibility, such as early abortion of hybrid embryo/endosperm (Weaver 1958; Pundir 1972).

Recently, there has been much interest in the transfer of genes in crops through genetic manipulation of cells and tissue culture, and through somatic hybridization by the fusion of protoplasts (see Bajaj 1994). Most of these techniques are in the process of being developed, and it will be quite some time before they are available to cotton breeders. However, some of the old and rather simple techniques, such as the culture of embryos and ovules, have already contributed a great deal to obtaining interspecific hybrids in a number of oilseed crops (Bajaj 1990). Thus, in this chapter, work on interspecific hybridization in cotton through embryo rescue/culture is discussed.

2 Interspecific Hybridization Through Embryo Rescue Methods

To date, numerous investigations have been undertaken utilizing embryo rescue/culture techniques to bring about hybridization amongst various species of *Gossypium*, both wild and cultivated, diploid and tetraploid. The literature is summarized in Table 2, and details primarily based on published work are discussed in the following pages.

2.1 Hybridization Among Diploid *Gossypium* Species, *G. arboreum*, *G. anomalum*, *G. herbaceum*, and *G. stocksii*.

(Gill and Bajaj 1984a,b, 1986, 1991, 1996)

Experiments were initiated by the authors involving two wild and two cultivated diploid species, and the following crosses were made:

G. arboreum × *G. stocksii*
G. arboreum × *G. herbaceum*
G. arboreum × *G. anomalum*
G. herbaceum × *G. stocksii*

In all four above-named interspecific crosses, hybrids were obtained by preventing the degeneration of embryos by treating the flowers/young bolls with growth regulators, and then culturing the immature rescued embryos/ovules.

Effect of Growth Regulators on Boll Retention. The daily application of a mixture of growth regulators (GA_3 50 mg/l + NAA 100 mg/l) to the flowers immediately after crossing considerably enhanced their retention and ensured further development in various crosses (Table 3).

In the controls (untreated), boll retention after 15 days of pollination was only 11.5% in *G. herbaceum* × *G. stocksii*, 14.8% in *G. arboreum* × *G. anomalum*, and 19.1% in *G. arboreum* × *G. stocksii*. Shedding started after 4 days of pollination, and abscission occurred at a point where the pedicel is joined to the stalk. However, when growth regulators were applied, boll retention was increased to 55.8% in *G. herbaceum* × *G. stocksii*, 62.8% in *G. arboreum* × *G. stocksii*, and 68.3% in *G. arboreum* × *G. anomalum*. The best response was observed in the cross *G. arboreum* × *G. anomalum*, where retention increased from 14.8 to 68.3%. The base of the pedicel became rather thick and strong, and abscission was prevented.

Culture of Hybrid Embryos. The hybrid embryos of three crosses, i.e., *G. arboreum* × *G. stocksii*, *G. arboreum* × *G. anomalum*, and *G. herbaceum* × *G. stocksii* cultured on MS + indoleacetic acid (1.5 mg/l) + kin (0.5 mg/l) + casein hydrolysate (250 mg/l) and 3% sucrose and incubated in the dark showed an overall increase in size during the first 2 weeks. The cotyledons enlarged, the radical underwent elongation, and penetrated the medium in most cases.

Table 2. Wide hybridization in cotton through embryo and ovule culture

Species crossed	Ovule/embryo	Results/remarks	Reference
		Diploid × diploid	
G. arboreum × *G. stocksii*	Embryo	Hybrid plants	Gill and Bajaj (1984a,b)
G. arboreum × *G. anomalum*			
G. herbaceum × *G. stocksii*			
G. arboreum × *G. stocksii*	Embryos	Backcrosses of hybrid plants	Gill and Bajaj (1986)
G. arboreum × *G. anomalum*	Embryos 15 DAP	Hybrids with several desirable combination were obtained	Gill and Bajaj (1991)
G. herbaceum × *G. stocksii*	Ovule 3 DAP Embryo 15 DAP	Hybrids were late-flowering, self-sterile and showed semicreeping habit	Gill and Bajaj (1996)
		Tetraploid × diploid	
G. hirsutum × *G. herbaceum*	Ovule	Hybrids	Stewart and Hsu (1978)
G. hirsutum × *G. trilobum*	Embryos from 15–25 DAP ovules	Cytogenetic studies in the hybrids	Unbeek and Stewart (1985)
G. hirsutum × *G. arboreum*	Ovule	Hybrid plants	Gill and Bajaj (1987)
G. sturtianum × *G. hirsutum*	Ovule 2 days after fertilization	Introgresion of glanded plant/glandless seed	Altman et al. (1987) Mergeai et al. (1995)
G. hirsutum × *G. harknessi*	Embryo 15 DAP	Hybrid plants	Bojvic (1992)
		Tetraploid × tetraploid	
G. hirsutum × *G. barbadense*	Ovule	Hybrid plants	Rafaat et al. (1984)

Table 3. Effect of growth regulators (GA_3 50mg/l + NAA 100mg/l) on boll retention in various interspecific crosses in *Gossypium*. (Gill and Bajaj 1984b)

Crosses	Control			Treated		
	No. of flowers crossed	No. of bolls shed	Retention (%)	No. of flowers crossed	No. of bolls shed	Retention (%)
1. *G. arboreum* × *G. stocksii*	21	17	19.1	86	32	62.8
2. *G. arboreum* × *G. anomalum*	27	23	14.8	102	35	68.3
3. *G. herbaceum* × *G. stocksii*	26	23	11.5	68	30	55.8

Table 4. In vitro growth response of the hybrid embryos cultured (15 days after pollination) on MS + IAA (1.5mg/l) + CH (250mg/l) + kin (0.5mg/l) + sucrose 3%. (Gill and Bajaj 1984a)

Cross	No. of embryos cultured	No. of plants formed	Plant formation (%)
1. *G. arboreum* × *G. stocksii*	31	22	71
2. *G. arboreum* × *G. anomalum*	74	51	69
3. *G. herbaceum* × *G. stocksii*	39	19	48

Upon exposure to light, the cotyledons turned green, the first leaf appeared in 30–35 days, and the embryos developed into small seedlings. In cases where the embryos did not produce seedlings, either the radical failed to penetrate the medium or the root stopped growing after a few days.

The response of the cultured embryos varied in different crosses (Table 4). The best growth response was observed in *G. arboreum* × *G. anomalum*, although the number of embryos producing seedling was slightly more in *G. arboreum* × *G. stocksii*. The *G. arboreum* × *G. anomalum* embryos exhibited fast growth and most of them produced seedlings in 30–35 days, whereas in *G. arboreum* × *G. stocksii*, it took at least 35 days. In *G. herbaceum* × *G. stocksii*, only 48% embryos produced healthy seedlings. Seedlings with first pair of leaves were obtained in 35 days.

The plants were transferred into pots at the two- to three-leaf stage with well-developed roots. There was 80–90% survival of potted plants, which were then grown in the greenhouse and field (Fig. 1). The hybrids grew normally; *G. arboreum* × *G. anomalum* hybrids showed vigorous growth, and attained almost double the height of either parent. This hybrid also produced a large number of flowers as compared to the parents as well as to other hybrids. The hybrids in all three crosses were perennial in nature and remained green throughout the winter season (Nov.–Feb.). The hybrids of *G. arboreum* × *G. anomalum* flowered throughout the year. The *G. herbaceum* × *G. stocksii* hybrids, however, produced only a few flowers in winter; *G. arboreum* × *G. stocksii* did not flower in winter at all, but resumed flowering in summer.

Effect of Light and Dark. The embryos incubated in the dark for the first 15 days, followed by exposure to light, yielded normal seedlings. The effect of

Fig. 1A–E. Production of interspecific hybrids in cotton through embryo culture. **A** *G. herbaceum* × *G. stocksii* embryo (15 DAP) after 35 days of culture of MS + IAA (1.5 mg/l) + kin (0.5 mg/l) + CH (250 mg/l), showing balanced root and shoot formation. **B** (*G. arboreum* × *G. herbaceum*) × *G. somalense* embryo (15 DAP) after 35 days of culture on MS + IAA (1.5 mg/l) + kin (0.5 mg/l) + CH (500 mg/l). **C** A *G. arboreum* × *G. stocksii* embryo-derived plant after 45 days of transfer into the pot. **D** (*G. arboreum* × *G. herbaceum*) × *G. somalense* plant after 5 months of transfer into the pot. **E** A glasshouse-grown *G. arboreum* × *G. anomalum* plant at the time of flowering (total time from embryo culture to this stage eight months). (Gill and Bajaj 1984b)

light vs. dark on percent plantlet formation from the embryos of *G. arboreum* and *G. herbaceum* is shown in Table 5. When the embryos were kept in the dark, most of them exhibited normal growth, and produced vigorous seedling. These embryos increased in size during the first 2 weeks when kept in the dark,

Table 5. Effect of light and dark treatments on the in vitro growth response of cotton embryos cultured (15 days after pollination) on MS + IAA (1.5 mg/l) + CH (250 mg/l) + kin (0.5 mg/l) + sucrose 3%. (Gill and Bajaj 1984b)

Species	Treatment	No. of embryos cultured	No. of plants formed	Plant formation (%)
G. arboreum	Dark (15 days)	36	27	75.0
	Light	32	14	43.7
G. herbaceum	Dark (15 days)	33	20	60.6
	Light	34	13	38.2

then turned green upon exposure to light, and produced plantlets. However, when the embryos were exposed to light from the very beginning, they turned green at a very early stage.

The radicle failed to elongate, and did not penetrate the medium in most of the cases. The hypocotyl was thin, and consequently the seedlings produced were also comparatively weak.

2.1.1 Gossypium arboreum × G. stocksii (Gill and Bajaj 1984a,b)

In this cross, normally the bolls started to shed after 4 days of pollination, and retention after 15 DAP was only 19%. However, on treatment with a mixture of NAA (100 mg/l) + GA (50 mg/l) it increased to 62.8% (see Table 3). The hybrid embryos were cultured on MS + IAA (1.5 mg/l) + kin (0.5 mg/l) + CH (250 mg/l) + 3% sucrose, and incubated in the dark for 2 weeks. After exposure to light, the cotyledons turned green, and in 35 days seedlings were formed, which on transfer to soil (Fig. 1C) grew further. These hybrids are perennials, and did not flower in winter, but resumed flowering in summer. A comparison of the leaves, flowers, and bracts is shown in Fig. 2. Note the differences in shape and size of the leaves, and shape and number of teethlike structures in the bracts.

2.1.2 Gossypium arboreum × G. anomalum (Gill and Bajaj 1991)

Gossypium anomalum is a wild diploid species found in Africa, and it belongs to the B genome (Endrizzi et al. 1985). This species is known to contribute genes for fiber quality, bacterial blight, jassids (Santhanam 1967), and narrow bract (imparts resistance to boll worms). The *anomalum* cytoplasm has also been found to interact with genomes of cultivated species, resulting in cytoplasmic male sterility (Meyer 1974). *Gossypium arboreum* is one of the two cultivated diploid species, and is grown widely in India and other Asian countries. Conventional methods of crossing these two species result in poor yield of hybrid plants due to lack of fertilization, premature boll shedding, and failure of seeds to germinate (Marappan and Santhanam 1962). In order to

Fig. 2A–C. Comparison of leaves, flowers, and bracts in *G. arboreum*, *G. stocksii*, and the embryo-derived hybrid. **A** Leaves of *G. arboreum*, *G. arboreum* × *G. stocksii*, and *G. stocksii* (*left to right*). Note the difference in lobe shape and size. **B** Flowers of *G. arboreum*, F_1, and *G. stocksii* (*left to right*). **C** Bracts of *G. arboreum*, F_1, and *G. stocksii* (*left to right*). Note the difference in shape and number of teethlike structures. (Gill and Bajaj 1984b)

overcome these incompatibility barriers, in vitro culture of rescued hybrid embryos was employed (Gill and Bajaj 1991).

Premature shedding of cross-pollinated bolls are prevented by the application of a solution of NAA (100 mg/l) and GA (50 mg/l) for 5 days after pollination (DAP). The hybrid embryos (15 DAP) cultured on modified MS medium supplemented with growth regulators developed into plants. In the field (Fig. 3), the hybrid plants showed vigorous growth and they resembled *G. anomalum* for most of the characters (Fig. 4). The fertility of F_1 plants was very low and they produced only a few seeds (Fig. 5).

Utilization of Hybrids in Backcrossing (Gill and Bajaj 1991). The F_1 plants predominantly exhibited characters of the wild parent, as is expected, since wild plant characteristics are usually dominant (Mehboob Ali et al. 1964). The *G. arboreum* × *G. anomalum* hybrid showed particularly vigorous growth. The hybrids flowered after 5–6 months of transfer into the field, and all were perennial and self-sterile. The *G. arboreum* × *G. anomalum* hybrid produced a large number of empty bolls. Hybrids involving *G. stocksii* had smooth stems, showed creeping habit, and were not weak, as observed by Mehboob Ali et al. (1964). In hybrids involving *G. arboreum*, red petal color and red

Fig. 3. Embryo-derived hybrid plants (*Gossypium arboreum* × *G. anomalum*) 6 months after transfer to the field (total time taken from embryo culture to flowering state is 11 months) being examined by Dr. Manjeet S. Gill (*left*) and Professor Y.P.S. Bajaj (*right*). (Bajaj and Gill 1985)

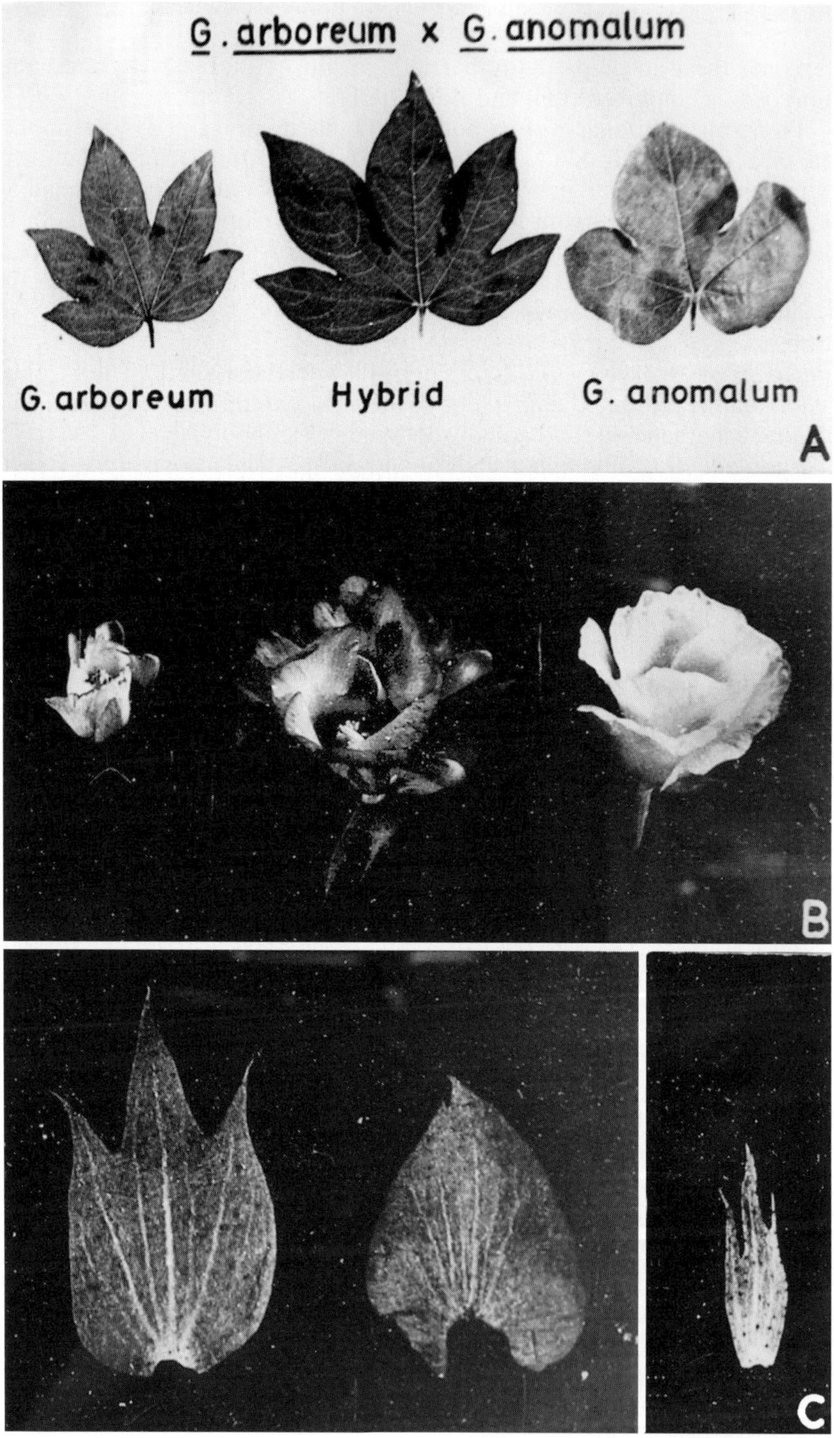

Fig. 4A–C. Comparison of leaves, flowers, and bracts of *G. arboreum*, *G. anomalum*, and their hybrids. **A** Leaves from (*left to right*) *G. arboreum*, hybrid, and *G. anomalum* photographed at the flowering state; note the broader and longer lobes of the hybrid leaves. **B** Flowers of *G. arboreum*, hybrid, and *G. anomalum*; note the dark color and larger size of the flowers. **C** Bracts of *G. arboreum*, hybrid, and *G. anomalum*; note the larger size of the bract from the hybrid plant. (Gill and Bajaj 1991)

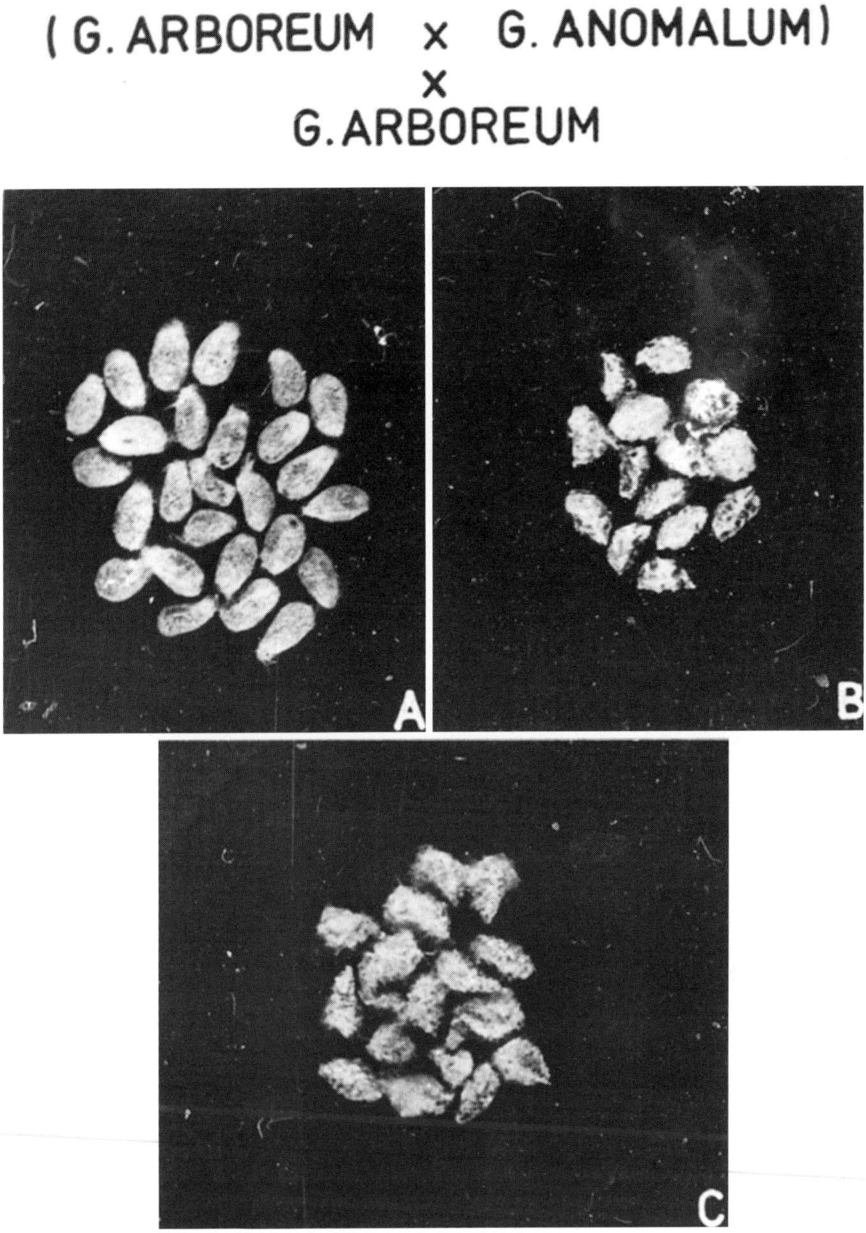

Fig. 5A–C. Comparison of seed shape and size of *G. arboreum*, *G. anomalum*, and backcross seeds obtained by pollinating the embryo-derived hybrid with *G. arboreum* pollen. **A** *Gossypium arboreum* seeds. **B** *G. anomalum* seeds. **C** *G. arboreum* × *G. anomalum* hybrid seeds obtained after pollinating the F_1 (obtained as a result of embryo culture – the hybrid was sterile) with *G. arboreum* pollen. (Gill and Bajaj 1991)

stem were dominant. In all the hybrids, flower size and shape, bract size and shape, and pedicel length resembled the wild species.

Efforts to backcross in F_1 plants with their respective cultivated parent met with limited success. The *G. arboreum* × *G. anomalum* hybrid produced some seeds when pollinated with *G. arboreum* pollen or vice versa. Seeds set on the hybrid plant possessed strong and fine fiber, as was also reported by earlier workers (Silow 1941; Afzal et al. 1945; Kalyanaraman and Santhanam 1955; Iyengar et al. 1957). In the backcross progeny, segregation was observed for plant color, leaf shape and size, bract shape and size, growth habit, and petal color.

In the BC_1 generation, all plants except one resembled *G. arboreum* in morphological characters; they had smaller bolls but with strong fiber. Bract size was variable. One plant was dark green and resembled *G. arboreum*, but had smaller flowers and was totally sterile. About 250 backcrosses were made. Some desirable combinations with longer fiber, more ginning output, higher seed weight, and bacterial blight resistance were obtained in the BC_2S_2 generation (Table 6; Gill and Bajaj 1991). Only one plant in the backcross progeny produced seeds upon self-pollination. However, second backcross generation seeds were obtained from all the plants after pollinating with *G. arboreum* pollen or vice versa (Gill and Bajaj 1986).

2.1.3 Gossypium herbaceum × G. stocksii (Gill and Bajaj 1996)

Gossypium herbaceum is one of the two A genome diploid cultivated species which is grown in some parts of central India, especially as a rainfed crop. *G. stocksii* is a wild diploid species belonging to the E genome and is found in West Asia. This species has a very good root system and is tolerant to drought (Sikka and Joshi 1960). Thus, hybridization between these two species is desirable to develop drought-tolerant varieties in *G. herbaceum*. The attempts to produce interspecific hybrids by the conventional methods are often unsuccessful due to the formation of underdeveloped abortive seeds. Thus, embryo rescue studies have been conducted (Gill and Bajaj 1996), and the results are shown in Table 7 and Figs. 6 and 7.

The hybrid embryos (15 DAP) and ovules (3 DAP) were cultured on MS + IAA (1.5 mg/l) + kinetin (0.5 mg/l) + CH (250 mg/l) + 3% sucrose. The hybrid plants were transferred to the field and exhibited most of the characters of the pollen parent, i.e., *G. stocksii*. The hybrids were late-flowering, self-sterile, and showed a semicreeping habit.

2.2 Hybridization Amongst Diploid × Tetraploid Species

2.2.1 Gossypium arboreum × G. hirsutum (Gill and Bajaj 1987)

Gossypium hirsutum ($2n = 52$) and *G. arboreum* ($2n = 26$) are the two species which occupy a major portion of the area under cotton in India. *G. hirsutum*

Table 6. Morphological characters of the two parents, *G. arboreum* × *G. anomalum* F_1 and the BC_2S_2 progenies. (Gill and Bajaj 1991)

Characters	*G. arboreum*	*G. anomalum*	F_1	BC_2S_2
Stem	Reddish brown, hairy, av. plant ht. 170 cm	Green, extremely hairy, av. plant ht. 170 cm	Slightly reddish, hairy av. plant ht. 170 cm	Green or reddish; hairy, plant ht. 140–190 cm
Leaf	Five-lobed with pointed ends, reddish, hairy	3–5-lobed, green hairy	5–7-lobed with pointed ends, red, hairy	3–7-lobed, color green or reddish, hairy
Flowers	Red, small size bract with shallow teeth, broad, do not flare back, calyx unlobed, pedicel 1–2 cm long, anthers deep red	Light pink, large size, bract deeply toothed, 3 teeth, very narrow, do not flare back. Calyx 5-lobed. Almost sessile flower (pedicel 2–3 mm), anthers very light pink	Red, large size, bract 3-toothed, breadth intermediate flare back. Calyx 5-lobed. Pedicel about 1 cm long, anthers orange-colored	Pink or red, size variable. Bract with shallow or deep teeth, bract width variable (very narrow to broad), in some plants the bracts flare back. Calyx 5-lobed. Sessile or 1–2 cm long pedicel. Anthers red, orange, or pink
Boll	Boll 3–4-loculed, brown color, boll weight 1.6 g	Bolls 3-loculed, green color. Boll weight 1.1 g	Bolls 3-loculed, light brown color. Boll weight 1.2 g	Bolls 3–4-loculed, green or brown color. Boll weight 1.4–2.3 g
Seed	Blackish seeds with long fiber, 100 seed wt. 4.3 g	Light brown seeds with small fibers, 100 seed wt. 2.2 g	Light brown seeds with small fibers, 100 seed wt. 2.3 g	Blackish or brownish seeds, 100 seed weight 4.7–5.9 g
Fiber length	15.6 mm	Very short fiber	Very short fiber	17.1–22 mm
Ginning outturn	35.7%	–	–	35–39.5%
Bacterial blight grade	3.2	0	0	0–2.5

Table 7. In vitro response of the hybrid embryos (15 DAP) and ovules (3 DAP) of a cross *Gossypium herbaceum* × *G. stocksii* cultured on MS + IAA (1.5 mg/l) + kinetin (0.5 mg/l) + CH (250 mg/l) + sucrose 3%. (Gill and Bajaj 1996)

Explant	No. of cultures raised	No. germinating	Germination (%)
Embryos	39	19	48
Ovules	72	16	22.2

types generally yield more, possess longer fiber of superior quality, and are resistant to diseases such as *Fusarium* wilt, whereas *G. arboreum* can better tolerate environmental stresses and is comparatively resistant to various insect pests. Efforts to combine the desirable characters of these two species have not been encouraging, due to abortion of the hybrid embryo/endosperm in the early stages of seed development (Weaver 1958; Pundir 1972; Stewart and Hsu 1978). Thus, tissue culture methods have been incorporated with a view to increasing genetic variation (Bajaj and Gill 1985, 1986).

Cultivars of *Gossypium hirsutum* (F414 and LH372) and *G. arboreum* (G27 and LD 133) were grown in the experimental fields of the Punjab Agricultural University at Ludhiana and reciprocal crosses between these species were made in the field. A mixture of growth regulator solution (gibberellic acid 50 mg/l + naphthalene acetic acid 100 mg/l) was applied in cotton wool at the base of the pedicel of the pollinated flowers on the day of pollination and the day after pollination. The developing bolls (cross-pollinated as well as selfed) were excised 3 days after pollination (DAP), sterilized by dipping them in 70% ethyl alcohol, and subjected to flaming. The ovules were scooped out of the bolls aseptically in a laminar flow chamber and cultured on agar-solidified MS medium or in liquid SH medium (Stewart and Hsu 1978) with or without growth regulators. The cultures were transferred to fresh medium every 4 weeks. The germinating ovules were transferred to a medium containing half the concentration of MS salts and 1% sucrose. The culture conditions and other details were as provided earlier (Gill and Bajaj 1984a).

Effect of Growth Regulators. In both parents there was normal development of the bolls; however, after interspecific pollination, the flowers generally abscised within 3 days. The application of a solution of GA_3 (50 mg/l) and NAA (100 mg/l) to the flowers immediately after cross-pollination and on the following day considerably enhanced their retention (Table 8).

In the controls (untreated) the flower retention with 3 days of pollination was only 6.4 to 17.6% in *G. hirsutum* × *G. arboreum*. When *G. arboreum* was used as the female parent, all the bolls were shed after 3 days. The process of shedding started on the second day after pollination, and by the third day the bolls were either completely shed or they fell on slight disturbance. However, with the application of growth regulators, depending on the cultivar, the

Fig. 6A–D. Plantlet formation from hybrid embryos of *G. herbaceum* × *G. stocksii*, and their transfer into pots and the field. **A** Embryo (15 DAP) after 35 days of culture on MS + IAA (1.5 mg/l) + kin (0.5 mg/l) + CH (250 mg/l), showing good root and shoot formation. **B** Hybrid plant 45 days after transfer into a pot. **C** Hybrid plant after 3 months of transfer in the field; note the bushy nature of the hybrid. **D** Hybrid plant at the flowering stage after 5 months of transfer to the glasshouse. (Gill and Bajaj 1996)

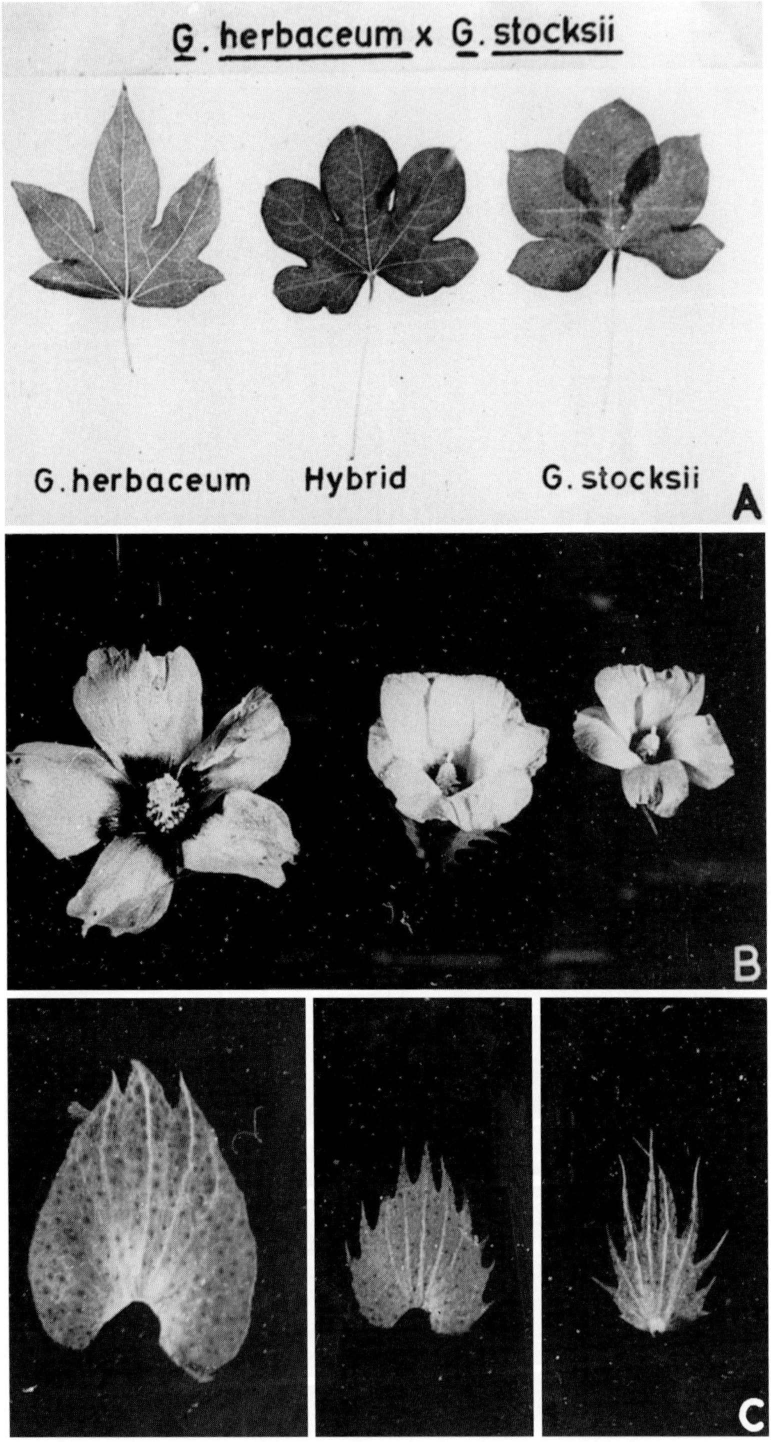

G. herbaceum x G. stocksii
G. herbaceum
Hybrid
G. stocksii
A
B
C

Table 8. Effect of growth regulators (GA_3 50 mg/l + NAA 100 mg/l) on boll retention in interspecific crosses in *G. hirsutum* × *G. arboreum* and their reciprocals. (Gill and Bajaj 1987)

Cross	Treated			Control		
	No. of flowers crossed	No. of bolls shed	Retention (%)	No. of flowers crossed	No. of bolls shed	Retention (%)
G. hirsutum × *G. arboreum*						
F 414 × G 27	71	20	71.8	36	30	16.7
F 414 × LD 133	62	13	79.0	15	14	6.6
LH 372 × G 27	56	24	57.1	31	29	6.4
LH 372 × LD 133	34	16	52.9	17	14	17.6
G. arboreum × *G. hirsutum*						
G 27 × F 414	28	11	60.7	11	11	0
LD 133 × F 414	16	9	43.7	17	17	0

Table 9. Percentage of ovules giving seedlings and the callus growth response of excised ovules (3 days after pollination) of *G. arboreum* and *G. hirsutum* cultured on various media (figures in parentheses indicate the number of ovules cultured). (Gill and Bajaj 1987)

Medium (mg/l)	*G. arboreum* germination (%)	Callus[a] development	*G. hirsutum* germination (%)	Callus development
MS basal	17.6 (125)	+	8.3 (96)	–
MS + CH 250	29.2 (96)	+	22.2 (99)	–
MS + CW 7%	21.2 (80)	+	25.4 (110)	–
MS + CH 250 + IAA 0.5 + kin 0.2	22.2 (108)	+	23.8 (122)	–
MS + CH 250 + IAA 0.5 + kin 0.5	23.7 (93)	+	25.0 (120)	–
MS + CH 250 + IAA 1 + kin 0.2	28.1 (121)	+	44.4 (133)	+ +
MS + CH 250 + IAA 1 + kin 0.5	20.3 (103)	+ + +	24.1 (108)	+ +
MS + CH 250 + IAA 1.5 + kin 0.5	37.5 (112)	+ + +	18.8 (117)	+ +

[a] – No callus; + little callus; + + good callus; + + + profuse callus.

retention varied between 52.9 and 79% in *G. hirsutum* × *G. arboreum*, and 43.7 and 60.7% in the reciprocal crosses.

Culture of Hybrid Ovules. The hybrid ovules were cultured on MS agar as well as on SH liquid medium (Tables 9, 10). On basal SH liquid medium there was no visible growth or germination, the ovules gradually sank to the bottom and turned brown. When IAA (0.5 or 1 mg/l) was added to this medium, there was profuse callusing (Fig. 8A–C), but no germination even after 4 months of culture.

◄

Fig. 7A–C. Comparison of leaves, flowers, and bracts in the cross *G. herbaceum* × *G. stocksii*. **A** Leaves of *G. herbaceum*, *G. herbaceum* × *G. stocksii*, and *G. stocksii* (*left to right*). Note the difference in shape and size of the lobes and the pedicel length. **B** Flowers of *G. herbaceum*, *G. herbaceum* × *G. stocksii*, and *G. stocksii* (*left to right*). Note the difference in size. **C** Bract of *G. herbaceum*, *G. herbaceum*, × *G. stocksii*, and *G. stocksii* (*left to right*). Note the difference in shape and size. (Gill and Bajaj 1996)

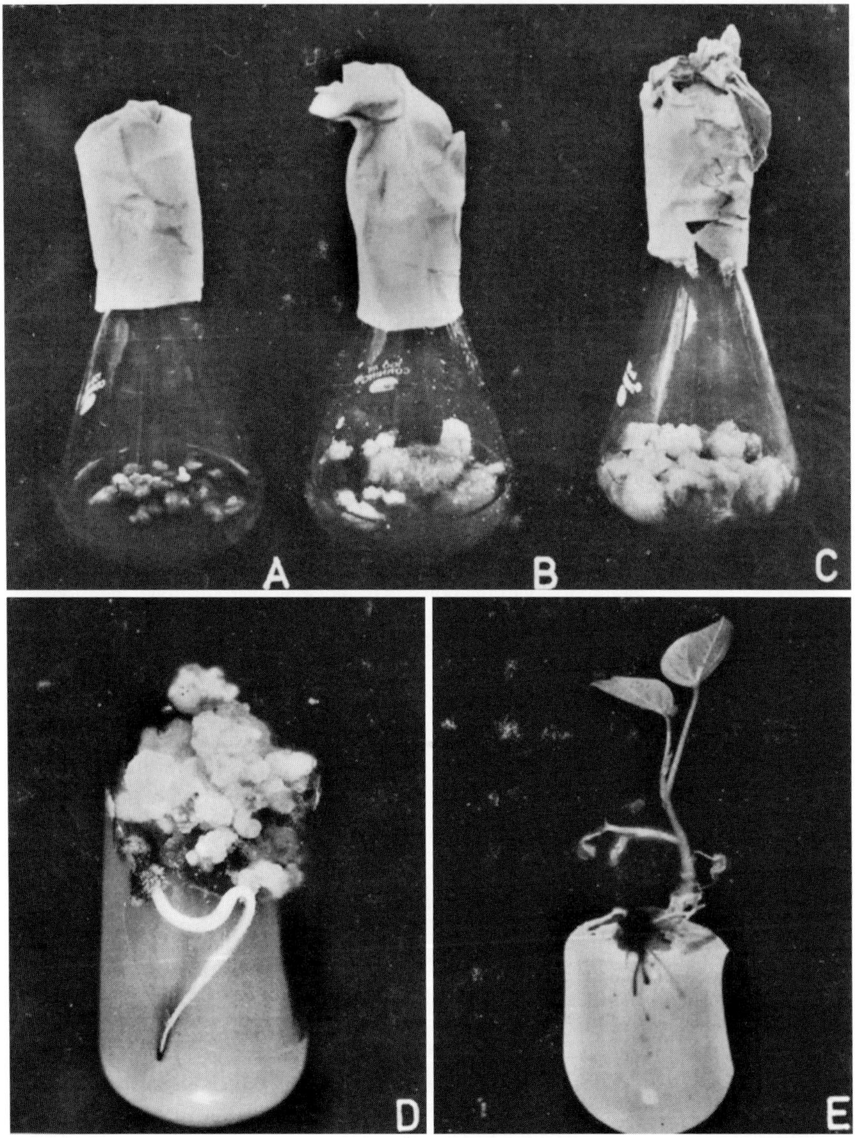

Fig. 8A–E. In vitro growth of hybrid ovules of *Gossypium hirsutum* × *G. arboreum*, and *G. arboreum* × *G. hirsutum* cultured 3 days after pollination on various liquid media containing IAA (0.5 mg/l). **A** After 14 days of culture. **B** Same, after 21 days of culture; note callus formation in some of the ovules. **C** Same, after 28 days of culture showing profuse proliferation. **D** *G. arboreum* × *G. hirsutum* ovule (3 DAP) after 55 days of culture on agar-solidified MS + IAA (1.5 mg/l) + kin (0.5 mg/l) + CH (250 mg/l) showing germination as well as callus. **E** 12-week-old culture showing a seedling derived from 3 DAP ovule of *G. hirsutum* × *G. arboreum* cultured on MS + IAA (1 mg/l) + kin (0.2 mg/l) + CH (250 mg/l) after transfer to half-strength MS + 1% sucrose. (Gill and Bajaj 1987)

Table 10. Percentage germination and callus growth response of hybrid ovules cultured on various media (data based on 100 ovules cultured in each experiment). (Gill and Bajaj 1987)

Media (mg/l)	*G. hirsutum* cv. F 414 × *G. arboreum* cv. G 27		*G. arboreum* cv. G 27 × *G. hirsutum* cv. F 414	
	Germination (%)	Callus[a]	Germination (%)	Callus
MS + CH 250 + IAA 1 + kin 0.2	27.5	+	10.7	+ +
MS + CH 250 + IAA 1.5 + kin 0.5	14.3	+	34.7	+ + +
			0	
			0	

[a] – No callus; + little callus; + + good callus; + + + profuse callus.

Table 11. Variability in chromosome number in hybrid callus cells obtained from embryos of a cross *Gossypium hirsutum* (2n = 52) × *G. arboreum* (2n = 26). Data based on 477 dividing cells. (Bajaj and Gill 1985)

Chromosome no.	No. of cells studied	Percent
25	7	1.47
26–29	70	14.67
30–33	105	22.01
34–37	161	33.75
38–41	123	25.79
42–45	7	1.47
46	4	0.84

On MS agar medium, the parental as well as the hybrid ovules underwent callusing and their responses were similar. However, the germination of hybrid ovules was slower in comparison to their respective parents, and no germination was observed in the absence of IAA and kin, whereas the parent ovules germinated even on basal medium. *G. hirsutum* × *G. arboreum* ovules produced more vigorous seedlings than the reciprocal hybrid.

The seedling from the in vitro-germinated ovules (Fig. 8E) were transferred to a fresh medium (half-strength MS with 1% sucrose) and placed in an illuminated room, where they continued further growth. The seedlings from *G. arboreum* × *G. hirsutum* crosses had a lower survival rate (40%) than the seedlings from the reciprocal crosses. The cytological examination showed a triploid chromosome number (2n = 3x = 39) in the root tips of hybrid plantlets.

Hybrid Embryo-Derived Callus. The embryos of a cross *G. hirsutum* × *G. arboreum*, cultured on Smith et al. medium (see Stewart and Hsu 1978) proliferated to form callus. This was then subcultured periodically on MS + NAA (2 mg/l) + BAP (0.5 mg/l). The callus was creamy-white, soft, and friable. Cytological examination of 1-year-old callus showed a wide range of chromosome number, varying from less than 26 to more than 45 (Table 11). However, the majority of the cells belonged to the category 26–41. Cells with less than 26 and more than 45 chromosomes were rare, i.e., 1.4 and 0.84%, respectively. No polyploids were observed.

The formation of hybrid callus which shows greater variations can be utilized to generate environments for enhancing chromosome breakage and reunion events (Larkin and Scowcroft 1981). Such a hybrid callus may thus provide means for obtaining genetic exchange in the interspecific hybrids. The plants from hybrid callus are likely to show an enhanced frequency of genetic variation. This is an area which needs to be explored, and which may open wide vistas for future plant breeding programs.

2.2.2 Gossypium hirsutum × G. trilobum (Umbeck and Stewart 1985)

Hybrids were obtained from immature embryo (from 15–25 DAP ovules) culture of a cross *G. hirsutum* (male) × wild diploid *G. trilobum*. In the hybrid triploid plant (39 chromosomes) almost all the bivalents were allosyndetic and showed a high frequency of chiasmata formation. The chromosome pairing indicated a very close homeology between *G. trilobum* DC ex. Skov (D_8) genome and the D subgenome of *G. hirsutum*. Cytogenetic analysis revealed no quadrivalent chromosome configurations.

2.2.3 Gossypium sturtianum × G. hirsutum (Altman et al. 1987; Mergeai et al. 1995)

The introgression of the glanded-plant/glandless seed trait from *G. sturtianum* ($2n = 26$) into *G. hirsutum* using conventional breeding is not successful beyond the first backcross (BC_1). Thus, ovule culture was used to overcome this fertility barrier (Altman et al. 1987). Pollination of BC_1 plants in the field followed by ovule culture produced 14 mature BC_2 progeny with partial female fertility. One BC_2 had no glands while in culture, but was fully glanded after the first true leaf emerged. From this glandless-seed plant, 8 BC_3 plants, and 125 BC_4 plants were obtained. Five BC_4 seeds were completely glandless.

Recently, a system of randomly amplified polymorphic DNA (RAPD) markers has been developed to facilitate the transfer of *G. sturtianum* genes into *G. hirsutum* genome by hybridization and beckcrossing (see Chap. II.4, this Vol.).

2.2.4 Gossypium hirsutum × G. harknessi (Bojinov 1992)

Gossypium harknesii has good resistance to *Verticillium*, in combination with drought resistance and fiber quality properties; however, it is incompatible with *G. hirsutum*. Stoilova and Savova (1986) reported obtaining young bolls through conventional crossing, but they shed between 6 and 12 days after pollination. However, application of growth regulators (NAA 100 mg/l + GA_3 50 mg/l) once a day for 15 days after pollination helped to retain 50% of the bolls on the plant, and about 26% of the ovules contained normal embryos. In culture, these embryos underwent normal cotyledons development, and the primary root penetrated deep into the medium. Cultures first incubated in the dark and then transferred to a 16/8-h photoperiod grew well. After 35–40 days the first true leaves were observed. The obtained regenerants developed further and had good fertility (more than 50% normal seed); but no further details were given.

2.3 In Vitro Fertilization and Interspecific Hybridization

In vitro fertilization, followed by culture of immature ovules, has also been experimented with some success. Rafaat et al. (1984) fertilized *G. hirsutum* in vitro with pollen of *G. barbadense*, and obtained hybrids. Later, Liu et al. (1992) brought about in vitro fertilization between *G. hirsutum* and *G. arboreum*. They studied the effects of various media, temperatures, relative humidity, and shake culture, and obtained a few hybrid seed/plants. These plants had stronger vegetative growth than their parents, but were highly sterile. The shapes of leaves and flowers were somewhat intermediate to those of their parents.

3 Summary

Interspecific hybrids in cotton can be successfully achieved through the culture of excised embryo/ovules. Normally, in many crosses, especially between diploids and tetraploids, the hybrid embryos abort at a very early stage. However, this has been delayed by the application of growth hormones, and when grown to a stage where the embryo is few celled, the ovules could be excised and cultured on agar medium to form hybrids. The hybrids (triploids) between *G. hirsutum* and *G. arboreum* were sterile. The pollen from triploid plants was used to pollinate *G. hirsutum* and *G. arboreum* and a few shriveled seeds were obtained.

Hybrids between diploid and diploid could be easily obtained. In all the hybrids, flower size and shape, bract size and shape, and pedicel length resembled the wild species. The hybrid *G. arboreum* × *G. anomalum* showed particularly vigorous growth and produced some seeds when pollinated with *G. arboreum* pollen or vice versa. However, second backcross generation seeds were obtained from all the plants.

References

Afzal M, Sikka SM, Rehman A (1945) Cytogenetic investigations in some *arboreum-anomalum* crosses. Indian J Genet 5:82–91

Altman DW, Stelly DM, Kohel RJ (1987) Introgression of the glanded-plant and glandless-seed trait from *Gossypium sturtianum* Willis into cultivated upland cotton using ovule culture. Crop Sci 27:880–884

Bajaj YPS (1990) Wide hybridization in legumes and oilseed crops through embryo, ovule, and ovary culture. In: Bajaj YPS (ed) Biotechnology in agriculture and forestry, vol 10. Legumes and oilseed crops I. Springer, Berlin Heidelberg New York, pp 3–37

Bajaj YPS (1994) Biotechnology in agriculture and forestry, vol 27. Somatic hybridization in crop improvement I. Springer, Berlin Heidelberg New York

Bajaj YPS, Gill MS (1985) In vitro induction of genetic variability in cotton (*Gossypium* spp). Theor Appl Genet 70:363–368

Bajaj YPS, MS Gill (1986) Biotechnology of cotton improvement. In: Crocomo J et al. (eds) Biotechnology of plants and microorganisms. Ohio State Univ Press, Columbus, pp 118–151

Berger J (1969) The world's major fibre crops, their cultivation and manuring. Conzett and Huber, Zurich

Bojinov B (1992) Interspecific hybrids *Gossypium hirsutum* × *G. harknessi* Brang. obtained through embryo culture. 2nd Natl Conf In vitro techniques applied to crop plants. Plovdiv, November 5, 1992

Endrizzi JE, Turcote EL, Kohel RJ (1985) Genetics, cytology and evolution of *Gossypium*. Adv Genet 23:271–375

Fryxell PA (1984) In: Kohel RJ, Lewis CF (eds) Cotton. Agronomy No 24. Am Soc Agron, Madison, pp 27–56

Fryxell PA (1992) A revised taxonomic interpretation of *Gossypium* L. (Malvaceae). Rheedea 2:108–165

Gerstel DV, Phillips LL (1958) Segregation of synthetic amphiploids in *Gossypium* and *Nicotiana*. Cold Spring Harb Symp Quant Biol 13:225–238

Gill MS, Bajaj YPS (1984a) In vitro production of interspecific hybrids in cotton. Curr Sci 53:102–103

Gill MS, Bajaj YPS (1984b) Interspecific hybridization in the genus *Gossypium* through embryo culture. Euphytica 33:305–311

Gill MS, Bajaj YPS (1986) In vitro production of interspecific hybrids in *Gossypium* and their utilization in backcrossing. Z Pflanzenzuecht 96:174–176

Gill MS, Bajaj YPS (1987) Hybridization between diploid (*Gossypium arboreum*) and tetraploid (*Gossypium hirsutum*) cotton through ovule culture. Euphytica 36:625–630

Gill MS, Bajaj YPS (1991) In vitro hybridization between *Gossypium arboreum* and *G. anomalum* and their backcross progeny. Plant Tissue Cult 1(2):101–108

Gill MS, Bajaj YPS (1996) Hybridization between *Gossypium herbaceum* and *Gossypium stocksii* through embryo and ovule culture. Cell Res (Beijing)

Hutchinson JB, Silow RA, Stephens SG (1947) The evolution of *Gossypium* and the differentiation of the cultivated cottons. Oxford Univ Press, London, 160 pp

Iyengar NK, Santhanam V, Leela M (1957) Further studies in *arboreum-anomalum* crosses. Indian Cotton Grow Rev 12:369–375

Kalyanaraman SM, Santhanam V (1955) A note on the performance of some interspecific hybrids involving wild species of *Gossypium* L. *Arboreum-anomalum* crosses. Indian Cotton Grow Rev 136–140

Larkin PJ, Scowcroft WR (1981) Somaclonal variation – a novel source of variability from cell cultures for plant improvement. Theor Appl Genet 60:197–214

Liu C, Shun J, Liu J (1992) In vitro interspecific fertilization, embryo development and formation of hybrid seedlings between *Gossypium hirsutum* and *G. arboreum*. Euphytica 60:79–88

Marappan PV, Santhanam V (1962) Breeding behaviour of some *arboreum-anomalum* crosses. Indian Cotton Grow Rev 16:24–30

Mehboob Ali S, Krishnayya VV, Satyanarayana Murthy K (1964) A note on an interspecific hybrid in cotton (*Gossypium arboreum* race *indicum* × *G. stocksii*). Indian Cotton Grow Rev 18:107–108

Mergeai G, Bi Irie V, Dujardin P, Baudoin JP (1995) Introgression of glanded-plant and glandless-seed trait from *G. sturtianum* Willis into tetraploid cotton plants. Proc Beltwide Cotton Improvement Conf, San Antonio, pp 513–514

Meyer VG (1974) Interspecific cotton breeding. Econ Bot 28:56–60
Percival AE, Kohel RJ (1990) Distribution, collection, and evaluation of *Gossypium*. Adv Agron 44:225–256
Prentice AN (1972) Cotton with special reference to Africa. Longmann, London, pp 68–76
Pundir NS (1972) Experimental embryology of *Gossypium arboreum* and *G. hirsutum* and their reciprocal crosses. Bot Gaz 133:7–26
Rafaat M, Rossignol L, Demarly Y (1984) Interspecific hybrid of *Gossypium hirsutum* L. × *G. barbadense* L. via in vitro fertilization and ovule culture. Z Pflanzenzuecht 93:137–146
Santhanam V (1967) Breeding procedure for cotton. Tech Bull (AGIR) Series No 10. ICAR, New Delhi
Sikka SM, Joshi AB (1960) Breeding. In: Cotton in India – a monograph. Indian Central Cotton Committee, Bombay, pp 137–235
Silow RA (1941) The comparative genetics of *Gossypium anomalum* and the cultivated Asiatic cottons. J Genet 42:259–358
Stewart JM, Hsu CL (1978) Hybridization of diploid and tetraploid cottons through in-ovulo embryo culture. J Hered 69:404–408
Stoilova A, Savova N (1986) Studies on growing and development of wild cotton relatives and the possibilities to their hybridization with cultivated *Gossypium hirsutum* L. Crop Sci (Bulg) 5:5–8
Umbeck PF, Stewart JMcD (1985) Substitution of cotton cytoplasm from wild diploid species for cotton germplasm improvement, Crop Sci 25:1015–1019
Weaver JB Jr (1958) Embryological studies following interspecific crosses in *Gossypium*. II. *G. arboreum* × *G. hirsutum*. Am J Bot 45:10–16

II.2 In Vitro Fertilization in Cotton

C. Liu[1] and Y.P.S. Bajaj[2]

1 Introduction

For interspecific hybridization in various crops, in addition to sexual hybridization through embryo rescue and somatic hybridization by protoplast fusion, intraovarian pollination and in vitro fertilization techniques can be used to produce hybrids in incompatible crosses. The technique of in vitro pollination/fertilization involves growing unpollinated ovaries/flowers/ovules on synthetic media, and then sprinkling pollen over them, or onto the cut end of the style. This technique has been used for seed production in a number of plant species (see Bajaj 1979; Rafaat et al. 1984; Zenkteler 1990; Kranz and Loerz 1993).

Gossypium hirsutum (2n = 52) and *G. arboreum* (2n = 26) are the two main species of cotton grown in India and China. *G. hirsutum* types generally yield more, possess longer fiber of superior quality, and are resistant to diseases such as *Fusarium* wilt, whereas *G. arboreum* can better tolerate environmental stress and is comparatively resistant to various insect pests. Efforts to combine the desirable traits of these two species have not been encouraging, due to abortion of the hybrid embryo/endosperm in the early stages of seed development (Weaver 1958; Pundir 1972). Thus, in vitro culture methods have been incorporated with a view to increasing genetic variability in cotton (Bajaj and Gill 1985).

In an earlier study (Gill and Bajaj 1987), hybrids were obtained between *G. arboreum* and *G. hirsutum* by the use of in vitro ovule culture. Early abortion of the embryo was prevented by repeated treatment of the flowers, immediately after pollination, with a solution of GA and NAA. Ovules excised 3 days after pollination and cultured, eventually germinated. However, most of the hybrids were sterile. Thus the present chapter, which is primarily based on an earlier publication (Liu et al. 1992), describes in vitro fertilization successfully applied to obtain interspecific hybrids. With a view to increasing the frequency of hybrid formation, various factors such as the effect of various media, temperature, relative humidity, and the effect of calyx and petals were studied.

[1] Department of Agronomy, Huazhong Agricultural University, Wuhan 430070, China
[2] Biotechnology in Agriculture and Forestry, A 135, New Friends Colony, New Delhi 110065, India

Biotechnology in Agriculture and Forestry, Vol. 42
Cotton (ed. by Y.P.S. Bajaj)

2 Methodology

Gossypium hirsutum L. cv. 3247 was fertilized in vitro with pollen of *G. arboreum* L. cv. Shuixianzhongmian, and hybrids were obtained by the following three steps; (1) in vitro fertilization, (2) embryo culture, and (3) growth of hybrid seedlings.

Five kinds of media were used:

M1 – Pollen germination medium, based on that of Taylor (1972), but contained 500 mg/l each of proline, lysine, serine, and glutamine.
M2 – Double fertilization medium, nearly the same as that of Beasley (1973).
M3 – Embryo development medium, nearly the same as that of Beasley, but 1 mg IAA, 0.5 mg NAA, and 0.5 mg kinetin per liter were added.
M4 – Seedling formation medium, same as that of Schenk and Hildebrandt (1972), but 500 mg/l activated charcoal was added.
M5 – Green seedling growth medium, same as M4 but without sucrose.

1. *In Vitro Fertilization.* Emasculation was carried out at 17.00 h on the day before pollination. After emasculation, the following three experiments were done, (1) retaining both sepals and petals, (2) petals removed but sepals retained, and (3) sepals and petals both removed. The emasculated flowers were excised and their stalks were inserted in dishes containing M1 (Fig. 1). The next day at 10.00 h, the stigmata were smeared with M1, then pollen of *G. arboreum* were sprinkled. During the next 48 h the pedicels were soaked in M2 till fertilization was accomplished. The method of Williams et al. (1982) was used for the study of pollen/measurement of pollen tube length.

2. *Embryo Culture.* The flowers were taken out of the dishes, 48 h after pollination, then sterilized by treating them with mercuric chloride (0.1%) for 8 min, washed three times with sterilized water, and subjected to flaming with ethyl alcohol for about 30 s. The ovaries were then opened, and the ovules cultured on M3 medium (Figs. 2, 3).

3. *Growth of Seedlings.* The embryos were excised from the cultured ovules, 56 days after pollination, and inoculated on M4 medium (Fig. 4). When the seedlings had five or six leaves, they were transferred to medium M5. After 1 month, they were transplanted into pots, where they grew into mature plants (Fig. 5).

3 Results and Discussion

Although in vitro fertilization resulted in the formation of seeds, their frequency was rather low. Thus, with a view to improving this aspect, various factors affecting fertilization/culture were studied; the results are summarized below, and shown in Figs. 1–27 and Tables 1–5.

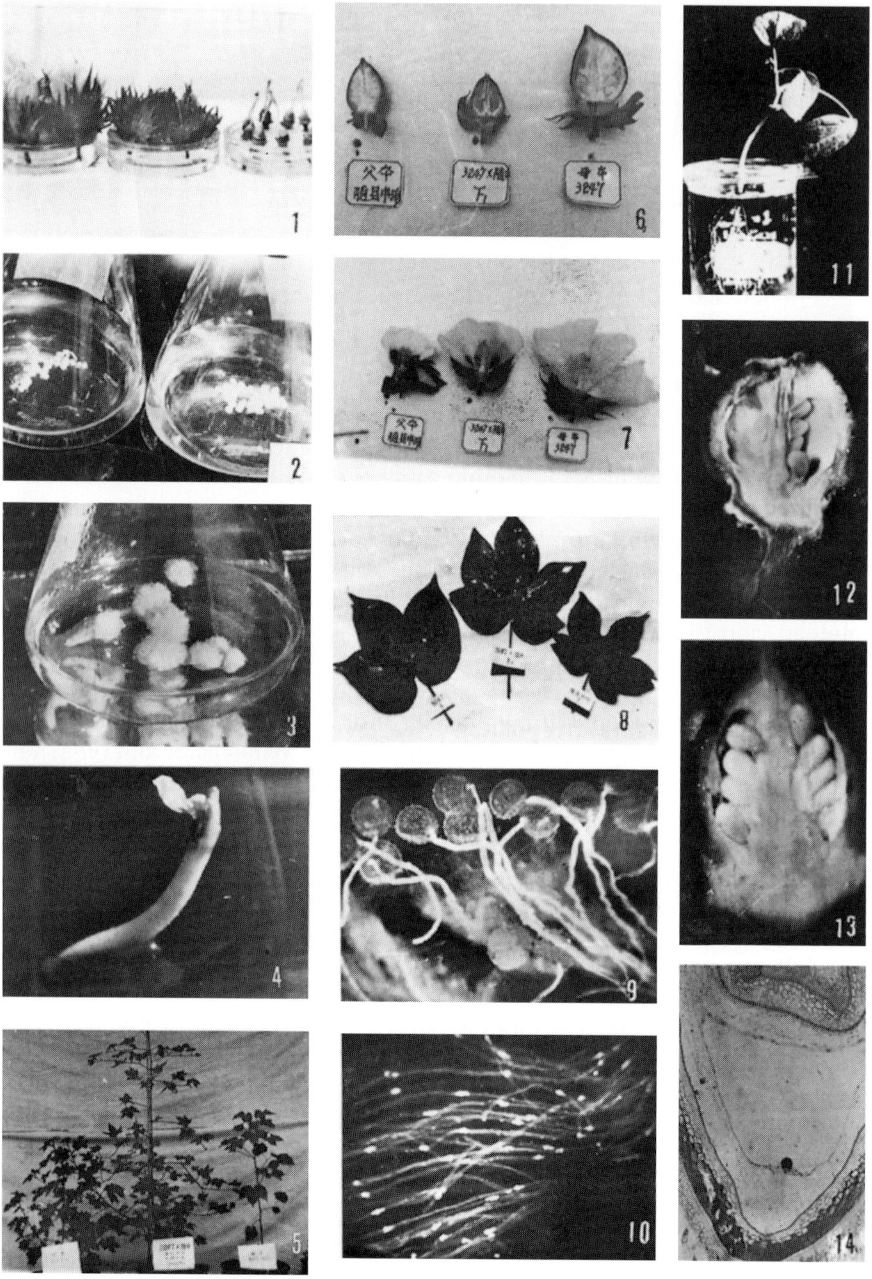

3.1 Effect of Temperature and Relative Humidity (RH)

In a preliminary experiment it was observed that in cotton flowers pollinated in vitro, although the ovary grew vigorously, the ovules were black or brown (Fig. 12). This was caused by high temperature and RH. Generally, temperature and RH were higher than 30 °C and 80%, respectively, when cotton was in bloom in Wuhan, China. Temperature and RH in test tubes were even higher, and proved to be unsuitable for in vitro fertilization. Thus, experiments were designed to overcome these problems.

The optimal RH was 65% (Table 1), the pollen germinating well (Figs. 9, 10), and the pollen tubes reaching the base of the style. The ovules developed better, and their mean width was 1.5 mm (Fig. 13). When RH increased, the fertilization percentage decreased rapidly. The inhibitive effects of a high RH were stronger when the temperature was raised to 32–35 °C (Table 2). Pollen germinated much less, ovules became black and shriveled, and no fertilization occurred. To conclude, the optimal RH and temperature for in vitro fertilization were 65% and 26–30 °C, respectively.

3.2 Effect of Various Media/Shake Culture

Smearing the stigmata with M1 medium promoted pollen germination/in vitro fertilization (Tables 3, 4). The smeared pollen grain germinated earlier, the germination ratio was higher, and the pollen tubes were longer than in the control. Ovules developed better because of better fertilization. Multicellular embryos were observed on the 6th day after pollination (Fig. 20), whereas in the control most egg cells degenerated on the 3rd day after pollination (Fig. 21).

Figs. 1–14. In vitro fertilization in ♀ *Gossypium hirsutum* × ♂ *G. arboreum*. **1** The emasculated excised flowers of *G. hirsutum* inserted in M2 medium (*left* with sepals and petals; *middle* with sepals but no petals; *right* both sepals and petals removed. **2** Ovule separated from ovaries at 48 h after pollination were cultured on M3 medium (*left* control – emasculated but not pollinated; *right* hybrid ovules). **3** Hybrid ovules cultured on M3 medium for 16 days. **4** Young seedlings from mature embryos cultured on M4 medium for 5 days. **5** Vigorous vegetative growth of a hybrid plant (*left* paternal plant; *middle* F_1 hybrid; *right* maternal plant). **6** Poor reproductive growth of hybrid plant (high sterility), Seeds were found in the parent's ovaries at 5th day after pollination, but the ovaries of the hybrid plants were shriveled and empty (*left* paternal plant; *middle* F_1 plant; *right* maternal plant). **7** Flowers of hybrid plant and its parents (*left* paternal plant; *middle* F_1 plant; *right* maternal plant). **8** Leaves of hybrid plant and its parents (*left* maternal plant; *middle* F_1 plant; *right* paternal plant). **9** Microscopic observations of *G. arboreum* pollen germinating luxuriantly on the pistils of *G. hirsutum* at RH 65%, 5 h after pollination. **10** Microscopic observation of pollen tubes at RH 65%, 12 h after pollination. **11** Young seedlings grown on M5 medium. **12** Vertical section of an ovary after 48 h pollination, showing black and shriveled ovules (RH 85%, temperature 32–35 °C). **13** Vertical section of the hybrid ovary after 48 h pollination showing normal growth of ovules (RH 65%, temperature 26–30 °C). **14** Early degeneration of the endosperm cells; the core of the ovule is empty

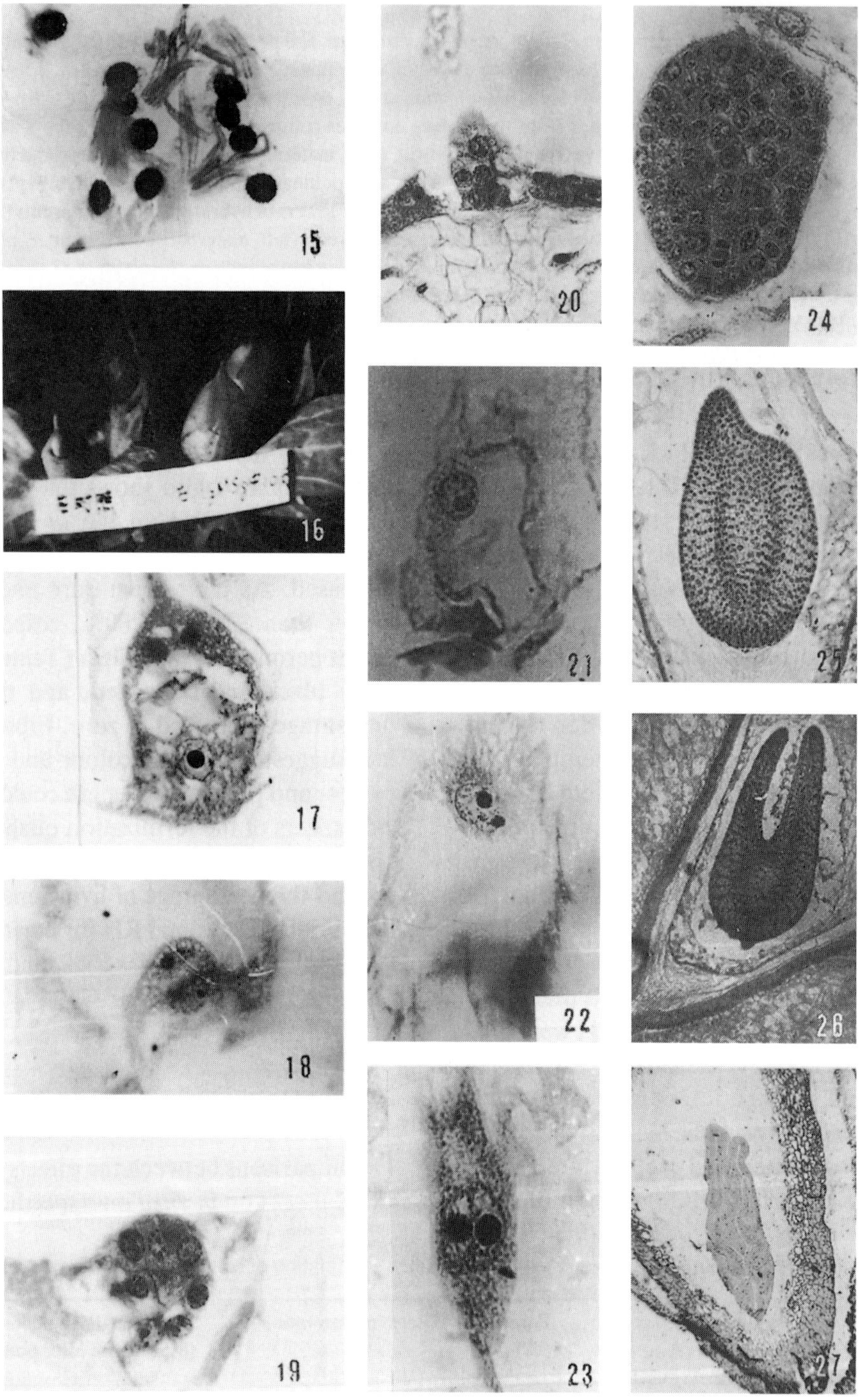

Table 1. Effects of RH on in vitro interspecific fertilization (temperature = 26–30 °C)

RH (%)	Ovule growth (48 h after pollination)		Mean length of pollen tubes (29 h after pollination) (mm)	Pollen germination ratio (%)	Fertilization (%)
	Color	Mean width (mm)			
45	White	1.10	4.15	–	
65	Milky white	1.50	21.50	100	74.0
75	White	1.00	7.65	80	9.7
85	Faintly brown	0.75	2.05	10	0
100	70% brown + 30% black	0.65	0.80	1	0

–, Not recorded.

Table 2. Effects of RH on in vitro interspecific fertilization (temperature = 32–35 °C)

RH (%)	Ovule growth (48 h after pollination)		Mean length of pollen tubes (29 h after pollination) (mm)	Pollen germination ratio (%)	Fertilization (%)
	Color	Mean width (mm)			
45	–	–	2.25	10	1.2
65	White	1.10	12.20	90	29.0
75	60% + slightly yellow + 30% brown + 10% white	0.96	5.30	60	2.0
85	Brown	0.70	1.15	1	0
100	Black	0.50	0.21	0.1	0

Shake cultures enhanced the growth of pollen tubes. The growth of pollen tubes influenced the fertilization ratio positively, and the first vigorous hybrid seedlings were obtained by shake culture (Fig. 4).

3.3 Effect of Sepals and Petals

The beneficial effect of sepals on the development of fruit/seeds in the in vitro-cultured ovaries was reported as early as 1966 (Bajaj 1966). In the present

◄

Figs. 15–27. 15 Ungerminated *G. arboreum* pollen grains under high temperature (32–35 °C) and high RH (85–90%). **16** Ovary (22 mm long) cultured on M2 medium for 11 days. **17** Hybrid zygote on the 3rd day after pollination. **18** Bicellular embryo after 4 days pollination. **19** Small sphere-shaped embryo consisting of over ten cells after 6 days pollination. **20** Multicellular embryo after 6 days pollination (the stigma was smeared with M1 medium). **21** Control of **20** (showing the degenerating egg cells after 3 days pollination). **22** The sperm of *G. arboreum* going towards the egg cells after 49 h pollination. **23** The sperm of *G. arboreum* going towards polar cell after 32 h pollination. **24** Sphere-shaped embryo after 10 days pollination. **25** Heart-shaped embryo after 14–16 days pollination. **26** Torpedo-shaped embryo after 16–20 days pollination. **27** Embryo after 25 days pollination (curving cotyledon)

Table 3. Effects of smearing the stigmata with M1 medium on pollen germination of cotton

Treatment	Replicates	Pollen grains germination ratio (%)	Mean length of pollen tubes (8.5 ± 0.5 h after pollination) (mm)
	1	100	21.00
Smearing	2	100	19.30
	3	95	9.70
	1	76	3.10
No smearing	2	70	1.24
	3	75	1.40

Table 4. Effects of smearing the stigmata with M1 medium on in vitro interspecific fertilization of cotton

Treatment	Size of ovaries (2 days after pollination) (mm)		Size of ovules (2 days after pollination) (mm)		Appearance of tomenta on ovules	Characteristics of young embryos
	Width	Length	Width	Length		
Smearing	11.0	20.0	1.4	1.6	Short tomenta	Multicellular embryos on the 6th day after pollination
No smearing	9.0	18.0	1.1	1.2	No tomenta or difficult to find	Beginning of unfertilized ovule's degeneration on the 3rd day after pollination

study, the retention of both petals and sepals resulted in larger ovules and higher percent formation (Table 5).

3.4 Hybrid Plants

Embryos excised from cultured ovules, and inoculated on M4 medium (Fig. 4) developed into plantlets. When finally transplated into soil, they matured. Hybrid plants had stronger vegetative growth than their parents, but were highly sterile. The shape of the flowers and leaves of the hybrids was intermediate between their parents (Figs. 5–8).

4 Summary and Conclusion

In vitro fertilization was carried out between ♀ *Gossypium hirsutum* and ♂ *G. arboreum*, and hybrid seedlings were obtained. The cut end of the pedicel of

Table 5. Effects of petals and sepals of maternal plants on in vitro interspecific fertilization

Treatment	Mean width of ovules (2 days after pollination) (mm)	White, plump ovules (2 days after pollination (%)	Percentage of living embryos (56 days after pollination)		
			No. of ovules	No. of embryos	Living embryos (%)
Petals + sepals	1.3	71	1328	25	1.88
Sepals but no petals	0.9	32	1052	2	0.19
No petals and no sepals	1.0	29	667	1	0.15

the emasculated excised flowers was inserted in a synthetic medium in dishes. The flowers were then pollinated, and fertilization was accomplished. The effect of various factors, such as temperature, relative humidity, media, petals, and sepals, on in vitro fertilization, embryogenesis, and seedling formation was studied. The important factors were temperature and RH. The optimum RH for in vitro fertilization was 65%, and a suitable temperature range 26–30 °C. The pollen germination ratio decreased rapidly at RH 80%. When temperature and RH were higher than 32 °C and 80%, respectively, the fertilization rate was zero.

There was a positive effect of sepals and petals of the maternal flowers on in vitro fertilization and ovule growth, and shake culture enhanced pollen-tube growth, and eventually fertilization. The processes of double fertilization and embryo development were also studied cytologically.

Although normal development of viable hybrid seeds and plants has been realized in a number of agricultural crops by the culture of embryo, ovule, and ovaries (See Bajaj 1990), the procurement of viable seed from in vitro-fertilized ovules, especially in cases related to pollen incompatibility, is an area which has great significance in wide hybridization programs.

Acknowledgment. This chapter is primarily based on a paper published in Euphytica in 1992. The authors are grateful to the Kluwer Academic Publishers Dordrecht for kindly granting permission to use data and photographs.

References

Bajaj YPS (1966) Growth of *Hyoscyamus niger* ovaries in culture. Phyton 23:57–62

Bajaj YPS (1979) Test-tube fertilization and development of maize (*Zea mays* L.) plants. Indian J Exp Biol 17:475–478

Bajaj YPS (1990) Wide hybridization in legumes and oilseed crops through embryo, ovule, and ovary culture. In: Bajaj YPS (ed) Biotechnology in agriculture and forestry, vol 10. Legumes and oilseed crops I. Springer, Berlin Heidelberg New York, pp 3–37

Bajaj YPS, Gill MS (1985) In vitro induction of genetic variability in cotton (*Gossypium* spp.). Theor Appl Genet 70:363–368

Beasley CA, Ting IP (1973) The effects of plant growth substances on in vitro fibre development from fertilized cotton ovules. Am J Bot 60:130–139

Gill MS, Bajaj YPS (1987) Hybridization between diploid (*Gossypium arboreum*) and tetraploid (*Gossypium hirsutum*) cotton through ovule culture. Euphytica 36:625–630

Kranz E, Loerz H (1993) In vitro fertilization with isolated, single gametes results in zygotic embryogenesis and fertile maize plants. Plant Cell 5:739–746

Liu Chengzhi, Shun J, Liu J (1992) In vitro interspecific fertilization, embryo development and formation of hybrid seedlings between *Gossypium hirsutum* and *G. arboreum*. Euphytica 60:79–88

Pundir NS (1972) Experimental embryology of *Gossypium arboreum* and *G. hirsutum* and their reciprocal crosses. Bot Gaz 133:7–26

Refaat M, Rossignol L, Demarly Y (1984) Interspecific hybrid of *Gossypium hirsutum* L. × *G. barbadense* L. via in vitro fertilization and ovule culture. Z Pfanzenzücht 93:137–146

Schenk RU, Hildebrandt AC (1972) Medium and techniques for induction and growth of monocotyledonous and dicotyledonous plant cell cultures. Can J Bot 50:199–204

Taylor RM (1972) Germination of cotton pollens on an artificial medium. Crop Sci 12:243–244

Weaver JB Jr (1958) Embryological studies following interspecific crosses in *Gossypium* II. *G. arboreum* × *G. hirsutum*. Am J Bot 45:10–16

Williams EG, Bruce Knox R, Rouse JL (1982) Pollination subsystems distinguished by pollen tube arrest after incompatible interspecific crosses in *Rhododendron* (Ericaceae). J Cell Sci 53:255–277

Zenkteler M (1990) In vitro fertlization and wide hybridization in higher plants. Crit Rev Plant Sci 9:267–279

II.3 In Vitro Culture of Cotton Ovules

J.E. MELLON[1]

1 Introduction – Early Studies

In vitro culture of developing cotton ovules is a significant achievement in the continuing investigation of cotton and its agronomic properties. Pioneering work on in vitro growth of cotton ovules met with limited sucess, as ovule growth was abnormal and fiber development did not occur (Joshi 1960). Later, the effects of hormones, casein hydrolysate, and yeast extract on the development of the cotton embryo and integuments were reported (Joshi and Johri 1972). Beasley and coworkers began development of the cotton ovule culture system in 1971 (see Beasley 1977) in order to conduct investigations on fiber development. Eid et al. (1973) were successful in culturing 10-day postanthesis ovules on MS medium (Murashige and Skoog 1962), but the resulting embryos did not develop normally. Beasley and Ting (1973) subsequently developed a phytohormone-supplemented medium (BT) derived from the MS medium which supported the development of the ovules in a manner close to that observed in vivo. The BT medium could be utilized for both fertilized (Beasley and Ting 1973) and unfertilized (Beasley and Ting 1974) ovules. While this medium supported normal growth of the ovule and fiber during the first 2 weeks of development, the embryos did not continue to develop normally, and subsequently degenerated. Stewart and Hsu (1977) resolved this problem by inclusion of an ammonium ion amendment to the medium; by making this medium modification, 50% of the ovules produced physiologically mature embryos. They transferred 10-week-old cotton ovules to a root-shoot medium and obtained healthy plants derived from individual ovules. The procedural details of in vitro cotton ovule culture have been reviewed (Beasley 1977, 1984). A more recent review of ovule methodology provides historical insight as well as possible future avenues of research for this system (Beasley 1992). Uses of ovule culture in cotton biotechnology have also been reviewed (Stewart 1991).

The cotton ovule culture system was developed primarily for the investigation of fiber physiology, and the technique has proved to be a valuable tool for this purpose. Once a medium was available which allowed for the complete

[1] USDA/ARS, Southern Regional Research Center, P.O. Box 19687, New Orleans, Louisiana 70179-0687, USA

Biotechnology in Agriculture and Forestry, Vol. 42
Cotton (ed. by Y.P.S. Bajaj)

developmental sequence from fertilized ovule to mature plant, the cotton ovule system was quickly utilized in hybridization studies involving source germplasms from diverse backgrounds. In addition, this culture technique has been used to study secreted proteins and polysaccharides of cotton origin, as well as physiological investigations of interactions between cotton and mycotoxigenic fungi. This chapter will look at each of these applications of the cotton ovule culture system in more depth, followed by a summary of what has been accomplished by use of this investigative tool in the past 25 years.

2 Applications of the Cotton Ovule Culture System

2.1 Ovule Culture and Hybridization Studies

The utilization of the entire *Gossypium* germplasm to incorporate desirable traits such as pest resistance, male sterility, and fiber quality into cultivated cotton has been hampered by incompatibility between the Old World (Asian) diploid species and the New World tetraploid species using traditional breeding techniques. Adopting the liquid medium used to culture self-fertilized *G. hirsutum* ovules, Stewart and Hsu (1978) developed a medium which supports the normal development of interspecific ovules to maturity by modifying the phytohormone regime. They noted with some crosses that the resulting cultured ovules produced excessive amounts of callus tissue (Stewart and Hsu 1978). Use of the in vitro ovule culture technique circumvents many of the problems which cause interspecific incompatibility. Thus, the entire *Gossypium* germplasm should now be available for improvement of cultivated tetraploid cotton varieties.

Bajaj and Gill (1985) carried this work forward by demonstraing that callus tissue derived from interspecific hybrid ovule cultures contains a variable number of chromosomes [1n to polyploidy (>6n)]. They suggested that this culture-induced variability might offer opportunities to augment genetic variability in commercial cotton varieties. Work with crosses of the incompatible species *G. hirsutum* and *G. arboreum* demonstrated that a series of five different media are necessary for development of fertilized ovules to healthy plants (Thengane et al. 1986). Additional work has shown that following pollination in incompatible interspecific crosses, immediate treatment of the flower with a phytohormone solution (gibberellic acid and naphthalene acetic acid) prevents early abortion of the embryo (Gill and Bajaj 1987). When fertilized ovules are excised at 3 days postanthesis and transferred to a solid medium (agar-based) supplemented with casein hydrolysate, indoleacetic acid, and kinetin, the enclosed embryos germinate to form hybrid plants. However, these plants (triploid) are sterile. A chromosome analysis of interspecific-derived pollen mother cells (*G. hirsutum* × diploid species) was made possible by the use of ovule culture to obtain the hybrid plants (Menzel et al. 1982). In vitro culture of 2-day postanthesis ovules has been successfully used to recover viable embryos with duplication deficiency products from heterozygous translocations in *G. hirsutum*, although the selected phytohormone regime is not optimal for this application (Contolini and Menzel 1987). However, the ovule-culture technique reduces the time required to obtain viable seedlings by 50%.

Introgression of a glanded plant (leaf, stem, floral structures) with the glandless seed trait (from *G. sturtianum*) has been hampered by fertility barriers. Embryo rescue through ovule culture has successfully overcome the barriers associated with pentaploid interspecific hybrids (Altman et al. 1987). This research produced germplasm which shows promise for developing cotton with higher levels of insect resistance coupled with improved nutritional characteristics (less gossypol). Altman (1988) compared in vitro-culture techniques (ovule and embryo) with traditional hybridization methods, incorporating a postanthesis flower application of phytohormone to obtain viable plants from interspecific crosses between *G. hirsutum* and various diploid *Gossypium* species. This work concluded that the traditional methods are superior to the in vivo methods in crosses utilized in the study. It is clear that the phytohormone flower treatment greatly improves boll retention. However, in most cases, the in vitro seed (viable) production is still poor. In addition, the ovule medium (Altman et al. 1987) used contained no phytohormones. Perhaps inclusion of phytohormones in the medium would have improved the yield of viable plants derived from the ovule-culture method. Ovule culture still appears to be a valuable technique when attempting crosses using diploid germplasm as the maternal cotton parent. A summary of the successful crosses obtained from tetraploid and diploid cotton species through ovule culture is included in a review of in vitro fertilization and embryo rescue in higher plants (Stewart 1981). The cumulative investigations of Bajaj and Gill using ovule culture for hybridization work have been reviewed as part of a discussion of micropropagation methods (Bajaj and Gill 1992; see also Chap. II.1, this Vol.)

In summary, it is apparent that cotton ovule culture methodology has provided a valuable tool to circumvent problems encountered in hybridization of tetraploid and diploid cotton species. The phytohormone regime required for successful development of ovule to whole plant would appear to be critical and may vary with each parental germplasm set. Thus, the full potential of ovule culture for improvement of commercial cotton germplasm remains to be tapped.

2.2 Ovule Culture and Fiber Development

Since the in vitro ovule-culture system was developed primarily to investigate fiber physiology, it is not surprising that many investigators have used this tool for that purpose. It has been a primary tool to determine the physiological effects of boron deficiency on ovule/fiber development. Ovules cultured in vitro produce callus and undergo necrosis in response to boron deficiency (Birnbaum et al. 1977). Exposure of cotton ovules to 6-azauracil produces symptoms similar to boron deficiency. The study suggests that boron deficiency interferes with UDP-glucose synthesis by stimulating UDP-glucose pyrophosphorylase, thereby retarding fiber growth. Incorporation of UDP-[^{14}C]-glucose into fibers developing on in vitro-grown ovules is significantly reduced in the case of boron deficiency (Dugger and Palmer 1980). A study utilizing radiolabeled orotic acid found that boron deficiency inhibits fiber uptake of this substrate (Wainwright et al. 1980). This physiological condition also shunts the substrate into RNA away from pyrimidine pools and results in a smaller incorporation of orotic acid into UDP-glucose. A study of [^{14}C]-glucose incorporation into cotton fibers offers support for the role of boron. Boron deficiency appears to center on the conversion of glucose to the wall-building substrate UDP-glucose (Dugger and Palmer 1985).

A thorough study was undertaken by Meinert and Delmer (1977) to compare biochemical parameters during development of cotton fibers derived from bolls or in vitro-cultured ovules. The temporal expression of weight, length, cellulose deposition, neutral sugar composition, uronic acid composition, and protein composition is similar for both types of fibers. The predominant precursor for both cellulose and β-1,3-glucans in developing cotton fibers is UDP-glucose, as demonstrated in fibers derived from bolls and ovule cultures (Carpita and Delmer 1981). These studies demonstrate the validity of using in vitro ovule cultures as a model for the developing cotton fiber.

The timing sequence for fiber differentiation was deduced by a study conducted with the cotton ovule system (Graves and Stewart 1988). Epidermal cells on the preanthesis ovule differentiate into fiber initials 2 to 3 days before anthesis and remain latent until stimulated by the phytohormones IAA and GA_3. Fiber initials at up to 2 days preanthesis would readily begin fiber synthesis if provided with the appropriate phytohormone regime (Graves and Stewart 1988). A separate study (Davidonis 1989) extends the time of fiber determination (potential) to 9-day preanthesis ovaries, using a modified MS medium (agar-based). However, the less mature the ovaries are, the fewer ovules produce fiber and the more produce callus. Callus production is not observed in vivo. It would seem, then, that the differentiation period for fiber initiation occurs 2 to 3 days before anthesis, in order to obtain a complete induction of fibers on the developing seed, as occurs in the boll.

The ovule-culture technique has been used in a few instances to characterize cotton mutations. Cultured ovules from a temperature-sensitive mutant produce fibers on a medium at 30 °C, but fail to do so at 32 °C (Beasley and Egli 1977). Unfortunately, the SMA2 and SMA4 stocks of *G. arboreum* which are deposited in the USDA cotton germplasm collection no longer exhibit the temperature-sensitive fiber formation reported by Beasley and Egli. It would appear that the germplasm utilized in their work has been lost (B.A. Triplett, pers. comm.). A developmental mutant which produces shorter-length fibers, termed Ligon-lintless (isogenic mutant of wildtype Texas Marker-1), has been described in cotton ovule, as well as suspension cultures (Triplett et al. 1989). The authors do not attribute developmental differences to deficiencies in the phytohormones GA_3 or IAA, since they are provided in the medium. However, in view of a proposed role for abscisic acid (ABA) in limiting fiber length, it would be interesting to examine the ABA levels in the mutant and wild-type cultures.

A comparison of fibers derived in vivo to those derived from ovule cultures using gel permeation chromatography yields valuable infomation concerning the polymer molecular weights and degree of branching at different stages of fiber development (Triplett and Timpa 1995). Molecular-weight profiles reveal that fibers grown in culture for 21 days appear more similar to 30-day postanthesis fibers (in vivo) than to mature cotton fibers. Polymer branching frequency for the ovule fibers is intermediate between 30-day postanthesis and mature fibers (in vivo). This technique is novel because it allows the determination of fiber polymer molecular weights without the normally observed polymer degradation associated with extraction from the cell walls.

Cotton ovule cultures have been used as a model system to investigate the role of cytoskeletal elements in fiber development. Microtubule arrays change orientation in cotton fibers in a developmentally regulated manner; the newly deposited wall micro-fibrils parallel the microtubules (Seagull 1986). A quantitative study confirmed that changes in the microtubule array occur during the shift from primary to secondary wall formation in the fiber (Seagull 1989). Young cotton fibers ($<$10 dpa) produce ordered wall microfibrils. Arrays of microtubules, containing various lengths and orientations, are observed close to the plasmalemma, parallel to the innermost layer of wall microfibrils. In fibers undergoing secondary wall deposition ($>$16 dpa), the microfibrils are paralleled by microtubules which are longer, more ordered, and more abundant than those observed in young fibers (Seagull 1989). Perturbation of cotton fiber cytoskeletal elements by various disrupting agents (cholchicine, oryzalin, trifluralin, taxol, cytochalasins B and D) indicates that microtubule populations during interphase are heterogeneous and differ in their susceptibility to these agents (Seagull 1990). Microtubule orientation changes suggest that microfilaments may play a role in developmental regulation of microtubule orientation. A study utilizing quantitative electron microscopic methods suggests the existence of three major stages in microtubule patterns during fiber development: (1) a transition between fiber initiation and elongation with microtubules developing a shallow-pitched helical orientation; (2) a transition between primary and secondary wall synthesis with microtubules distinctly shifting to a steeply pitched helical orientation; and (3) early secondary wall synthesis with a fourfold increase in microtubule number. Microfibril orientations change in a manner similar to microtubules, but careful comparison of those different elements reveals distinct differences between them (Seagull 1992). Delineation of the role of cytoskeletal elements in cotton fiber development will require further investigative efforts.

Cultured cotton ovules have been demonstrated to be an excellent model system for investigating effects of cycling warm-cool temperatures (34/15 °C) on development of fibers (Haigler et al. 1991). When observed under cross-section, fibers grown under these conditions demonstrate the presence of growth rings. The cool temperature effects on fiber development appear to be fiber-/ovule-specific events and are not dependent on whole-plant physiology (Haigler et al. 1991). Using a cycling temperature regime with cultured ovules, respiration in the ovule/fiber increases between 18 and 34 °C and remains level up to 40 °C. Cellulose synthesis in the fiber, on the other hand, increases above 18 °C and reaches a plateau between 28 and 37 °C, followed by a decrease at 40 °C (Roberts et al. 1992). Thus, respiration and cellulose (fiber) synthesis are affected differently by cool temperatures. By using scanning electron microscopy and fiber length measurement in the cotton ovuleculture system, there appear to be three phases of fiber elongation: initiation, early elongation (up to 0.5 mm), and later elongation. Initiation and early elongation can be independently delayed by cycling cool temperatures; late elongation proceeds in a temperature-independent manner (Xie et al. 1993). In vitro ovule cultures have played a pivotal role in defining the physiological effects of cool temperatures on cotton fiber development.

Secondary cell walls of the green lint type of cottons reveal concentric osmiophilic layers in the walls, separated by cellulosic material (Ryser et al. 1983). These osmiophilic layers contain suberin, with the predominant monomer being 22-hydroxy-docosanoic acid. Ultrastructural studies reveal a periodic deposition of these layers which resemble the ultrastructural features of fibers exposed to cool-warm cycling temperature regimes. However, green lint ovules cultured in vitro at constant temperature display periodic suberin layers in the fiber secondary wall (Ryser et al. 1983). Thus, this structural feature is not low-temperature-induced. The wax component of the green fibers contains caffeic acid esterfied to ω-hydroxy fatty acids along with glycerol (Schmutz et al. 1993). Further, when 2-aminoindan-2-phosphonic acid, an inhibitor of phenylalanine ammonia lyase, is added to in vitro ovule cultures of the green lint mutant (at the beginning of secondary wall formation), the fibers remain white and the caffeic acid derivative is absent. Ultrastructural analysis of inhibitor-treated fibers by electron microscopy reveals that the suberin layers normally present in the secondary fiber wall are replaced by empty space (Schmutz et al. 1993). These results suggest that cinnamic acid derivatives (along with glycerol) are integrated into the suberin layer and may provide a critical structural role in the polymer or may attach the phenolic polymer to the cellulosic portion of the secondary wall.

Protoplast preparations have been successfully derived from in vitro-cultured ovules in order to investigate various parameters of fiber-wall genesis (Gould et al. 1986). The wall material formed in this system more closely resembles callose (β-1,3-glucans) than the in vivo-associated β-1,4-glucans of cellulose. Suspension cell cultures derived from cotton ovule callus display developmental changes associated with fiber growth in vivo, including extended length to width cell dimensions and secondary walls (Trolinder et al. 1987). Cotton ovule cultures have been used to help determine the role of seed position in a developing boll with respect to resource competition resulting from assimilate partitioning in the cotton plant (Davidonis and Hinojosa 1994). When culture-grown fibers are examined, differences in fiber length seen with plant-grown boll material largely disappear due to the absence of resource competition.

Evidence obtained with cotton ovule cultures suggests that callose synthesized concurrently with cellulose may act as a reservoir for glucose which can subsequently be incorporated into cellulose (Francey et al. 1989). Using [^{14}C]-glucose feeding experiments with in ovulo-grown fibers (in vitro), glucose is incorporated into cellulose at twice the level observed for callose. Pulse-chase experiments demonstrate that there is turnover in the callose glucan material with possible reincorporation of glucose into cellulose. The inhibitor 2,6-dichlorobenzonitrile significantly inhibits substrate incorporation into cellulose, but not callose. This inhibitor does not inhibit cytoplasmic enzymes involved in sugar metaboeism, suggesting that it operates at the plasma membrane of the fiber, perhaps with cellulose synthase itself (Francey et al. 1989).

Cotton ovules in culture secrete a mixture of polysaccharides into the surrounding medium, consisting mainly of pectins and a xyloglucan (Buchala et al. 1989). These polysaccharides are consistent with primary cell wall synthesis and are probably derived from callus tissue associated with the cotton ovules.

A number of studies have employed inhibitors in conjunction with ovule cultures to investigate cotton fiber development. The most effective and specific inhibitor of cellulose synthesis yet described is 2,6-dichlorobenzonitrile (Montezinos and Delmer 1980). Use of less potent inhibitors has yielded better understanding of the fiber developmental process. When AMO-1618, a growth retardant, is added to ovule cultures along with IAA, both ovule dry weight and fiber production are reduced; when GA_3 is added with IAA and AMO-1618, only a small reduction in fiber production is observed, but a large decrease in ovule dry weight occurs (Dhindsa 1978a). The effects of AMO-1618 are apparently caused by its inhibition of the gibberellin biosynthetic pathway, reducing the endogenous GA_3 pool. Gibberellin appears to be largely responsible for ovule growth, while IAA is primarily responsible for fiber growth. Malate synthetic enzyme activities (phosphoenolpyruvate carboxylase, malate dehydrogenase, glutamic oxalacetic transaminase) are increased in unfertilized ovule cultures in the presence of IAA and GA_3 (Dhindsa 1978b). Malate concentration drives turgor pressure, which is associated with fiber growth. Abscisic acid (ABA) inhibits the increased levels of these enzymes, presumably by counteracting the hormonal response of GA_3. Recent work suggests that auxin (IAA) is the critical phytohormone which stimulates increased production of fiber-specific PEP carboxylase (Ramsey et al. 1996). It has been suggested that ABA plays a direct role in fiber growth by exerting a regulatory effect on fiber elongation (Nayyar et al. 1989). This role for ABA is supported by results which show fluridone, an inhibitor of ABA biosynthesis, causing an increase in ovule culture fiber growth (Kaur et al. 1990). Treatment of 2- to 6-day postanthesis cotton ovule cultures with tunicamycin, an inhibitor of lipid-associated glycosyl transfer, inhibits fiber elongation and secondary wall deposition. Treatment of 6- to 11-day postanthesis cultures with monensin, a Golgi function inhibitor, inhibits secondary wall deposition but not fiber elongation (Davidonis 1993a). A tentative role for ethylene in cotton fiber development has been suggested, but the case for this phytohormone's direct involvement is still unclear (Davidonis 1993b).

Inhibitor experiments have very real limitations and, as such, should be interpreted with considerable caution. First, when investigating a complex system such as a developing cotton ovule, the question of adequate inhibitor uptake into the tissue of interest must be considered. If the biochemical mechanism and site of action of the inhibitor are known, then a straightforward interpretation of the physiological response may be possible. However, observed inhibitor effects may be secondary or the response may be a general one, involving many interacting processes.

2.3 Ovule Cultures and Peroxidase

During an investigation of ovule-specific proteins, it was discovered that a soluble peroxidase activity (guaiacol-based spot assay) is secreted into medium of 5- to 35-day-old cotton ovule cultures (Mellon 1986). Peroxidase activity increases 200-fold over the 30-day period of the study. The activity is thermostable to 60 °C and displays a pH optimum between 5.5 and 6. The ovule peroxidase is susceptible to periodate treatment, but is resistant to protease (pronase) digestion (Mellon 1986). The data are consistent with peroxidase as a glycoprotein.

Electrophoretic analysis of cotton ovule medium indicates a relatively simple protein profile with the presence of only two major proteins with molecular weights of 30 and 56 kDa (Mellon and Triplett 1989). In vivo labeling of cotton ovule cultures with [^{35}S]-methionine revealed that all secreted proteins detected with Coomassie or silver visualization agents are radiolabeled. Thus, all the ovule-secreted proteins, including peroxidase, are de novo synthesized (Mellon and Triplett 1989). It was subsequently determined that the 56-kDa protein is native peroxidase and the 30-kDa protein is a degradation product of the peroxidase. Thus, the predominant protein secreted into the ovule culture medium is peroxidase.

Cotton ovule peroxidase has been isolated from culture medium and purified about 150-fold to apparent homogeneity (Mellon 1991). The molecular weight of the purified enzyme is 56 kDa by denaturing gel electrophoresis, confirming the value obtained in the previous electrophoreis study, and 55 kDa by gel filtration chromatography, suggesting the enzyme exists as a monomer in native conditions. The pH optimum is 6, confirming that the pH optima obtained with crude preparations (5 to 6) are derived from a mixture of isoforms. The isoelectric point of the purified enzyme is 4.4. The absorption spectrum of cotton ovule peroxidase displays a major peak at 400 nm and a minor peak at 280 nm, confirming the presence of a heme moiety (400 nm) in the molecule. Amino acid analysis of this peroxidase confirms its acidic nature and reveals its similarity to anionic peroxidases from tobacco and potato. Its glycoprotein nature is confirmed by neutral sugar analysis, yielding 12 to 14% sugar by weight, mainly in the form of galactose and mannose (Mellon 1991).

Plant anionic peroxidases, as a group, tend to be cell wall-associated. Since maximum peroxidase secretion rates in culture correspond closely with seed coat lignification (20 to 35 days postanthesis), cotton ovule peroxidase probably functions in cottonseed coat lignification.

2.4 Ovule Culture as a Model System for Plant-Fungal Interactions

The cotton ovule culture system (fertilized) has been utilized as a model system for investigation of interactions between cotton and the toxigenic fungus *Aspergillus flavus*. The fungus can infect developing seeds and produce the potent carcinogen, aflatoxin. One technical problem which is associated with cotton concerns the morphology of the cotton fruit (boll). The boll is divided into three to five sections (locules), each containing six to eight developing seeds (ovules). Each locule is surrounded by a tough, fibrous protective tissue, the carpellary membrane. In order to inoculate developing bolls with fungal conidia, the carpellary membrane must first be penetrated. There is a real possibility that a host wound response will be encountered which is distinct from the fungal stress response and may complicate interpretation of the data. Thus, ovule cultures offer the substantial benefit of avoiding a wound response. Another feature of cotton bolls which can cause technical problems concerns the presence of lysigenous glands on the outer wall surface which contain numerous terpenoid materials. These terpenoids, in addition to exhibiting antiinsect and antimicrobial properties, are phytotoxic to the host plant's own developing seeds. Direct exposure of seeds may occur when the boll wall is penetrated during the process of fungal inoculation. Again, use of ovule culture circumvents this technical complication.

Cotton ovule cultures (30 days old) inoculated with *A. flavus* conidia accumulate peroxidase (activity) in levels up to 20-fold greater than noninoculated controls (Mellon 1988). Culture filtrates of axenic cultures of test fungi display no peroxidase activity. An infection time course with *A. flavus* on cultured cotton ovule tissue shows a rapid secretion of peroxidase into culture medium from 2 to 6 days of incubation, reaching an activity plateau after 6 days. Peroxidase profile analysis by isoelectric focusing electrophoresis reveals distinct differences between noninoculated (control) and inoculated ovule cultures. Control cultures contain several anionic isoforms, one major and three minor forms. The inoculated cultures contain two anionic isoforms, with the major form corresponding to a minor species in control cultures and the minor form corresponding to the major control isoform (Mellon 1988). This anionic isoperoxidase, which is elicited by fungal stress, appears to account for much of the increased peroxidase activity observed in challenged ovule cultures. The purification and characterization of this isoperoxidase is described in the previous section. It is present at very low levels in noninoculated controls. Lignification of the developing seedcoat represents a considerable mechanical and chemical defense barrier to potentially pathogenic microorganisms. Thus, involvement of an inducible isoperoxidase in a seed coat lignification function is consistent with a host defense role.

The response range of cotton ovule cultures to fungal challenge includes the production of low molecular weight stress metabolites. Incubation of cultured ovule tissue (without medium, 30 days in culture) with *A. flavus* for 7 days produces a distinct hypersensitive reaction, as evidenced by an intense darkening of the host tissue (Mellon 1988). Inoculated ovule tissue produces a complex mixture of extractable stress metabolites. Five of these components demonstrate antifungal activity on a thin layer plate bioassay, using *Cladosporium cladosporioides* as the screen organism. Two of these materials were subsequently identified as lacinilene C and lacinilene C7-methyl ether. Both are oxidation products of the cadalene family of sesquiterpenes known to be potent antimicrobial metabolites (Mellon 1988). Crude extracts derived from *A. flavus*-inoculated cotton ovule tissue inhibit aflatoxin production in axenic cultures of *A. flavus* by as much as 93% (Mellon 1992).

Formation of antifungal stress metabolites by developing cottonseed tissues is consistent with exposure to insect mechanical damage and confirms the existence of multiple defense strategies in the cotton plant. Such chemical defense mechanisms are often localized in peripheral tissues regularly exposed to interaction with potential pathogens. Pigment glands in the boll wall and the carpel tissue surrounding the developing seed constitute important defense barriers for the cotton plant. Potential for antifungal metabolite biosynthesis in the developing seed suggests an evolutionary exposure of internal host tissues to potentially harmful entities, necessitating the availability of further defense measures.

3 Summary and Conclusion

Cotton ovule cultures offer some considerable advantages when compared to intact plant systems. Manipulation of the developing seeds for different experimental procedures is much easier when using in vitro cotton ovule cultures than with in vivo bolls. Control of environmental conditions is more easily maintained with ovule cultures than with intact plants. Isolation of low molecular weight natural products and macromolecules is more simply accomplished with ovule cultures, especially if the molecule of interest is secreted into the surrounding medium. Elimination of problems associated with cotton glands and carpel tissue of the boll enhances the ease with which purification procedures are effected. By selecting a favorable set of culture conditions for the production of cotton ovule peroxidase, the predominant protein secreted into culture medium is a single isoform, considerably simplifying the purification procedure (Mellon 1991). This procedure offers many advantages over purification of proteins from whole-tissue homogenates, with their accompanying problems of complex protein mixtures and release of deleterious agents (phenolics, terpenes). Fortunately, the physiological development of cotton seeds in the in vitro ovule system appears to parallel boll development in many respects.

There are some disadvantages associated with the ovule culture system. As with any in vitro system, there is a potential for artifacts which can reduce the system's value. Any potential artifacts must be carefully considered in the experimental design in order to effectively evaluate a culture system's advantages vs. its liabilities. The developing cotton ovule is a complex organ, consisting of multiple integrated tissues. One would predict that uptake of experimental materials such as inhibitors and other metabolites might be a more complex situation than encountered with homogenous tissues. Indeed, this situation was found to be the case when incubating cotton ovule cultures with [^{35}S]-methionine in order to investigate de novo synthesis of secreted proteins. Ovule cultures required incubation times on the order of 10 to 20 times those needed by the near-homogenous suspension culture systems. The researcher should consider whether the in vitro system adequately simulates the in vivo system of interest. In the case of cotton fiber development, the ovule culture system would seem to fulfill these requisites.

The in vitro cotton ovule culture system was developed as a tool for the investigation of fiber growth, and it is in this field of research that it has been most utilized. It has proven to be an excellent tool in the arsenal of procedures available to the cotton fiber researcher. One of the goals of cotton fiber research is to incorporate fiber improvements, in both quality and structural (strength) characteristics. The ovule culture system will probably continue to be an important tool for this goal, as cotton fiber research is being actively pursued throughout the cotton-growing regions of the world. In general, it is a good model system for investigation of the developmental biology of cotton fiber. The system has also proved valuable in circumventing infertility problems associated with cotton hybridization work, and may well continue to play a role in this research area.

While cotton ovule cultures have received limited use a model system for cotton-microorganism interactions, they offer a valuable tool for the investigation of ovule interactions with different organisms, including bacteria, fungi, and insects. The system also offers much potential for the investigation of ovule-secreted components, both low molecular weight materials and macromolecules. The cotton ovule culture system is also an excellent tool for regulation and function studies of a specific stress-induced protein, peroxidase.

References

Altman DW (1988) Exogenous hormone applications at pollination for in vitro and in vivo production of cotton interspecific hybrids. Plant Cell Rep 7:257–261

Altman DW, Stelly DM, Kohel RJ (1987) Introgression of the glanded-plant and glandless-seed trait from *Gossypium sturtianum* Willis into cultivated upland cotton using ovule culture. Crop Sci 27:880–884

Bajaj YPS, Gill MS (1985) In vitro induction of genetic variability in cotton (*Gossypium* spp.). Theor Appl Genet 70:363–368

Bajaj YPS, Gill MS (1992) Micropropagation of cotton (*Gossypium* species). In: Bajaj YPS (ed) High-tech and micropropagation III. Biotechnology in agriculture and forestry, vol 19. Springer, Berlin Heidelberg New York, pp 483–504

Beasley CA (1977) Ovule culture: fundamental and pragmatic research for the cotton industry. In: Reinert J, Bajaj YPS (eds) Applied and fundamental aspects of plant cell, tissue, and organ culture. Springer, Berlin Heidelberg New York, pp 160–178

Beasley CA (1984) Culture of cotton ovules. In: Vasil IK (ed) Cell culture and somatic cell genetics of plants, vol 1. Academic Press, New York, pp 232–240

Beasley CA (1992) In vitro cotton ovule culture: a review. In: Proc Cotton Fiber Cellulose: Structure, Function and Utiliztion. National Cotton Council of America, Memphis, TN, pp 65–89

Beasley CA, Egli E (1977) Fiber production in vitro from a conditional fiberless mutant of cotton. Dev Biol 57:234–237

Beasley CA, Ting IP (1973) The effects of plant growth substances on in vitro fiber development from fertilized cotton ovules. Am J Bot 60:130–139

Beasley CA, Ting IP (1974) Effects of plant growth substances on in vitro fiber development from unfertilized cotton ovules. Am J Bot 61:188–194

Birnbaum EH, Dugger WM, Beasley CA (1977) Interaction of boron with components of nucleic acid metabolism in cotton ovules cultured in vitro. Plant Physiol 59:1034–1038

Buchala AJ, Roulin S, Meier H (1989) Polysaccharides in the culture medium of cotton (*Gossypium hirsutum* L.) ovules cultured in vitro. Plant Cell Rep 8:25–28

Carpita NC, Delmer DP (1981) Concentration and metabolic turnover of udp-glucose in developing cotton fibers. J Biol Chem 256:308–315

Contolini CS, Menzel MY (1987) Early development of duplication-deficiency ovules in upland cotton. Crop Sci 27:345–348

Davidonis GH (1989) Fiber development in preanthesis cotton ovules. Physiol Plant 75:290–294

Davidonis GH (1993a) Cotton fiber growth and development in vitro: effects of tunicamycin and monensin. Plant Sci 88:229–236

Davidonis GH (1993b) A comparison of cotton ovule and cotton cell suspension cultures: response to gibberellic acid and 2-chloroethylphosphonic acid. J Plant Physiol 141:505–507

Davidonis GH, Hinojosa O (1994) Influence of seed location on cotton fiber development in planta and in vitro. Plant Sci 103: 107–113

Dhindsa RS (1978a) Hormonal regulation of cotton ovule and fiber growth: effects of bromodeoxyuridine, amo-1618 and *p*-chlorophenoxyisobutyric acid. Planta 141:269–272

Dhindsa RS (1978b) Hormonal regulation of enzymes of nonautotrophic CO_2 fixation in unfertilized ovules. Z Pflanzenphysiol 89:355–362

Dugger WM, Palmer RL (1980) Effect of boron on the incorporation of glucose from udp-glucose into cotton fibers grown in vitro. Plant Physiol 65:266–273

Dugger WM, Palmer RL (1985) Effect of boron on the incorporation of glucose by cotton fibers grown in vitro. J Plant Nutr 8:311–325

Eid AAH, DeLange E, Waterkeyn L (1973) In vitro culture of fertilized cotton ovules. I. The growth of cotton embryos. Cellule 69:361–371

Francey Y, Jaquet JP, Cairoli S, Buchala AJ, Meier H (1989) The biosynthesis of β-glucans in cotton (*Gossypium hirsutum* L.) fibres of ovules cultured in vitro. J Plant Physiol 134:485–491

Gill MS, Bajaj YPS (1987) Hybridization between diploid (*Gossypium arboreum*) and tetraploid (*Gossypium hirsutum*) cotton through ovule culture. Euphytica 36:625–630

Gould JH, Palmer RL, Dugger WM (1986) Isolation and culture of cotton ovule epidermal protoplasts (prefiber cells) and analysis of the regenerated wall. Plant Cell Tissue Organ Cult 6:47–59

Graves DA, Stewart JM (1988) Chronology of the differentiation of cotton (*Gossypium hirsutum* L.) fiber cells. Planta 175:254–258

Haigler CH, Rao NR, Roberts EM, Huang JY, Upchurch DR, Trolinder NL (1991) Cultured ovules as models for cotton fiber development under low temperatures. Plant Physiol 95:88–96

Joshi PC (1960) In vitro growth of cotton ovules. Symp Plant Embryology. Council of Scientific and Industrial Res, New Delhi, pp 199–204
Joshi PC, Johri BM (1972) In vitro growth of ovules of *Gossypium hirsutum*. Phytomorphology 22:195–209
Kaur K, Nayyar H, Basra AS, Malik CP (1990) Stimulation of enzymes of non-photosynthetic C_4 metabolism in cultured cotton ovules by fluridone. Acta Physiol Plant 12:3–6
Meinert MC, Delmer DP (1977) Changes in biochemical composition of the cell wall of the cotton fiber during development. Plant Physiol 59:1088–1097
Mellon JE (1986) Some characteristics of peroxidase secreted by cotton ovule cultures. Plant Cell Rep 5:338–341
Mellon JE (1988) Use of ovule culture as a model for cotton-*Aspergillus flavus* interactions. In: Brown JM (ed) Proc Beltwide Cotton Prod Res Conf, National Cotton Council of America, Memphis, TN, pp 35–37
Mellon JE (1991) Purification and characterization of isoperoxidases elicited by *Aspergillus flavus* in cotton ovule cultures. Plant Physiol 95:14–20
Mellon JE (1992) Inhibition of aflatoxin production in *Aspergillus flavus* by cotton ovule extracts. J Am Oil Chem Soc 69:945–947
Mellon JE, Triplett BA (1989) De novo synthesis of peroxidase in cotton ovule culture medium. Physiol Plant 77:302–307
Menzel MY, Hasenkampf CA, Stewart JM (1982) Incipient genome differentiation in *Gossypium*. III. Comparison of chromosomes of *Gossypium hirsutum* and Asiatic diploids using heterozygous translocations. Genetics 100:89–103
Montezinos D, Delmer DP (1980) Characterization of inhibitors of cellulose synthesis in cotton fibers. Planta 148:305–311
Murashige T, Skoog F (1962) A revised medium for rapid growth and bio-assays with tobacco tissue cultures. Physiol Plant 15:473–497
Nayyar H, Kaur K, Basra AS, Malik CP (1989) Hormonal regulation of cotton fibre elongation in *Gossypium arboreum* L. in vitro and in vivo. Biochem Physiol Pflanz 185:415–421
Ramsey P, Menke S, Zeiher C (1996) Stimulation of PEP carboxylase by auxin and gibberellin in cultured cotton ovules. Plant Physiol 111(S):114
Roberts EM, Rao NR, Huang JY, Trolinder NL, Haigler CH (1992) Effects of cycling temperatures on fiber metabolism in cultured cotton ovules. Plant Physiol 100:979–986
Ryser U, Meier H, Holloway PJ (1983) Identification and localization of suberin in the cell walls of green cotton fibres (*Gossypium hirsutum* L., var. Green Lint). Protoplasma 117:196–205
Schmutz A, Jenny T, Amrhein N, Ryser U (1993) Caffeic acid and glycerol are constituents of the suberin layers in green cotton fibres. Planta 189:453–460
Seagull RW (1986) Changes in microtubule organization and microfibril orientation during in vitro cotton fiber development: an immunofluorescent study. Can J Bot 64:1373–1381
Seagull RW (1989) Changes in microtubule arrays during cotton fiber development. In: Bailey GW (ed) Proc 47th Annu Meet Electr Micros Soc Am, San Francisco Press, San Francisco, pp 760–761
Seagull RW (1990) The effects of microtubule and microfilament disrupting agents on cytoskeletal arrays and wall depositon in developing cotton fibers. Protoplasma 159:44–59
Seagull RW (1992) A quantitative electron microscopic study of changes in microtubule arrays and wall microfibril orientation during in vitro cotton fiber development. J Cell Sci 101:561–577
Stewart JM (1981) In vitro fertilization and embryo rescue. Environ Exp Bot 21:301–315
Stewart JM (1991) Biotechnology of cotton. CAB International, Wallingford, pp 21–22
Stewart JM, Hsu CL (1977) In-ovulo embryo culture and seedling development of cotton (*Gossypium hirsutum* L.). Planta 137:113–117
Stewart JM, Hsu CL (1978) Hybridization of diploid and tetraploid cottons through in-ovulo embryo culture. J Hered 69:404–408
Thengane S, Paranjpe SV, Khuspe SS, Mascarenhas AF (1986) Hybridization of *Gossypium* species through in-ovulo embryo culture. Plant Cell Tissue Organ Cult 6: 209–219
Triplett BA, Timpa JD (1995) Characterization of cell-wall polymers from cotton ovule culture fiber cells by gel permeation chromatography. In Vitro Cell Dev Biol 31:171–175

Triplett BA, Busch WH, Goynes WR Jr (1989) Ovule and suspension culture of a cotton fiber development mutant. In Vitro Cell Dev Biol 25:197–200
Trolinder NL, Berlin JD, Goodin JR (1987) Differentiation of cotton fibers from single cells in suspension culture. In Vitro Cell Dev Biol 23:789–794
Wainwright IM, Palmer RL, Dugger WM (1980) Pyrimidine pathway in boron-deficient cotton fiber. Plant Physiol 65:893–896
Xie W, Trolinder NL, Haigler CH (1993) Cool temperature effects on cotton fiber initiation and elongation clarified using in vitro cultures. Crop Sci 33:1258–1264

II.4 Use of Random Amplified Polymorphic DNA (RAPD) Markers to Assist Wide Hybridization in Cotton

G. Mergeai[1], I. Vroh Bi[1,2], J.P. Baudoin[1], and P. du Jardin[2]

1 Interest of RAPD Markers to Assist Interspecific Breeding Programs

In recent years, molecular marker-assisted selection has been integrated into several plant breeding programs to select traits of agronomical importance. Isozymes were initially sought for this purpose, but their use was hindered by the rather small number of loci and alleles available for analysis. The development of molecular biology has resulted in alternative DNA-based procedures for the detection of polymorphism. These include restriction fragment length polymorphism (Tanksley et al. 1989), random amplified polymorphic DNA (Williams et al. 1990), and cleaved amplified polymorphic sequences (Akkaya et al. 1992).

Although RFLPs were widely used in genetic studies concerning cotton (Brubaker and Wendel 1994; Reinisch et al. 1994) and other crop species they promise to be a cost-effective diagnostic technology for plant breeding. This technique uses a single short oligonucleotide primer (10 bp) of arbitrary DNA sequence and polymerase chain reaction (PCR)-mediated amplification of random fragments from genomic DNA. The main advantages of the RAPD technique include the ease and rapidity of analyses: it does not require preparation of genomic or cDNA clones, restriction enzyme digestion of DNA, Southern blotting, or hybridization with radioactively labeled probes (data are directly scored from agarose gels). Other advantages concern the use of universal sets of primers that are commercially available, its requirement for minimal substrate DNA and negligible primer quantity, and its lower cost. All these characteristics make RAPD analysis particularly well adapted for detecting polymorphic situations in unknown germplasm where no preexisting set of RFLP clones is available (Heun and Helentjaris 1993). Because of its simplicity, RAPD analysis can be performed in a moderately equipped laboratory, and since only a small amount of tissue is required, plants in a breeding program can be screened at the seedling stage. Moreover, because RAPD

[1] Unité de Phytotechnie des Régions intertropicales, Faculté Universitaire des Sciences agronomiques, B-5030 Gembloux, Belgium,
[2] Unité de Biologie végétale, Faculté Universitaire des Sciences agronomiques, B-5030 Gembloux, Belgium

Biotechnology in Agriculture and Forestry, Vol. 42
Cotton (ed. by Y.P.S. Bajaj)

analyses require much smaller amounts of DNA and less pure DNA than RFLP analysis, the RAPD method is particularly attractive to genetists working with species recalcitrant to nucleic acid extraction, such as cotton (Multani and Lyon 1995; Tatineni et al. 1996).

Clear and reproducible detection of polymorphims has been demonstrated using RAPD in many plants including *Dastica*, *Yucca*, and *Helianthus* (Fritsch et al. 1993), corn (Heun and Helentjaris 1993), wheat (Devos and Gale 1992; Vierling and Nguyen 1992), barley (Tinker et al. 1993), rice (Heun et al. 1994), and others (Hu and Quiros 1991).

Since its discovery, the RAPD technique has been applied in plants for the construction of genetic linkage maps (Reiter et al. 1992; Tulsieram et al. 1992; Kesseli et al. 1994; Rowland and Levi 1994), estimation of genetic relationships (He et al. 1992; Vierling and Nguyen 1992; Halward et al. 1991, 1992), tagging disease resistance traits (Michelmore et al. 1991), identification of cultivars (Hu and Quiros 1991; Quiros et al. 1992; Klein-Landhorst et al. 1991; Wilde et al. 1992), parentage determinations (Welsh et al. 1991), population genetics (Van Hausden and Bachmann 1992), analyzing interspecific hybrids and exotic germplasm utilization (Arnold et al. 1991; Baird et al. 1992; Veli-Matti Roka et al. 1992; McCoy and Echt 1993).

Given all the putative advantages of the RAPD analysis, we have chosen this technique to investigate the potential of molecular markers to monitor the introgression of diploid *Gossypium* genomes into upland cotton. This study was realized in the frame of a three-species breeding program involving the main cultivated amphididiploid, *G. hirsutum*, two wild diploid American cottons (*G. thurberi* and *G. raimondii*), and a diploid wild species endemic to Australia (*G. sturtianum*). Like all the other Australian cottons, *G. sturtianum* produces gossypol-free seeds, while gossypol glands are present in the rest of its aerial parts. Our goal is to transfer this glandless-seed and glanded-plant trait into cultivated cotton in order to transform it into a true food crop whilst preserving one of its natural defense mechanism (the gossypol glands) against insect pests. Because direct crossing strategies between *G. hirsutum* and *G. sturtianum* followed in the past by other laboratories were not successful to obtain the introgression of the researched character into upland cotton (Dilday 1986; Altman et al. 1987; Koto 1989; Rooney et al. 1991), we created two different trispecific hybrids by bridge crosses in order to confront all the chromosomes of *G. sturtianum* to the genome of *G. hirsutum* in a tetraploid structure allowing recombination (Mergeai 1992). In this context a system of random amplified polymorphic DNA (RAPD) markers was developed to facilitate the transfer of *G. sturtianum* genes into *G. hirsutum* genome by hybridization and backcrossing.

2 Materials and Methods

2.1 Plant Material

Two trispecific allotetraploids, including the Australian diploid species *G. sturtianum* ($2C_1$), have been produced using either *G. thurberi* ($2D_1$) or *G. raimondii* ($2D_5$) as bridge species (Mergeai et al. 1995). These hybrids are respectively designated by the initials TSH for G405 (*G. thurberi* × *G. sturtianum* × *G. hirsutum*: $D_1C_1D_hA_h$) and HRS for G376 (*G. hirsutum* × *G. raimondii* × *G. sturtianum*: $A_hD_hC_1D_5$). Figure 1A,B shows the crossing scheme followed to obtain these genotypes. Both trispecific hybrids were backcrossed by three *G. hirsutum* ($2[AD]_1$) glanded varieties originating from Africa (C2, Stam F and NC8) and one US glandless cultivar (LPB5). Twenty seven BC_1 plants obtained by backcrossing each trispecific hybrid to *G. hirsutum* were evaluated to test the suitability of RAPDs for detecting the amount of introgression from wild diploid species into *G. hirsutum*. Genomic DNA was extracted from all the genotypes involved in the breeding program. To obtain the BC_1 plants analyzed in this program it was necessary to treat the flowers immediately after pollination with a solution of gibberellic acid and naphthalene acetic acid (Gill and Bajaj 1987), to cultivate the seeds in vitro for 2 weeks on the medium of Stewart and Hsu (1977) and to graft most of them

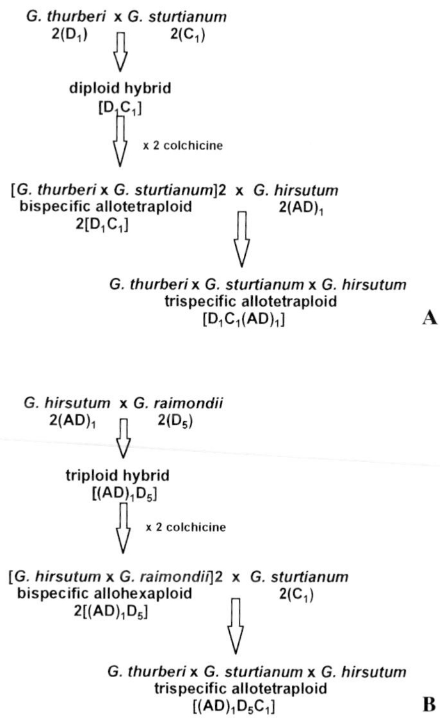

Fig. 1. A Creation scheme of the *G. thurberi-G. sturtianum-G. hirsutum* (TSH) tispecific hybrid (G405). **B** Creation scheme of the *G. hirsutum-G. raimondii-G. sturtianum* (HRS) trispecific hybrid (G 376)

on vigorous *G. hirsutum* seedlings. All the plants used in the experiments were grown in a greenhouse in Gembloux. Young leaves, when available, were preferably used for DNA extraction.

2.2 DNA Extraction and Amplification

The detailed protocol followed for DNA extraction has been published elsewhere (Vroh Bi et al. 1996). This protocol is adapted from the standard CTAB method of Murray and Thompson (1980), and includes incubation of the tissue homogenate with activated charcoal. This original modification proved to significantly improve the subsequent PCR amplification of the extracted DNA.

Using a Techne PHC-3 DNA thermal cycler, we optimized RAPD amplification conditions proposed by Williams et al. (1990). The main parameters studied to adapt the conditions to cotton DNA amplification are the concentration of template DNA, primers, $MgCl_2$, and dNTPs, the annealing temperature, the number of cycles, the duration of the first denaturation, different sources of polymerase, and the design of premixes to avoid variations between different reactions (Vroh Bi et al. 1997).

Seventy five decamers from Operon Technologies Inc. (Alameda, California, USA) were tested. Amplification reaction volumes were 25 µl, each containing 50 mM KCl, 10 mM Tris-HCL (pH 9), 2 mM MgCl2, 200 µM each of dATP, dCTP, dGTP, and dTTP (Pharmacia), 0.8 µM random primer, 1 U Taq Polymerase (Pharmacia), and 50 ng template DNA. Reaction mixtures were overlaid with mineral oil and exposed to the following conditions: 94 °C for 5 min, followed by 45 cycles of 1 min at 94 °C, 1 min at 34 °C, 2 min at 72 °C, ending with 7 min 30 s. at 72 °C, and then kept at 15 °C till electrophoresis. The RAPD products were resolved on 1.8% agarose gel and visualized by ethidium bromide staining.

All reactions were repeated twice, and only reproducible bands were considered.

2.3 Data Analysis

The data for RAPD analysis were scored from photographs of the ethidium bromide stained agarose gels.

Bands were considered to be the same if they occurred at exactly the same position on the electrophoresis gel. Band sizes for each genotype were determined by comparison with lambda and Phi X 174 DNA, respectively, digested by HindIII and HaeIII. For each genotype and for all the informative primers used, a RAPD bands presence (1)/absence (0) profile was recorded into a matrix. A cluster analysis based on UPGMA (unweighted pairgroup method with arithmetical mean) was then performed with systat version 5.2 software (Systat 1992). RAPDs are identified by Operon kit, primer, and fragment size (i.e., OPB-01-1500).

3 Results and Discussion

3.1 Selection of RAPD Markers Suitable for Introgression Monitoring in Cotton

The main objective of our research was to assess the suitability of the RAPD technique for monitoring the introgression of wild diploid species characters into *G. hirsutum*. For this purpose, 75 RAPD primers on four *G. hirsutum* cultivars and three diploid species implicated in the creation and backcrossing of TSH and HRS trispecific hybrids were tested. The aim of this preliminary investigation was to select primers exhibiting polymorphism between the parents of two trispecific hybrids in order to detect species-specific RAPD markers that could be used to trace the introgression of wild diploid DNA segments in the genome of *G. hirsutum* (identification of RAPD markers that are present in the trispecific hybrids and their ascendants). Among the 75 primers used, 6 (8%) primers did not amplify detectable products, 23 (30.7%) failed to amplify in one or two parents and gave clear amplification in the remaining parents, 46 (63.3%) succeeded in amplifying in all the parents. Of these 46 informative primers, 30 were tested twice to generate the results showed in this study. The sequences and the number of bands generated by the 30 primers are given in Table 1.

These primers produced 375 different RAPD bands, among which only 36 (9.6%) were shared by all the hybrids parents, the 339 (90.4%) other fragments were polymorphic.

The number of bands amplified per primer ranged from 6 (OPC-10, OPC-15 and OPC-17) to 20 (OPB-18) with a mean value of 12.5 bands per primer. These values are rather high for RAPD amplification compared to the average numbers of amplified bands recorded in other crops: (3 fragments in *Triticum turgidum* L.; Joshi and Nguyen 1993), 4.3 fragments in *Solanum tuberosum* L. (Masuelli et al. 1995), 6.7 in corn (Heun and Helentjaris 1993), and 9 in coffee (Orozco-Castillo et al. 1994).

The large size of the *Gossypium* genomes, 2246 Mbp for *G. hirsutum* (Arumunganathan and Earle 1991), the high frequency of repetitive DNA sequences (Walbot and Duke 1976), and the quality of the DNA produced by the improved extraction method are possible explanations for the relatively high numbers of amplified products obtained.

The amplified fragment sizes varied from 160 to 3200 base pairs (bp). The most reproducible types of RAPD fragments, as might be expected, were those of the greatest intensity in any one reaction. "Minor" fragments seemed to possess the greatest propensity for irreproducibility, as was also found in corn (Heun and Helentjaris 1993) and canola (Marshall et al. 1994). Species-specific RAPD fragments which were not present in both reaction repetitions were eliminated. On the average, the great majority of the amplified fragments produced by the 30 selected primers were reproducible and the interpretation of the banding pattern was straightforward. RAPD markers unique to the different parents that are also present in TSH and HRS hybrids were easily

Table 1. Informative primers and respective number of polymorphic bands generated

Primers code	Primers sequences 5′-3′	No. of RAPD bands generated	No. of polymorphic[a] bands
OPB-01	GTTTCGCTCC	10	10
OPB-02	TGATCCCTGG	12	11
OPB-03	CATCCCCCTG	12	11
OPB-04	GGACTGGAGT	13	10
OPB-05	TGCGCCCTTC	17	17
OPB-06	TGCTCTGCCC	9	8
OPB-08	GTCCACACGG	14	14
OPB-10	CTGCTGGGAC	15	14
OPB-18	CCACAGCAGT	20	20
OPC-01	TTCGAGCCAG	9	7
OPC-03	GGGGGTCTTT	16	13
OPC-04	CCGCATCTAC	13	13
OPC-05	GATGACCGCC	13	11
OPC-08	TGGACCGGTG	11	11
OPC-09	CTCACCGTCC	17	15
OPC-10	TGTCTGGGTG	6	6
OPC-11	AAAGCTGCGG	15	9
OPC-12	TGTCATCCCC	11	8
OPC-13	AAGCCTCGTC	16	15
OPC-14	TGCGTGCTTG	17	15
OPC-15	GACGGATCAG	6	5
OPC-16	CACACTCCAG	12	11
OPC-17	TTCCCCCCAG	6	3
OPC-18	TGAGTGGGTG	10	10
OPC-19	GTTGCCAGCC	8	7
OPD-03	GTCGCCGTCA	17	17
OPD-11	AGCGCCATTG	13	12
OPD-13	GGGGTGACGA	14	14
OPD-15	CATCCGTGCT	14	14
OPD-18	GAGACGCAAC	9	8
	Total	375 (100%)	339 (90.40%)

[a] A band is considered as polymorphic when it is not shared by all the parents of the trispecific hybrids.

recognized (Fig. 2A). This allowed the confirmation of the composition of our trispecific hybrids. The use of RAPD markers to detect alien gene introgression or to confirm the composition of interspecific hybrids has also been reported in other crop species (Arnold et al. 1991; Baird et al. 1992; McCoy and Echt 1993).

Besides the bands characteristic of only one genotype, some amplified products were shared by two, three, four, or more genotypes, and some bands were present in all genotypes but one.

Table 2 summarizes the results obtained concerning the identification of wild species-specific RAPD makers in the trispecies hybrids G405 and G376.

Given the fact that trispecies hybrids do not have full chromosome complements of the wild species used for their production, it is interesting to note

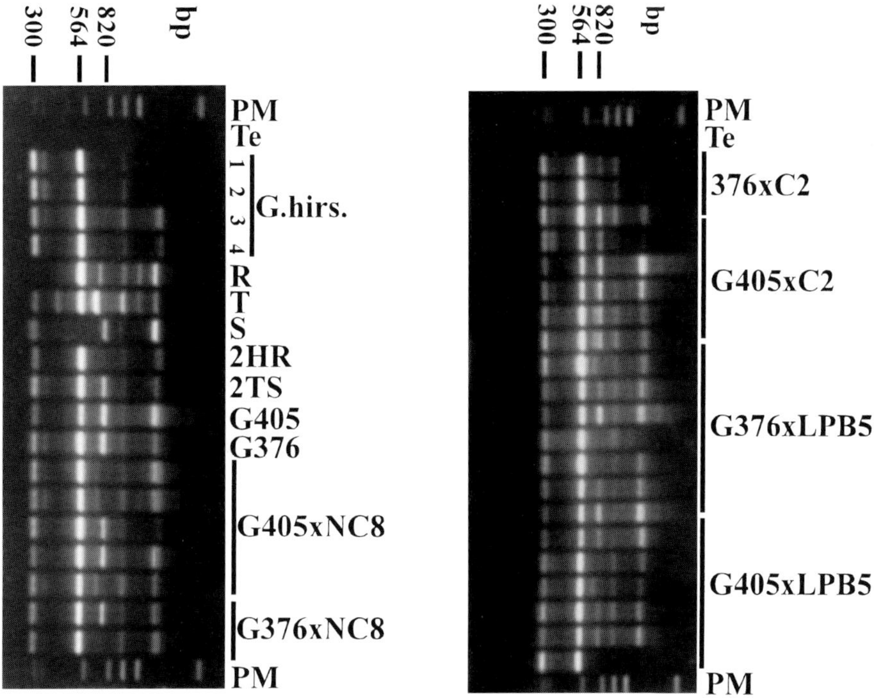

A B

Fig. 2A,B. RAPD bands generated using primer *OPC-16* (5′-CACACTCCAG-3′) in 1.8% agarose gels. The cultivated upland cotton *G. hirsutum* = *G. hirs.* (with four cultivars: *1 Stam F*; *2 NC8*; *3 C2*; *4 LPB5*); *R G. raimondii*; *T G. thurberi*; *S G. sturtianum*; *2HR* (*G. hirsutum* × *G. raimondii*) doubled; *2TS* (*G. thurberi* × *G. sturtianum*) doubled; Trispecies hybrids *G405* = (2TS × *G. hirsutum*) and *G376* = (2HR × *G. sturtianum*). The differents groups of backcrosses 1 are indicated by their crosses (*G405* × *NC8*, *G376* × *NC8*, *G376* × *C2*, *G405* × *C2*, *G376* × *LPB5*, and *G405* × *LPB5*). Each lane contains 12 µl of RAPD products; *PM* represents the molecular size marker (DNA of lambda/Hind III and PhiX 174/Hae III); *Te* is the negative control without DNA, and molecular weights are given in base pairs (bp). The 820-bp fragment is specific to the Australian wild diploid donor *G. sturtianum*; this fragment is present in the two trispecies hybrids, G405 and G376, and clearly segregates in the backcrosses 1. The polymorphism between all the genotypes under study is easily detectable

Table 2. Transmission of RAPD diploid wild species-specific markers to trispecific hybrids

Tested genotype	No. of specific RAPD markers in the parent genotypes	No. of specific RAPD markers in the TSH hybrid (G405)	No. of specific RAPD markers found in the HRS hybrid (G376)
G. sturtianum	49	18[a]	20[a]
G. raimondii	12	0	4
G. thurberi	13	4	0

[a] The two trispecies hybrids shared 18 bands specific to *G. sturtianum*.

that some *G. sturtianum*, *G. thurberi*, and *G. raimondii* markers are not present in the trispecies hybrids. The same kind of observation was made by McCoy and Echt (1993) on *Medicago* trispecies hybrids. Actually, one can speculate that the loci for markers not present in the trispecies hybrids are on chromosomes that were not transmitted to them. Another complementary explanation is given by Heun and Helentjaris (1993), who stated that many unexpected RAPD inheritance results could be attributed to the "competition" amongst fragments where one very intense product seemed to outcompete and overwhelm other expected fragments, even those of very high yields in inbred situations. This may also explain the reappearance in the BC_1 progeny of some RAPD markers that were identified as characteristic of the parent species and vanished in the trispecific hybrids. Indeed, as the presence of some fragments can interfere with the amplification of others, the absence of these latter from some genotypes does not prove the absence of the corresponding target sequences, which may "reappear" in segregating situations if the former fragments are allowed to segregate away.

To minimize the risk of interference between fragments for amplification, we took into consideration only the products less sensitive to the competition of other fragments, that means intense specific bands that were present in the parent species and in the trispecific hybrids. A total of 22 different markers of this kind were detected for *G. sturtianum* in both trispecific hybrids against 4 markers for *G. raimondii* and *G. thurberi*.

As the two wild diploid species belonging to genome D (*G. thurberi* and *G. raimondii*) are phyletically very close to the D_h subgenome of *G. hirsutum*, it is not surprising to observe for them lower numbers of specific RAPD markers than for *G. sturtianum*, which belongs to the more remote C genome. Specific RAPD bands detected in both D genome species seem to be more sensitive to the competition of other markers for amplification.

This higher sensitivity to competition may limit the usefulness of the RAPD markers to trace introgression of genes from diploid species considered as the most closely related to *G. hirsutum* compared to introgression of genes from a more distant genome such as *G. sturtianum*.

3.2 Assessment of Introgression in Upland Cotton Using RAPD Markers

The species-specific RAPD markers detected in our investigations were used to assess the introgression of cotton wild diploid DNA segments in the BC_1 plants produced by backcrossing the TSH and HRS trispecific hybrids with four *G. hirsutum* cultivars.

The distribution of the species-specific markers observed in each BC_1 plant are given in Tables 3, 4, and 5.

Figure 2A,B shows the RAPD markers produced with primer OPC-16 by all the genotypes compared in our work. RAPD markers specific to the wild diploid parents of the trispecific hybrids that are also present in some plants of their BC_1 progeny can be easily identified. This is, in particular, the case for the band OPC 16-820, which is specific to *G. sturtianum*. This fragment is present in both trispecific hybrids and in 17 introgressed genotypes out of the 27 BC_1 plants analyzed.

In the case of a totally randomly distribution of the chromosomes during meiosis, a 50% transfer of a RAPD wild species-specific marker observed in one of the trispecific hybrids to the BC_1 generation should be expected if this marker is present in only one chromosome and if intergenomic recombination and deletion are rare.

Table 3 shows the segregation of the *G. sturtianum*-specific markers in the backcross generation. All 22 *G. sturtianum*-specific markers selected from the analysis of the trispecific hybrids were found among the tested BC_1 plants. The number of *sturtianum*-specific markers per plant ranged from 7 to 20 (mean = 13 markers per plant).

The marker OPB-01-1500 was found in significantly less than 50% of the BC_1 plants, OPC-08-760 was significantly found in more than 50% of the BC_1. Among the rest of the RAPD markers, they were systematically observed in all the BC_1 plant and 17 fit the expected 1:1 Mendelian segregation ratio. As was noted by Masuelli et al. (1995) in the case of *Solanum*-interspecific hybrids, one can speculate that the loci for markers not present in the BC_1 plants are on chromosomes that were not transmitted from the trispecies hybrids.

On the contrary, the markers that are systematically present in the backcrossed progeny of the trispecific hybrids (OPC-04-1200, OPB-10-1078, and OPB-18-3200) should be carried on *G. sturtianum* chromosomes that are preferentially transmitted to the BC_1 plants. The average numbers of chromosome pairs in AC and DC diploid hybrids are respectively 8 and 5 (Endrizzi et al. 1985). This means that at least five chromosomes of the Australian diploid species present enough remnant homology to pair regularly with the chromosomes of the A_h and D_h subgenomes. These chromosomes with a higher pairing affinity certainly have more chance to be transmitted to the progeny of the trispecific hybrids than the others.

The presence of markers at more than one chromosome in the *G. sturtianum* genome might also explain the transfer frequency of RAPD markers significantly higher than 50%. Extensive gene duplication has been found within diploid cotton species from A and D genomes (Reinisch et al. 1994). Therefore, it is likely that some RAPD markers could be present at several unlinked loci. However, it is worthwhile to note that the number of markers that are transmitted with an abnormally high frequency is rather low and could be explained for most of them by the preferential pairing capacities of the chromosome that carries them.

With the detection of more *G. sturtianum*-specific RAPD markers it should be possible to identify synteny groups among them and assess their organization. The number of Australian species with chromosomes that are preferentially transferred to the BC_1 and the types of translocation, deletion, or insertion that occur in BC_1 should be revealed by coupling synteny groups analyses with GISH (genomic in situ hybridization) (Jacobsen et al. 1995).

All but one of the D species-specific RAPD markers fit the 1:1 segregation ratio (Tables 4, 5). No explanation can be given so far for this exception. The data recorded for the rest of the *thurberi*- and *raimondii*-specific markers

Table 3. Segregation of RAPD markers specific to *G. sturtianum* in the 27 BC_1 plants obtained from G376 and G405

Cross formulas and genotype nos. / Specific RAPD markers	G405 × NC8				G376 × NC8		G376 × C2			G405 × C2				
	5	6	9	7	3	4	11	S1	2	10	5	7	6	3
OPC04-1200	+	+	+	+	+	+	+	+	+	+	+	+	+	+
OPC08-840	−	+	+	−	−	+	+	−	−	−	−	+	+	+
OPC08-760	+	+	+	+	+	+	+	+	+	+	−	−	+	−
OPC13-1250	+	+	+	+	+	+	+	+	−	−	−	−	+	−
OPC13-800	−	−	−	+	−	−	+	+	−	+	+	+	+	−
OPC14-1300	+	+	−	−	−	−	+	−	+	−	−	+	+	+
OPC16-820	−	−	+	+	−	+	−	+	−	+	+	+	+	+
OPC18-740	+	+	−	−	−	+	+	−	−	−	+	−	+	+
OPC19-872	−	−	−	+	+	+		+	+	−	+	+	+	+
OPC19-500	+	−	+	−	−	−	−	−	−	−	+	+	+	−
OPB01-1500	+	+	−	+	−	−	+	−	+	+	−	+	−	−
OPB03-360	+	+	+	−	−	+	+	+	−	−	+	−	+	+
OPB03-180	−	−	+	+	+	−	+	+	+	+	−	−	+	+
OPB04-740	+	+	+	+	+	−	−	−	−	−	+	+	−	−
OPB10-1078	+	+	+	+	+	+	+	+	+	+	+	+	+	+
OPB18-3200	+	+	+	+	+	+	+	+	+	+	+	+	+	+
OPD03-690	+	−	+	+	+	−	+	−	−	−	−	+	+	+
OPD03-510	+	+	+	−	+	+	−	−	−	−	+	+	+	+
OPD03-271	−	+	+	−	−	−	−	+	−	−	−	+	−	−
OPD-220	+	+	+	+	+	−	+	+	−	+	−	+	−	−
OPD13-310	−	+	−	+	−	+	−	+	−	+	+	+	+	−
OPD18-690	+	−	+	+	+	+	−	+	+	−	+	+	−	−
Total of introgressed bands per backcross	15	15	16	15	12	13	14	14	9	10	13	17	17	12

[a] $\chi^2 > 3.85$ ($P < 0.05$) indicates that the frequency of transfer of the marker is significantly different from 50%.
[b] χ^2 values were not calculated for markers with presence or absence number <5.

G376 × LPB5							G405 × LPB5						segregation in backcrosses I	
17	18	6	11	10	14	5	6	8	1	10	12	4	Presence : absence	Chi-square
+	+	+	+	+	+	+	+	+	+	+	+	+	27:0	[a]
+	–	–	+	–	+	+	+	+	+	+	–	–	15:12	0.33
+	–	–	–	+	+	+	+	–	+	+	+	+	20:7	6.25[a]
+	+	+	–	+	–	–	+	–	+	+	+	+	18:9	3
+	+	+	–	–	–	+	+	–	–	+	+	+	15:12	0.33
–	+	+	+	+	–	–	–	–	+	–	+	–	13:14	0.03
+	–	+	+	–	–	+	+	+	–	–	+	+	17:10	1.06
+	–	–	+	+	+	+	+	–	–	+	+	+	16:11	0.925
–	–	+	+	+	–	–	+	–	–	+	+	+	16:11	0.925
+	–	–	–	–	–	–	+	+	–	–	+	–	10:17	1.06
–	–	+	–	–	–	–	–	–	–	–	–	–	8:19	4.48[a]
+	–	–	–	–	–	–	–	–	+	+	+	–	13:14	0.03
–	–	–	+	+	+	+	–	–	–	+	+	–	15:12	0.33
–	+	–	–	–	–	+	–	+	–	+	+	–	12:15	0.33
+	+	+	+	+	+	+	+	+	+	+	+	+	27:0	[b]
+	+	+	+	+	+	+	+	+	+	+	+	+	27:0	[b]
+	–	–	+	–	–	+	–	+	+	+	+	+	16:11	0.925
–	–	+	–	–	–	–	+	+	–	+	+	–	14:13	0.03
+	–	–	–	+	+	–	–	+	+	–	+	–	10:17	1.06
–	–	–	–	–	–	–	–	–	+	+	+	+	13:14	0.03
+	–	+	–	–	+	+	+	+	–	+	+	+	17:10	1.06
+	–	–	–	–	+	+	+	+	–	+	+	+	16:11	0.925
15	7	10	10	10	10	13	13	12	11	16	20	13		

Table 4. Segregation of RAPD bands specific to *G. raimondii* in the G376 BC_1 plants

Cross formulas and genotype nos. / Specific RAPD markers	G376 × NC8		G376 × C2			G376 × LPB5							Segregation in backcrosses I	
	3	4	11	S1	2	17	18	6	11	10	14	5	Presence : absence	Chi-square
OPB01-1130	–	+	+	–	+	–	+	+	–	+	–	–	6:6	0
OPB05-980	–	+	+	+	+	–	–	–	+	+	–	+	7:5	0.33
OPB05-872	–	–	–	+	+	–	+	–	–	–	–	–	3:9	3
OPC11-672	–	–	+	–	+	–	+	+	–	–	–	–	5:7	0.33
Total of introgressed bands per backcross	0	2	3	2	4	4	3	2	1	2	0	1		

Table 5. Segregation of RAPD bands specific to *G. thurberi* in the G405 BC_1 plants

Cross formulas and genotype nos. / Specific RAPD markers	G405 × NC8				G405 × C2					G405 × LPB5						Segregation in backcrosses I	
	5	6	9	7	10	5	7	6	3	6	8	3	10	12	4	Presence : absence	Chi-square
OPC13-460	−	−	+	−	−	−	−	−	−	−	+	+	−	+	+	5:10	1.66
OPC18-370	−	−	−	−	+	+	+	−	−	+	+	−	−	−	−	5:10	1.66
OPC18-1078	+	+	−	−	−	+	+	−	−	+	+	+	−	−	−	7:8	0.066
OPB10-496	−	−	−	−	+	−	−	−	−	−	−	−	−	−	−	1:14[a]	12.13[a]
Total of introgressed bands per backcross	1	1	1	0	2	2	2	0	0	2	3	2	0	1	1		

[a] $\chi^2 > 3.85$ ($P < 0.05$) indicates that the frequency of transfer of the marker is significantly different from 50%.

are consistent with the pairing affinity of D chromosomes with the D_h subgenome.

3.3 Prospects of Gene Mapping with RAPD Markers in the Progeny of Trispecific Hybrids

The backcross progeny should be either deficient for entire *G. sturtianum* chromosomes or for fragments of chromosomes, depending on whether homeologous recombination took place. In either case, groups of linked markers are expected to be absent from the genome of most individuals. Linkage could thus be established by the comparative testing of a relatively small number of backcross progeny (20–100, depending on the rate of chromosome loss). If no homeologous recombination is taking place, linkage groups will correspond to *G. sturtianum* chromosomes. If homeologous recombination takes place, linkage groups as defined by a single individual may cover only part of a chromosome. However, chromosomal groups could be defined by comparing individuals containing overlapping sets of missing markers.

If the backcross 1 progeny is tested phenotypically for the presence of genes of interest, positive and negative plants could be bulked and subjected to bulk segregant analysis, as described by Michelmore et al. (1991). For dominant genes, the markers absent in the negative plants should be linked to the gene of interest. Therefore, in order to facilitate the transfer of the glanded-plant and glandless-seed genes from *G. sturtianum* to *G. hirsutum*, we plan to analyze the first backcross progeny for *G. sturtianum* RAPD markers linked to the desired character by bulk segregant analysis.

3.4 Genetic Relationships Between Introgressed Tetraploid Cotton and Their Ascendants

One of the goals of our study was to investigate the efficiency of RAPD markers in determining accurately the genetic relationships between the BC_1 and their parents (*G. hirsutum*, *G. sturtianum*, *G. raimondii*, *G. thurberi*, and the trispecies G405 and G376).

Figure 3 represents the dendrogram generated by UPGMA analysis of the crosses G405 × NC8 and G405 × C2.

Three clusters can be observed. The first cluster comprises only *G. sturtianum*, showing clearly that this Australian wild diploid species of genome C is the most remote from the other genotypes studied. The American wild diploid *G. thurberi* (genome D) clusters also apart; it is situated between *G. sturtianum* and the remaining genotypes, which form the third cluster. This last cluster is divided into four subgroups; the first consists of the two cultivars, NC8 and C2, of *G. hirsutum* (genome AD), the second is the backcross 1 G405 × C2/10 that failed to integrate the group in which it was expected to be. With this exception, all the backcrosses derived from the same cross group together, G405 × NC8 backcrosses forming the third group and backcrosses G405 × C2

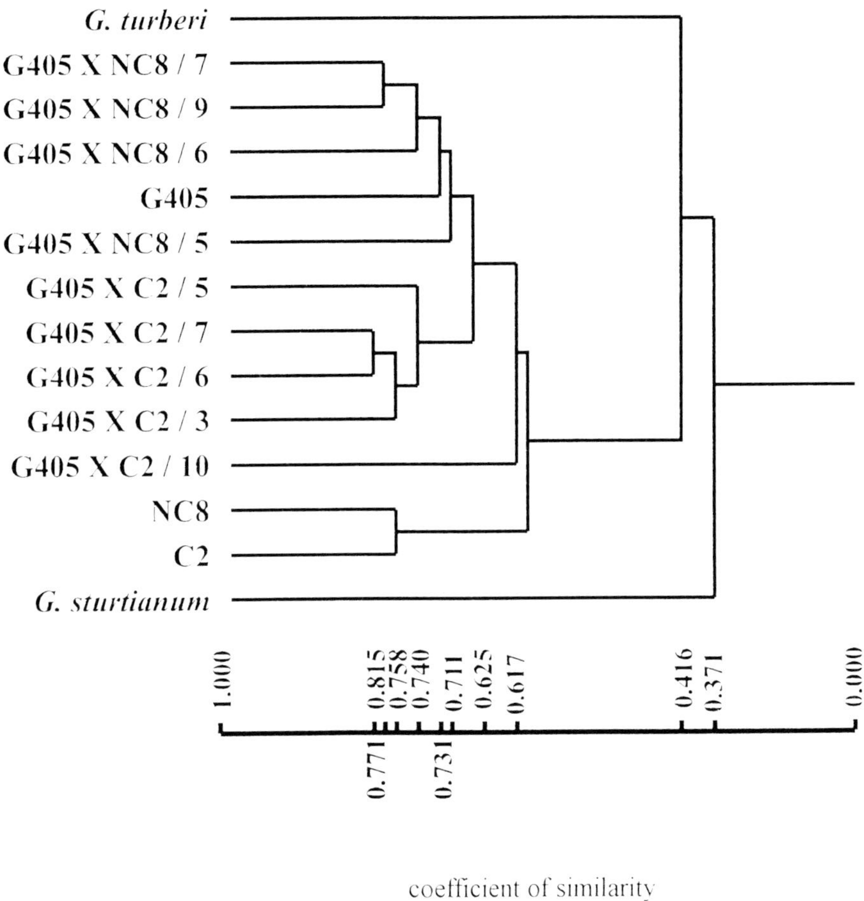

Fig. 3. Dendrogram obtained by UPGMA cluster analysis showing the genetic relationship between the backcrosses 1 and their parents used in the crosses G405 × NC8 and G405 × C2

the fourth. The placement of the trispecies G405 into the G405 × NC8 group instead of the G405 × C2 group is not surprising since G405 was itself generated by the cross (*G. thurberi* × *G. sturtianum* × *G. hirsutum* NC8); it is due to the fact that in a backcross 1 of G405 × NC8, the cultivated cotton NC8 is double-dosed whereas in a backcross 1 of G405 × C2, NC8 is only single-dosed.

In all crosses, the similarity between the wild species and the cultivated parent ranges from 31.30 to 39.20% and the similarity between the three upland cotton cultivars used varies from 78.3 to 78.70%. RAPD data also show that a large amount of genetic variation exists in the backcrosses 1, with a similarity ranging from 57.70 to 81.50% between them. This wide variation in the backcrosses may suppose that genetic events such as translocations, deletions, and homeologous recombinations frequently occur in such materials.

The composition of the trispecies hybrids G405 and G376 was confirmed by RAPD; these hybrids are more similar to the cultivated upland cotton than to the wild species parents *G. raimondi*, *G. thurberi*, and *G. sturtianum*.

These results are in close agreement with the current taxonomy of the *Gossypium* genus. They confirm the conclusions of previous research works (Multani and Lyon 1995; Tatineni et al. 1996) concerning the interest of RAPDs to determine genetic relationships within a diverse array of *Gossypium* germplasm.

4 Conclusion and Prospects

A system of random amplified polymorphic DNA (RAPD) markers has been developed to monitor the transfer of wild diploid cotton genes into *Gossypium hirsutum* by hybridization and backcrossing. It is shown here that RAPD markers, specific to wild diploid species, can be generated quickly and efficiently. As Klein-Lankhorst et al. (1991) demonstrated for potato, and McCoy and Echt (1993) for *Medicago*, the RAPD technique provides a highly effective means to fingerprint cotton species and interspecific hybrids. Moreover, our results demonstrate the ability of RAPD markers to reliably differentiate between *G. hirsutum* varieties and provide a molecular tool to assess introgression from several wild diploid species (*G. sturtianum*, *G. thurberi*, and *G. raimondii*) into *G. hirsutum*.

Given the important genetic similarities existing between the diploid species belonging to A and D cytotypes and both *G. hirsutum* subgenomes, RAPD seems to be more convenient to detect useful specific markers in species from phyletically more remote genomes (such as the C cytotype).

One obvious use of the markers we selected is to identify the BC_1 individuals introgressed for a useful character and which simultaneously show the minimum number of wild species RAPD markers.

It should also be possible to use them in a bulk segregant analysis according to the method developed by Michelmore et al. (1991) in order to identify markers associated with the glandless-seed and glanded-plant trait of *G. sturtianum*.

For homozygous parents, the gel pattern of RAPD products of interspecific hybrid should be the sum of the parental patters. However, we found a relatively high rate of exception to this prediction. The same kind of observation was made in the analysis of potato and alfalfa interspecific hybrids by Masuelli et al. (1995) and McCoy and Echt (1993). The appearance of products that are not present in the parents, or the disappearance of some parental products, could be explained by primer competition during annealing (Williams et al. 1990) or by inversion and deletion due to heterogenetic pairing in the intermediate tetraploid (for the HRS hybrid) or hexaploid (for the TSH hybrid) stages.

RAPD analysis is sensitive to minor changes in the amplification procedure because it uses short (10-mer primers). To limit this problem, Paran and

Michelmore (1993) proposed the sequencing of RAPD products and their conversions to normal PCR products by the use of longer primers. These markers constitute a type of sequence tagged site (STS, Olson et al. 1989) and have been termed sequence characterized amplified regions (SCARs).

As was observed by Masuelli et al. (1995) in the framework of an interspecific breeding program of potato, the conversion into SCARs of the RAPD primers we selected should improve the accuracy of the analysis of introgression from wild diploid species in subsequent generations. Furthermore, combination of primers in RAPD may provide additional species-specific markers.

Cluster analysis of genetic distances between the ascendants and the progeny of the trispecific hybrids shows phylogenetic relationships in agreement with the known lineage of the compared genotypes.

The clear and reliable detection of polymorphism among different upland cotton species, cultivars, and backcrosses makes us anticipate that RAPD markers will prove to be extremely useful in introgressing economic traits in cotton.

Acknowledgment. This research was supported by the Belgian Fonds de la recherche fondamentale collective.

References

Akkaya MS, Bhagwat AA, Cregan PB (1992) Length polymorphisms of simple sequence repeat DNA in soybean. Genetics 132:1131–1139

Altman DW, Stelly DM, Kohel RJ (1987) Introgression of the glanded-plant and glandless-seed trait from *Gossypium sturtianum* Willis into cultivated upland cotton using ovule culture. Crop Sci 27:880–884

Arnold ML, Buckner CM, Robinson (1991) Pollen-mediated introgression and hybrid speciation in Louisiana irises. Proc Natl Acad Sci USA 88:1398–1402

Arumunganathan K, Earle ED (1991) Nuclear DNA content of some important plant species. Plant Mol Biol Rep 9:208–218

Baird E, Lojkowska E, Ehlenfeldt MK, Kelman A, Helgeson JP (1992) Molecular characterization of inter- and intra-specific somatic hybrids of potato using random amplified polymorphic DNA (RAPD) markers. Mol Gen Genet 233:469–475

Beckman JS, Soller M (1986) Restriction fragment length polymorphism and genetic improvement of agricultural species. Euphytica 35:111–124

Brubaker CL, Wendel JF (1994) Reevaluating the origin of domesticated cotton (*Gossypium hirsutum*, Malvaceae) using nuclear restriction fragment length polymorphisms (RFLPs). Am J Bot 81(10):1309–1326

Devos KM, Gale MD (1992) The use of random amplified polymorphism DNA markers in wheat. Theor Appl Genet 84:194–200

Dilday RH (1986) Development of cotton plant with glandless seeds and glanded foliage and fruiting forms. Crop Sci 26:639–641

Endrizzi JE, Turcotte EL, Kohel RJ (1985) Genetics, cytology and evolution of *Gossypium*. Adv Genet 23:271–275

Fritsch P, Hanson MA, Spore CD, Pack PE, Rieseberg L (1993) Constancy of RAPD primer amplification stength among distantly related taxa of flowering plants. Plant Mol Biol Rep 11:10–20

Gill MS, Bajaj YPS (1987) Hybridization between diploid (*Gossypium arboreum*) and tetraploid (*Gossypium hirsutum*) cotton through ovule culture. Euphytica 36:625–630

Halward T, Stalker HT, LaRue E, Kochert G (1991) Genetic variation detectable with molecular markers among unadapted germplasm resources of cultivated peanut and related wild species. Genome 34:1013–1020

Halward T, Stalker HT, LaRue E, Kochert G (1992) Use of single-primer DNA amplifications in genetic studies of peanut (*Arachis hypogea*). Plant Mol Biol 18:315–325

He S, Ohm H, Mackenzie S (1992) Detection of DNA sequence polymorphisms among wheat varieties. Theor Appl Genet 84:573–578

Heun M, Helentjaris T (1993) Inheritance of RAPDs in F_1 hybrids of corn. Theor Appl Genet 85:961–968

Heun M, Murphy JP, Phillip TD (1994) A comparison of RAPD and isozyme analysis for determining the genetic relationships among *Avena sterilis* L. accessions. Theor Appl Genet 87:689–696

Hu J, Quiros CF (1991) Identification of brocolli and cauliflower cultivars with RAPD markers. Plant Cell Rep 10:505–511

Jacobsen E, De Jong JH, Kamstra SA, Van Den Berg PM, Ramanna MS (1995) Geomic in situ hybridization (GISH) and RFLP analysis for the identification of alien chromosomes in the backcross progeny of potato (+) tomato fusion hybrids. Heredity 74:250–257

Joshi CP, Nguyen HT (1993) Application of the random amplified polymorphic DNA technique for the detection of polymorphism among wild and cultivated tetraploid wheats. Genome 36:602–609

Kesseli RV, Paran I, Michelmore RW (1994) Analysis of a detailed genetic linkage map of *Lectuca sativa* (lettuce) constructed from RFLP and RAPD markers. Genetics 136:1435–1446

Klein-Lankhorst RM, Vermunt A, Weide R, Liharska T, Zabel P (1991) Isolation of molecular markers for tomato (*Lycopersicon esculentum*) using random amplified polymorphic DNA (RAPD). Theor Appl Genet 83:108–114

Koto E (1989) Tentative de transfert du caractère "retard à la morphogénèse des glandes à gossypol" I. Caractéristiques des hexaploides *G. hirsutum* × *G. sturtianum* et *G. hirsutum* × *G. australe*. lère Conférence de la recherche cotonnière africaine, Lomé, Togo, 31 janv.–2 fév., 1989, IRCT, Montpellier, Tome I:167–173

McCoy TJ, Echt CS (1993) Potential of trispecies bridge crosses and random amplified polymorphic DNA markers for introgression of *Medicago daghestanica* and *M. pironae* germplasm into alfalfa (*M. sativa*). Genome 36:594–601

Marshall P, Marchand MC, Lisieczko Z, Landry BS (1994) A simple method to estimate the percentage of hybridity in canola (*Brassica napus*) F_1 hybrids. Theor Appl Genet 89:853–858

Masuelli RW, Tanimoto EY, Brown CR, Comai L (1995) Irregular meiosis in a somatic hybrid between *Solanum bulbocastanum* and *S. tuberosum* detected by species-specific PCR markers and cytological analysis. Theor Appl Genet 91:401–408

Mergeai G (1992) New perspectives concerning the methodology to be used for introgression of the glanded-plant and glandless-seed character in cultivated cotton (*Gossypium hirsutum* L.). Coton Fibres Trop 47:113–119

Mergeai G, Vroh Bi I, du Jardin P, Baudoin JP (1995) Introgression of glanded-plant and glandless-seed trait from *G. sturtianum* Willis into tetraploid cotton plants. Proc Beltwide Cotton Improvement Conf, San Antonio, 6–7 Jan 1995, pp 513–514

Michelmore RW, Paran I, Kesseli RV (1991) Identification of markers linked to disease-resistance genes by bulked segregant analysis: a rapid method to detect markers in specific genomic regions by using segregating population. Proc Natl Acad Sci USA 88:9828–9832

Multani DS, Lyon BR (1995) Genetic fingerprinting of Australian cotton cultivars with RAPD markers. Genome 38:1005–1008

Murray M, Thomson WF (1980) Rapid isolation of high molecular weight plant DNA. Nucleic Acids Res 8:4321–4325

Olson M, Hood L, Cantor C, Botstein D (1989) A common language for physical mapping of the human genome. Science 245:1434–1435

Orozco-Castillo C, Chalmers KJ, Wangh R, Powell W (1994) Detection of genetic diversity and selective gene introgression in coffee using RAPD markers. Theor Appl Genet 87:934–940

Paran I, Michelmore W (1993) Development of reliable PCR-based markers linked to downy mildew resistance genes in lettuce. Theor Appl Genet 85:985–993

Quiros CF, Hu J, This P, Chevre AM, Delseny M (1992) Developmental and chromosomal localization of genome-specific markers by polymerase chain reaction in *Brassica*. Theor Appl Genet 82:627–632

Reinisch AJ, Dong Jian-min, Brubaker CL, Stelly DM, Wendel JF, Paterson AH (1994) A detailed RFLP map of cotton, *Gossypium hirsutum* × *Gossypium barbadense*: chromosome organization and evolution in a disomic polyploid genome. Genetics 138:829–847

Reiter RS, Williams JGK, Feldmann KA, Rafalski JA, Tingey SV, Sclonik PA (1992) Global and local genome mapping in *Arabidopsis thaliana* by using recombinant inbred lines and random amplified polymorphic DNAs. Proc Natl Acad Sci USA 81:8014–8018

Rooney WL, Stelly DM, Altman DW (1991) Identification of four *Gossypium sturtianum* monosomic alien addition derivatives form a backcrossing program with *G. hirsutum*. Crop Sci 31:337–341

Rowland LJ, Levi A (1994) RAPD-based genetic linkage map of blueberry derived from a cross between diploid species (*Vaccinium darrowi* and *V. elliottii*). Theor Appl Genet 87:863–868

Stewart J McD, Hsu CL (1977) In-ovulo embryo culture and seedling development of cotton (*Gossypium hirsutum* L.). Planta 137:113–117

Tanksley SD, Young ND, Pat AH, Bonierbale MW (1989) RFLP mapping in plant breeding – new tools for an old science. Bio/Technology 7:257–264

Tatineni V, Cantrell RG, Davis DD (1996) Genetic diversity in elite cotton germplasm determined by morphological characteristics and RAPDs. Crop Sci 36(1):186–192

Tinker NA, Fortin MG, Mather GE (1993) Random amplified polymorphic DNA and pedigree relationships in spring barley. Theor Appl Genet 85:976–984

Tulsieram LK, Glaubitz JC, Kiss G, Carlson JE (1992) Single tree linkage mapping in conifers using haploid DNA from megagametophytes. Biotechnolgy 10:686–690

Van Hausden AW, Bachmann K (1992) Genotype relationships in *Microseries elegans* (Asteraceae, Latcuceae) revealed by DNA amplification from arbitrary primers (RAPDs). Plant Syst Evol 179:221–233

Veli-Matti Rokka, Yong-Shen Xu, Jyri Kankila, Anja Kuusela, Seppo Pulli, Eija Pehu (1992) Identification of somatic hybrids of dihaploid *Solanum tuberosum* lines and *S. brevidens* by species-specific RAPD patterns and assessment of disease resistance of the hybrids. Euphytica 80:207–217

Vierling RA, Nguyen HT (1992) Use of RAPD markers to determine the genetic diversity of diploid wheat genotypes. Theor Appl Genet 84:835–838

Vroh Bi I, Harvengt L, Chandelier A, Mergeai G, du Jardin P (1996) Improved RAPD amplification of recalcitrant plant DNA by the use of activated charcoal during DNA extraction. Plant Breed 115:205–206

Vroh Bi I, du Jardin P, Mergeai G, Baudoin J-P (1997) Optimisation et application de la RAPD (random amplified polymorphic DNA) dans un programme de sélection récurrente chez le cotonnier (*Gossypium* spp.). Biotechnol Agron Soc Environ 1:142–150

Walbot V, Duke LS (1976) Developmental biochemistry of cotton seed embryogenesis and germination. 7. Characterisation of the cotton genome. J Mol Biol 101:503–536

Welsh J, Honeycutt RJ, McClelland M, Sobral BWS (1991) Parentage determination in maize hybrids using arbitrarly primed polymerase chain reaction (AP-PCR). Theor Appl Genet 82:473–476

Wilde J, Waugh R, Powell W (1992) Genetic fingerprinting of *Theobroma* clones using randomly amplified polymorphic DNA markers. Theor Appl Genet 83:871–877

Williams JGK, Kubelik AR, Livak KJ, Rafalski JA, Tingery SV (1990) DNA polymorphisms amplified by arbitrary primers are useful as genetic markers. Nucleic Acids Res 22:6531–6535

II.5 Heterosis in Cotton

D.G. Roupakias[1], E. Gouli-Vavdinoudi[1], M. Koutsika-Sotiriou[1], S. Galanopoulou-Sendouca[2], and A.S. Mavromatis[1]

1 Introduction

Heterosis is a term used to describe the phenomenon in which the performance of an F_1, produced by the crossing of two homozygous but genetically different individuals, is superior to that of the best parent. Since the time Shull (1914) introduced the term heterosis, breeders have made wide use of this phenomenon, and the list of crops in which heterosis has led to economic gains is ever-increasing (Sinha and Khanna 1975). In addition, the remarkable improvement in yield of maize during the 20th century has largely resulted from utilization of heterotic F_1 hybrids (Eberhart 1979). Thus, while plant breeders have been utilizing heterosis for the improvement of crop productivity, the biological basis of heterosis remains unknown. There are, at present, two major theories suggested to explain the genetic basis of heterosis, the dominance and the overdominance hypotheses. In the past, geneticists and breeders found it difficult to approach the dominance vs. overdominance controversy experimentally. Tsaftaris (1995) reviewed the work pertaining to heterosis using molecular analysis and concluded that, in maize, overdominace is the main genetic basis of heterosis. Later, however, Stuber (1995) suggested that the overdominance observed in maize was probably pseudooverdominance (i.e., nearby loci at which alleles having dominant or partially dominant advantageous effects are in repulsion-phase linkage). This, together with the finding that in rice heterosis is largely due to dominance (Xiao et al. 1995), indicates that dominance seems to be the genetic basis of heterosis.

Cotton is one of the oldest crops in which heterosis in interspecific crosses of *Gossypium hirsutum* × *G. barbadense* was observed almost 90 years ago (Cook 1909). Loden and Richmond (1951) reviewed the early work on heterosis in cotton and pointed out findings ranging from little or no heterosis in certain crosses to substantial amounts in others. Today, it is widely recognized that economically important heterosis resulting in higher lint yield and improved fiber properties exist in some interspecific and intraspecific hybrids

[1] Dept. of Genetics and Plant Breeding, School of Agriculture, Aristotelian University of Thessaloniki, 54006 Thessaloniki, Greece
[2] Lab. of Agronomy, School of Agriculture, University of Thessaly, 38334 Volos, Greece

Biotechnology in Agriculture and Forestry, Vol. 42
Cotton (ed. by Y.P.S. Bajaj)

(Davis et al. 1986; Thomson and Lukett 1988b; Galanopoulou-Sendouca and Roupakias 1995). Commercial utilization of heterosis, however, is currently limited in cotton by our inability to produce F_1 hybrids in an economically feasible way.

During the past 40 years or so, much effort has been devoted to studying three basic questions: (1) the level of heterosis within and between cotton species for a number of agronomic traits, (2) the genetic basis of heterosis in cotton, and (3) the identification of the most profitable way of utilization. In this chapter, heterosis in cotton in relation to these questions is discussed.

2 General Considerations

2.1 Occurrence and Manifestation of Heterosis

Heterosis is a widely documented phenomenon in diploid and polyploid organisms that undergo sexual reproduction. There is sufficient evidence today that heterosis can be observed in plants, animals, and microorganisms for a number of economically important characters (MacKey 1976). In plants, the phenomenon has been observed in both cross and self-pollinated crops, but its level of expression is highly variable. Thus, the original concept that "heterosis is the increase in size, yield, etc." was extended to cover inferiority (Powers 1944) and adaptability (Dobzhansky 1950). Yet, most textbooks of genetics and plant breeding describe heterosis as the manifestation of greater vigor in height, leaf area, growth, dry matter accumulation, and higher yield of the F_1 hybrid in comparison with its inbred parents (Allard 1960; Brewbaker 1964; Hageman et al. 1967; Fehr 1987). All these characters are considered to be quantitative, and the study of their inheritance became an important part of the quantitative genetics (Sprague 1983). Most physiologists, however, viewed these characters as a consequence of physiological stimulus due to heterozygosity (Tsaftaris 1990). A number of studies demonstrated that the heterotic F_1 seedlings exhibit higher growth rates than the inbred lines during the first weeks of germination (Luckwill 1937; Shrivastava 1972). It is well known today that the basic physiological activity during early seedling growth is the formation of enzymes, translocation, transformation, and utilization of stored food material in the seed, and then the building up of the active protoplasmic base for further physiological activity. At this stage, the heterotic effect is apparent in the acceleration of growth rate and in the building up of a differential level of metabolism which ultimately results in quantitative differences in size, yield, etc. However, no matter whether heterosis is manifested in quantitative or in physiological traits, the biological basis of heterosis remains unknown. The present-day explanations of heterosis are based on the following theories.

2.2 Present-Day Theories of Heterosis

The genetic basis of heterosis has been debated for more than 80 years and is still not well established. During this long period of intensive investigation of the phenomenon, two major hypotheses have been suggested to explain heterosis, the dominance and the overdominance hypotheses. The dominance hypothesis, proposed by Davenport (1908) and Bruce (1910), assumes that the contribution of dominant alleles by one parent cancels the deleterious effects of recessive alleles contributed by the other parent in the heterozygous F_1. The overdominance hypothesis (East 1908; Shull 1908) supposes that the heterozygous combination of the alleles at a given locus results in a value superior to the value of either homozygote at that locus.

To resolve the dominance vs. overdominance controversy experimentally is very important because it reflects essential differences for practical breeding. According to dominance, heterosis is fixable by inbreeding and therefore the development of inbred lines with all the desirable dominant alleles is possible through inbreeding and selection. It must be clarified, however, that in practical breeding, heterosis due to dominance is easily fixable only when the number of complementary loci controlling an agronomic trait is relatively small and not closely linked. If the number of favorable and unfavorable loci is large, the probability of recovering an F_2 segregant homozygous for the favorable dominant alleles at all loci is extremely small but not impossible. In contrast, heterosis due to overdominance is not fixable and, therefore, the breeder's goal should be the development of hybrid cultivars.

In the past, geneticists and breeders found it difficult to approach the dominance vs. overdominance controversy experimentally. Tsaftaris (1995) concluded that, in maize, overdominance is the main genetic basis of heterosis. Stuber (1995), however, suggested that the overdominance observed in maize was probably pseudooverdominance. Bingham et al. (1994) and Woodfield and Bingham (1995) also reported that complementation of favorable alleles in repulsion phase linkage must predominate in heterosis of tetraploid alfalfa. These reports, together with the finding that in rice heterosis is largely due to dominance (Xiao et al. 1995), indicate that dominance seems to be the genetic basis of heterosis. In some species, however, with a high load of deleterious genes where dominant and recessive alleles of closely linked loci occur in repulsion phase, pseudooverdominance is expressed (Crow 1952) which has been taken as overdominance.

2.3 Measurement of Heterosis

In most studies when hybrid vigor was first commercially utilized, heterosis was evaluated in comparison with the best parent (heterobeltiosis) (Shull 1952; Allard 1960). Later, Liang et al. (1972) estimated heterosis using the formula:

$$\text{Heterosis} = \frac{\left(\text{F}_1 - \text{better parent}\right)}{\text{better parent}} \times 100.$$

However, it is also common to estimate heterosis by using the midparent (MP) value (relative heterosis) (Walton 1971). In such a case, one could use the above formula by substituting the value of the better parent with the midparent value ($P_1 + P_2/2$). This estimate does, indeed, provide an expression of heterosis, but the first one is preferable, since no breeder would look for a hybrid plant poorer than one of its parents. Furthermore, in the practical breeding of self-pollinated crops, we look for hybrid plants with a higher value than the better-adapted commercial check cultivar. This expression of heterosis in comparison with the best local variety is termed useful heterosis (Meredith and Bridge 1972) and it has a direct economic meaning. It is clear, therefore, that in comparing levels of heterosis one should take into consideration the basis of comparison used in each case.

3 Intraspecific *G. hirsutum* Heterosis

3.1 Yield

Loden and Richmond (1951) reviewed the early work pertaining to heterosis in cotton and were of the opinion that "the list of intraspecific crosses in which an appreciable amount of heterosis has been exhibited would have been quite extensive if all the cases observed had been recorded and published". Indeed, since then, numerous studies from different areas have shown significant relative heterosis (for review see Meredith 1984). Yield expressed the highest heterosis in the studies reviewed and the 18% increase was mainly attributed to the two main yield components: boll number (13.5%) and boll weight (8.3%). From the breeding point of view, however, no breeder would look for a hybrid cultivar poorer than one of its parents. Thus, Jones and Loden (1951) were the first to report that the cotton seed yields from nine F_1s ranged from 0.8 to 47.0% higher than the best parent. Since that time, there are plenty of reports in which best-parent (BP) heterosis was recorded. The most impressive ones are shown in Table 1. White and Richmond (1963), taking into consideration that hybrid vigor is positively correlated with genetic divergence (Loden and Richmond 1951), conducted a study based on a diallel crossing system in which they used five genetically diverse cultivars. From the 15 different F_1 hybrids produced, only 2 yielded significantly more than their best parent, but they did not significantly outyield the best open-pollinated commercial check variety. Thus, it became clear that genetic divergence alone could not guarantee heterosis higher than the best commercial varieties. This is very important because a hybrid cultivar should be more profitable than the best available commercial variety in order to overcome the higher cost of seed production. The range of useful heterosis reported so far in intraspecific and interspecific cotton hybrids is shown in Table 2.

The question that is open here is related to the factors which have to be taken into consideration in an attempt to produce F_1 hybrids exhibiting useful

Table 1. Heterobeltiosis (BP heterosis) for yield, components of yield, and fiber properties of *G. hirsutum* and *G. barbadense* intraspecific hybrids

Species	Yield		Yield components						Fiber properties			Reference
	Seed	Lint	Lint (%)	No. of bolls	Boll weight	Seed Index	Seed/ boll	Lint index	Length	Strength	Micr. value	
G. hirsutum	–	19.0*	–	–	–	–	–	–	–	–	–	Hawkins et al. (1965)
G. hirsutum	–	8.5*	–1.3	–	–0.7	–1.2**	–	–	0.5**	1.1**	–1.3**	Meredith and Bridge (1972)
G. hirsutum		13.3*	–3.2*	–4.0	2.1	–0.8	–1.6	9.8**	–	–	–	El-Adl and Miller (1971)
G. hirsutum	11.9	13.5	–1.0	–5.8	1.0	2.1	1.7	–1.9	–	–	–	Marani (1963)
G. hirsutum	–	12.9	–1.3	–4.5	4.6	–2.5	4.2	–6.4	–	–	–	Wells and Meredith (1986)
Mean	11.9	13.4	–1.7	–4.8	1.8	–0.6	1.4	0.5	0.5	1.2	–1.3	
G. barbadense	32.7	18.9	–2.7	9.9	4.8	–2.7	2.5	4.9	–	–	–	Marani (1963)

*, ** Significant at the 0.05 and 0.01 levels, respectively.

Table 2. Useful heterosis for yield, components of yield and fiber properties of *G. hirsutum* intraspecific and *G. hirsutum* × *G. barbadense* intrespecific hybrids

Species	Commercial check variety	Yield		Yield components			Fiber properties			Reference
		Seed	Lint	Lint (%)	No. of bolls	Boll weight	Length	Strength	Micr. value	
G. hirsutum	Coker 100A	–	10.5*	1.0	–	8.5**	–1.7	–	0.25	Lee et al. (1967)
G. hirsutum	Coker 100 Wilt	–	6.7	–4.4	–	11.6	–0.8	1.6	–	Miller and Marani (1963)
G. hirsutum	Coker 100A	–	14.9	0.3	–	7.0	–1.1	–	0.2	Miller and Lee (1964)
G. hirsutum	Stoneville 7A	–	13.3	–3.6	24.6	–0.8	–	4.9	–7.4	Thomson (1971)
G. hirsutum	Stoneville 7A	–	15.4	–1.5	16.6	0.63	3.7	–	–7.1	Thomson and Luckett (1988b)
G. hirsutum	Deltapine 16	–	20.3	–3.6	10.1	13.4	4.4	–	12.3	
G. hirsutum	Sindos – 80	60.0	76.0	–1.7	–	20.8	1.1	–	10.8	Galanopoulou-Sendouca and Roupakias (1995)
Mean		60.0	22.4	–1.9	17.1	8.7	0.9	3.3	1.5	
G. h × G. b	Acala 1517 – 70	–	–5.8	–4.4	52.6**	–	–	–	–	Davis (1974)
G. h × G. b	Acala 1517 – 70	45.0*	29.5*	–	–	–	19*	23.9*	23.8*	Davis (1979)
G. h × G. b	Taskend – 6	13.0	–	–10.4	–	–29.6	18.8	–	7.7	Galanopoulou-Sendouca and Roupakias (1995)

*, ** Significant at the 0.05 and 0.01 levels, respectively.

heterosis. Genetic divergence is, of course, important, but not sufficient (White and Richmond 1963; Hawkins et al. 1965; Thomson 1971). In addition, it has been observed that the highest-yielding cultivars tend to produce the highest-yielding hybrids (Marani 1963; Miller and Lee 1964; Hawkins et al. 1965; Pathak and Kumar 1975; Thomson and Luckett 1988b). The level of heterosis expressed, however, seems to be affected by the growth environment. Thus, we have observed (Galanopoulou-Sendouca and Roupakias 1995) that the level of heterosis fluctuated from location to location and from year to year within the same location (Table 3). Similar data have also been presented by others (Turner 1953a; Hawkins et al. 1965; Lee et al. 1967; Meredith and Bridge 1972; Baker and Verhalen 1975; Galanopoulou-Sendouca 1988). Generally, the percentage of heterotic yield increase was found to be greater at low yield levels than at maximum yield levels (Table 3). The same has been reported by other workers (Miller and Lee 1964; Hawkins et al. 1965; Thomson 1971). Thus, it seems that, in areas with high yield levels, the probability of success of hybrid cultivars is reduced as compared to the lower-yielding areas with cool conditions and restricted vegetative growth. Nevertheless, one could suggest that the choice of parents for hybrid production should be based on per se performance of the cultivars in the area under interest, the genetic divergence, and general combining ability effects. Such parents could then be intercrossed and their hybrids should be evaluated in the area. Under such a program one should be able to isolate intraspecific F_1 hybrids with useful heterosis at both yield and quality properties (Thomson 1971; Meredith and Bridge 1972; Galanopoulou-Sendouca 1987).

Table 3. Useful heterosis (%) for lint yield over two locations in Greece and 4 years of two intraspecific and two interspecific cotton hybrids compared to the adapted cultivar Sindos 80. (After Galanopoulou-Sendouca and Roupakias 1995)

Location	Year	Intraspecific *hirsutum* hybrids		Interspecific hybrids *G. h.* × *G. b.*		Lint yield Sindos 80 (g/m^2)
		Thalia – 2 × Taskend – 6	PD × Taskend – 6	73468-3 × Pima S4	432 × Pima S4	
Sindos	1985	157	99	119	102	74
	1986	97	82	55	49	93
	1987	18	13	6	–10	160
	1988	30	27	–	5	64
	Mean	75.5	55.3	60	36.5	
Bardates	1985	29	28	–21	–10	116
	1986	23	15	–3	–20	152
	1987	36	47	–12	–20	108
	1988	31	77	–	+32	90
	Mean	29.8	41.8	–12	–4.5	

3.2 Yield Components

The concept of yield components has been used to clarify the nature of heterosis for yield. A simplified cotton yield model used so far is the one represented by bolls per unit area × seed cotton per boll × lint percentage (Davis 1978). The two most important yield components in *G. hirsutum* intraspecific hybrids are the number of bolls per unit area and boll weight, with an average useful heterosis of 17.1 and 8.7, respectively (Table 2). These values are close to those reported by Meredith (1984). The effect of lint percentage on the final yield heterosis seems to be of minor importance (average useful heterosis – 1.9%; Table 2). A similar but positive value has been reported by Meredith (1984). In addition to the major yield components, some investigators have reported significant heterosis for seed index (Marani 1963, 1967; Meredith 1979), lint index (Al-Rawi and Kohel 1969, 1970; Marani 1963, 1967, 1968b), and number of seeds per boll (Marani 1963, 1967; Al-Rawi and Kohel 1969, 1970). These components contribute to the increase in boll weight and therefore to yield heterosis.

Theoretically, therefore, the highest final yield would result from a large number of heavy bolls, but such is rarely ever the case. This is because the two major yield components are negatively correlated and big bolls are seldom developed on cotton plants with a large number of bolls (Turner 1953a). It is expected, therefore, that the greatest gain in yield improvement should come from exerting selection pressure on the number of bolls per square meter (Maner et al. 1971).

3.3 Fiber Properties

A considerable number of investigations have reported small but significant relative heterosis for fiber length (Meredith 1984). Our data also agree with these findings (Fig. 1D). Small and significant heterosis has also been reported for fiber strength (Miller and Marani 1963; Thomson and Luckett 1988b; Meredith 1990). Meredith (1990) also reported a small negative heterosis for micronaire value. In general, however, the fiber performance of an intravarietal *G. hirsutum* cross is expected to be relatively stable and similar to the midparental value (Davis 1978; Galanopoulou-Sendouca 1987). It has been found that the genetic control of the fiber properties is largely additive, although partial dominance has also been detected for upper-half mean length (Miller and Marani 1963; Verhalen and Murray 1967, 1969; Marani 1968a; Al-Rawi and Kohel 1970).

Since fiber properties are inherited in an additive manner, hybridization can only partially obscure the deficiencies of a poor parent. This fact alone tends to exclude much obsolete varietal material from consideration as possible parents for hybrids. Therefore, if commercial hybrid seed production becomes possible, breeders will need to prepare lines with the proper fiber attributes and disease and insect resistance, which, in addition, should combine well.

3.4 Other Traits

Davis (1978) reviewed extensively the heterotic expression of many agronomic characters in cotton hybrids. Thus, significant increases in seedling vigor have been reported (Harris and Loden 1954; Thomson 1971; Marani and Avieli 1973). In addition, Harris and Loden (1954) noted that hybrid vigor was evident in dry weight of stems, dry weight of leaves, dry weight of fruits, total dry weight of above-ground plant parts, number of squares, and number of fruits. Galal et al. (1966), in a study of heterosis in relation to accumulation of dry matter in upland cotton, observed that the hybrids began to diverge sharply from their parents when all plants were about 6 weeks old. The relative rates of dry matter accumulation continued to diverge for a period of about 3 weeks and then proceeded at comparable rates in both parents and hybrids to maturity. The hybrids thus maintained their vegetative superiority throughout the period of fruiting, resulting in plants larger than their parents, with a greater yield of fruit. The same conclusion had been reached by White and Richmond (1963) and Wells and Meredith (1986). Significant heterosis has also been reported for earliness (Miller and Marani 1963; Al-Rawl and Kohel 1969; Baker and Verhalen 1975; Wells and Meredith 1986; Galanopoulou-Sendouca 1987) and disease resistance (Shepherd 1974). We have also observed heterosis for disease resistance and specifically in *Verticillium* wilt (Fig. 1G). In general, therefore, in an F_1 hybrid, the breeder may incorporate a good yielding ability together with important attributes such as fiber quality, earliness, and disease resistance.

4 Intraspecific *G. barbadense* Heterosis

There are few reports on intraspecific crosses of *G. barbadense* (Marani 1963, 1967, 1968a,b). Heterosis was significant for both seed cotton yield and lint yield (23.3 and 24.2%, respectively) and it was mainly associated with an increase in the number of bolls produced. Heterosis for boll weight and number of seeds per boll was not significant, but there was a significant heterosis for lint index and seed index (2.9 and 1.2, respectively). A significant relative heterosis was also detected for fiber length (4%), while the fiber strength was closer to the value of the parent with the strongest fiber, but not significantly so. Finally, the micronaire value of the F_1 hybrids was reported to be intermediate between the parental values with a slight tendency toward coarseness.

5 Interspecific *G. hirsutum* × *G. barbadense* Heterosis

5.1 Yield

Striking hybrid vigor in interspecific crosses of *G. hirsutum* × *G. barbadense* was noted almost 90 years ago by Cook (1909). Some good results, however, in both yield and quality characters were reported in the early 1950s (Richmond and Lewis 1951; Patel and Patel 1952). This, together with the idea of producing hybrid cultivars in cotton, triggered the interest of the breeders for a detailed study of the interspecific hybrids. These studies documented that heterosis for yield of seed cotton and yield of lint in the interspecific crosses is very large (Table 4). This is also confirmed by our data (Table 3). The relative heterosis for seed cotton yield ranged from 60.8% (Ali and Lewis 1962) to 97% (Marani 1963), and for lint yield from 63% (Percy and Turcotte 1991) to 82% (Marani 1963). In addition, some of the hybrids examined in these studies exhibited best-parent heterosis for seed cotton yield, which ranged from 50% (Marani 1968c) to 87.5% (Marani 1967) and for lint yield from 26% (Marani 1968c) to 65.6% (Marani 1967; Table 4). Useful heterosis of the F_1 interspecific hybrid for seed cotton yield (45%) and for lint yield (29.5%) has also been reported (Davis 1979). Galanopoulou-Sendouca and Roupakias (1995), in a study of five interspecific hybrids, also observed positive useful heterosis for lint yield (4%), but in only one of two locations (Table 3). The authors attributed this difference to the excessive vegetative growth and later maturity observed at the second location (Vardates, Fig. 1B). Generally, it is recognized today that economically important heterosis for yield exists in some interspecific hybrids.

5.2 Yield Components

In *G. hirsutum* × *G. barbadense* hybrids the primary component of heterosis is the number of bolls produced per unit area (Table 4). The relative heterosis for number of bolls per square meter was significant in all studies reported so far and ranged from 68.5% (Marani 1968c) to 94.2% (Krishnaswami and Kothandaraman 1977). In contrast, relative heterosis for boll weight and number of seeds per boll was negative (Table 4). Thus, the boll weight in interspecific hybrids was lower than their respective *G. hirsutum* parent, but higher than that of *G. barbadense* parent, and did not contribute to heterosis in yield. Negative heterosis as compared to the *hirsutum* parent was also recorded for lint percentage. This was mainly due to the higher heterosis observed for seed index as compared to lint index.

5.3 Fiber Properties

The fiber of the interspecific hybrids was very long, fine, and strong. Hybrid fiber was generally similar to *G. barbadense* fiber, except for micronaire val-

Table 4. Relative heterosis and heterobeltiosis for yield, components of yield, and fiber properties of interspecific hybrids (*G. hirsutum* × *G. barbadense*)

Yield		Yield components						Fiber properties			Reference
Seed	Lint	Lint (%)	No. of bolls	Boll weight	Seed Index	Seed/boll	Lint Index	Length	Strength	Micr. value	
60.8	–	–8.0	–	11.0	23.8*	–	8.0	15.9*	28.9*	–	Ali and Lewis (1962)
(–0.7)[a]	–	(–17.6)	–	(–18.3)	(19*)	–	(–0.1)	(26.6*)	(8.4)	–	
–	64.7	–4.5	81.7**	–5.5	–	–	–	–	–	–	Davis (1974)
–	68.0*	–	–	–	–	–	–	13.3*	6.4	–22.7	Davis (1979)
83.6**	–	–	94.2**	–8.3	–	–	–	10.0	–6.5	–22.7	Krishnaswami and Kothandaraman (1977)
(63.7*)	–	–	(52.6**)	(–23.4)	–	–	–	(–4.6)	–	(–36)	
97.0**	82.0**	–7.0**	82**	–3.0	21**	15.2**	8.0**	–	–	–	Marani (1963)
(71.7**)	(54.5)	(–9.8)	(45.7)	(–30.5)	(16.5)	(–36.8)	(–1.3)	–	–	–	
91.2**	72.8**	–9.3**	79.5**	–6.9	20.6**	–17.7**	3.9	–	–	–	Marani (1967)
(87.5**)	(65.6**)	(–14.4**)	(32.4*)	(–30.5)	(18.1**)	(–36.5**)	(–6.6)	–	–	–	
79.3**	62.7**	–8.6**	68.5**	–0.8	20.5**	–13.3**	5.2**	–	–	–	Marani (1968c)
–	–	–	–	–	–	–	–	9.8**	12.2**	–16.1**	Marani (1968d)
–	–	–	–	–	–	–	–	(5.8)	(–2.0)	(–19.5)	

*, ** Significant at the 0.05 and 0.01 levels, respectively.

[a] Numbers in parenthesis indicate the percentage of heterobeltiosis.

ues. Micronaire values usually were lower than those of both parents. Thus, significant relative heterosis was found for both fiber length and fiber strength (Table 4). Relative heterosis for micronaire values, however, was negative. In a recent study Galanopoulou-Sendouca and Roupakias (1995) also observed that the interspecific hybrids had the good fiber properties of the *barbadense* parent. The range of useful heterosis for fiber length, micronaire value, and strength of the five interspecific hybrids studied is given in Fig. 1. Useful heterosis has also been reported by other workers (Table 2).

The low micronaire value observed in the interspecific hybrids, which was closer to the *barbadense* parent, may be due to inherent reduction in fiber diameter, inadequate cellulose deposition of the fiber wall, or a combination of the two factors (Davis 1979). In conclusion, the results obtained so far indicate that it is possible to combine good fiber qualities with a high degree of heterosis for yield in interspecific hybrids.

5.4 Other Traits

Although several investigators have reported excellent heterosis for yield and fiber properties in interspecific hybrids (*G. hirsutum* × *G. barbadense*), their commercial use is endangered if the parent cultivars are not selected on the basis of their plant type. This is because such crosses may produce tall and late hybrids. Significant relative heterosis for plant height has been reported by several workers (Marani 1967, 1968c; Davis 1974, 1979; Krishnaswami and Kothandaraman 1977). Heterosis has also been reported for maturity (Fig. 1B). Krishnaswami and Kothandaraman (1977) reported negative midparent heterosis for all the three maturity parameters, i.e., days to square, days to flowering, and days to first boll opening. The same authors reported that some hybrids were as early as their *G. hirsutum* parent. In general, however, the hybrids matured later than the *G. hirsutum* parent and earlier than the *G. barbadense* parent (Marani 1967, 1968c; Davis 1979; Galanopoulou-Sendouca 1987). Finally, we observed that heterosis for earliness of the interspecific hybrids varies significantly from location to location (Fig. 1B). This may indicate that the interspecific hybrids have a higher potential for commercial use in areas with early maturing conditions.

Davis (1978) reviewed the pest resistance potential of the interspecific hybrids, and indicated that there is a good prospect for pest resistance as compared to *G. hirsutum* cultivars. Indeed, Galanopoulou-Sendouca (1987) and Galanopoulou-Sendouca and Roupakias (1995) reported that the interspecific hybrids exhibited good resistance to *Verticillium* wilt (Fig. 1G).

Beyond the above agronomic characters, the interspecific hybrids exhibited a best parent heterosis for callogenesis in in vitro culture of anthers (Fig. 2). The same has been recorded for hypocotyl explants (A.G. Mavromates and D.G. Roupakias, unpubl.). In conclusion, heterosis in the interspecific hybrids seems to be a more general phenomenon, and may help to overcome difficulties related with the in vitro response of cotton cultivars.

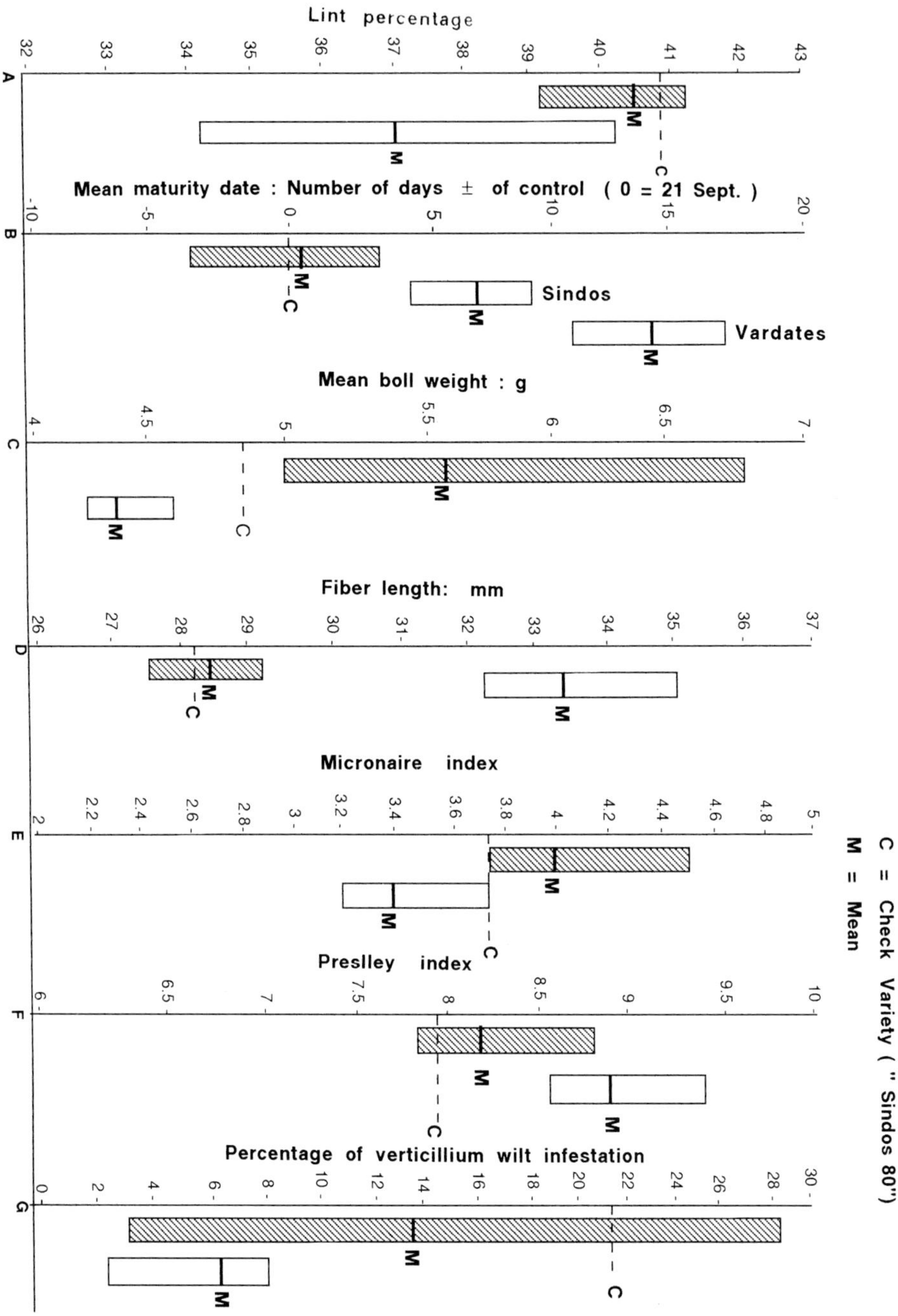

Fig. 1. Range of F_1 heterosis for lint percentage, earliness (mean maturity date), mean boll weight, fiber length, Pressley index, micronaire index, and percentage of *Verticillium* wilt infestation, in 14 intraspecific ▨ and 5 interspecific ▭ cotton hybrids tested at two locations (Sindos and Vardates) in Greece during the period 1985–88 together with the control variety (Sindos 80). (After Galanopoulou-Sendouca and Roupakias 1995)

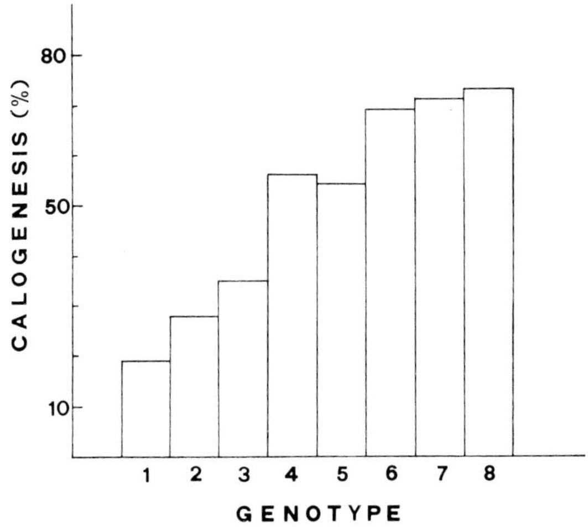

Fig. 2. Anther culture response to callogenesis of one *G. barbadense* cultivar (*1* Pima S4), three *G. hirsutum* cultivars (*2* Acala SJ-2; *3* 4S and *4* Coker 312), three intraspecific hybrids (*5* Acala SJ-2 × 4S; *6* 4S × Coker; *7* Coker × 4S), and one interspecific hybrid (*8* 4S × Pima)

6 Combining Ability and Gene Action

An understanding of the nature of combining ability and gene action in relation to yield and fiber properties is of great help in the choice of parents and in the planning of an appropriate breeding program. Thus, many researchers attempted to obtain estimates of variances of general and specific combining ability in cotton. Some of them are summarized in Table 5. The great variability in the results reported should not be surprising, since the studies were conducted by several researchers over a range of environments and with genetically diverse materials. These results clearly indicate that an appreciable amount of variance in cotton is mainly due to additive and dominance genetic effects.

7 Effects of Inbreeding

The effects of inbreeding were obvious from the earliest studies in maize (East 1908; Shull 1908). It was observed then that vigor and productiveness were reduced in maize with increasing homozygosity. This phenomenon is widely known today as inbreeding depression and it could be viewed as the opposite of heterosis. A usual way of measuring inbreeding depression is the average percent decrease of the F_2 from the F_1.

$$\text{Inbreeding depression } \% = \left(F_1 - F_2\right)/F_1 \times 100.$$

Table 5. General and specific combining abilities variance components for yield, yield components, and fiber properties

Yield	Yield components						Fiber properties			Reference
Lint	Lint (%)	No. of bolls	Boll weight	Seed index	Seed/ boll	Lint Index	Length	Strength	Micr. value	
3894** (–441)[a]	1500** (0.047)	– –	0.229** (0.068)	– –	– –	– –	0.002** (0.000)	0.008** (0.001**)	– –	Miller and Marani (1963)
–22 (272)	1637** (71)	– –	2.88** (2)	– –	– –	– –	379** (20)	360** (–9)	378** (37)	Lee et al. (1967)
1.77 (4.60)	0.58 (0.55)	– –	– –	– –	– –	– –	0.42** (0.11)	0.149 (0.044)	0.561 (0.277)	Baker and Verhalen (1975)
87 (3)	54 (27)	92 (3)	83 (8)	– –	– –	– –	– –	52 (8)	57 (21)	Thomson (1971)
56** (22**)	62** (20.5**)	87.5** (5.5**)	83** (10**)	– –	– –	– –	72.5** (14**)	70** (17**)	61** (17**)	Thomson and Luckett (1988a)
1787.49 (4705.36)	– –	64.26 (75.15)	0.23 (0.22)	0.14 (0.32)	6.72 (4.62)	0.09 (0.15)	– –	– –	– –	Pathak and Kumar (1975)
3393* (–484)	0.005 (0.144*)	920* (54)	0.051* (0.006)	0.571** (0.008)	1.25* (0.19)	0.031** (0.001)	– –	– –	– –	El-Adl and Miller (1971)

*, ** Significant at the 0.05 and 0.01 levels, respectively.
[a] Numbers in parenthesis refer to SCA.

Cotton is a self-pollinated crop with a range of cross-pollination dependent on the level of pollinating insects. In spite of the possibility, however, that high levels of cross-pollination may occur, cotton cultivars are not greatly divergent from self-pollinated species in their response to inbreeding depression. In general, cotton cultivars suffer from small but measurable inbreeding depression when subjected to continued self-pollination (Brown 1942; Simpson and Duncan 1953; Young and Murray 1966). Inbreeding depression, however, was more pronounced after the first selfing of the F_1 to produce the F_2 generation. Thus, Turner (1953b) in 21 F_1 hybrids of seven inbred lines reported that, over all hybrids, the F_2 generation lost all vigor for earliness of bloom, shedding rate, and boll size, but retained from 15 to 50% of the F_1 gain for boll production, seed per boll, seed cotton yield, and seedling vigor. The inbreeding depression for yield (the single character of greatest interest in cotton production), yield components, and fiber properties reported from seven different research groups conducted in different years is given in Table 6. On average, the F_2 generation retained almost 50% of yield heterosis recorded in the F_1 hybrids. This is in accordance with the expected value based on a 50% decrease in dominance from the F_1 to F_2. There were individual hybrids, however, where the inbreeding depression for yield was very low, and therefore these F_2 hybrids were superior in yield to well-established cultivars (Meredith 1990). The existence of such hybrids is very important because such F_2 are promising populations within which to select.

8 Utilization of Heterosis in Cotton

8.1 Hybrid Cotton

The data presented so far indicate that heterosis in cotton is well documented today. The question that has to be answered, however, is related to the identification of the most profitable, economically feasible, and scientifically sound way to utilize heterosis in cotton. Cotton is a largely self-pollinated crop with a low load of deleterious genes, as they are expressed by a relatively low inbreeding depression (Brown 1942; Simpson and Duncan 1953; Turner 1953b; Christidis 1955; Young and Murray 1966; Al-Rawi and Kohel 1969). In addition, heterosis in cotton is basically due to additive and dominant genes and therefore must be fixable. Indeed, El-Adl and Miller (1971), in a study including two inbred parental lines, their F_1 hybrid, and six lines selected for high yield after three cycles of recurrent selection in this population, reported that the mean yield of the six selected lines exceeded that of the F_1 by 5.5%. In addition, a large part of heterosis observed in cotton during the past 30 years has been fixed by breeders (Lee 1987).

Historically, however, heterosis is strongly affiliated with the use of hybrid cultivars as it has been developed in the model crop of maize. It was logical, therefore, for the cotton breeder to think that the most effective way to utilize

Table 6. Average inbreeding depression for yield, components of yield, and fiber properties of *G. hirsutum* intraspecific crosses

Yield	Yield components						Fiber properties			Reference
	Lint (%)	No. of bolls	Boll weight	Seed index	Seed/ boll	Lint index	Length	Strength	Micr. value	
5.8** (5.5)[a]	2.7**	6.0**	5.4**	7.9**	4.0**	–	–	–	–	Al-Rawi and Kohel (1969)
7.7** (14.0)	0.8**	–	–	–	–	–	1.1**	0.4	–2.0**	Baker and Verhalen (1975)
12.5* (12.1)	0.8	7.9	4.3	1.3	3.3	3.6	–	–	–	Meredith (1979)
17.4** (27.5)	1.2**	–	5.3**	–	–	–	1.7**	0.5	–	Miller and Marani (1963)
3.7** (11.8)	0.6**	–	3.5**	–	0.9**	–	0.3*	0.9*	–0.7*	Meredith (1990)
2.6 (19.4)	–	–	–	–	–	–	–	–	–	Christidis (1955)
11.3 (16)	0.3	–	1.7	–	–	–	0.7	–	2.4	Galanopoulou-Sendouca and Roupakias (1995)
Mean 8.7 (15.2)	1.1	7.0	4.0	4.6	2.7	3.6	1.0	0.6	–0.1	

*, ** Significant at the 0.05 and 0.01 levels, respectively.

[a] Numbers in parentheses indicate the corresponding yield heterosis observed in the F_1 hybrids.

heterosis in cotton was to follow the maize trend and produce hybrid cultivars. This tendency is strongly demonstrated in the statement made by Stroman (1961) "that although around the year 1920, results from experiments on the value of hybrid corn were not too encouraging, the majority of today's corn in the developed world is raised from hybrid seed". He made this statement because he wanted to emphasize the good future of hybrid cotton although the experimental results up to that date were not encouraging. He recognized, however, the differences that exist between corn and cotton in their reproductive systems (i.e., monoecious-monogenic vs. digenic, cross-pollination vs. self-pollination, wind-blown pollen vs. insect-borne, and several hundred seeds vs. 50 seeds at most per female flower), but this was not enough indication to drive him against hybrid cultivars. In contrast, it was believed that hybrid cultivars could have a future in cotton if an effective male sterility system could be identified and a method of pollen dispersion could be developed. Thus, in the following decades, the researchers devoted a large part of their endeavor towards the development of proper methods for the commercial production of hybrid cotton seed in an economically feasible way. For this to be accomplished, however, two major problems had to be solved successfully. Firstly, the removal of the male gametophyte from the perfect cotton flowers, and secondly, the transfer of pollen from the donor cultivar to the stigmas of the seed cultivar. Both problems could be faced very successfully by hand emasculation and pollination if the labor cost was not prohibitive. Indeed, farmers in India planted approximately 800000 ha in 1972 to hybrid cotton from seed produced by using male sterility and hand-pollination techniques (Srinivasan et al. 1972; Davis 1978) and in 1992 the area planted with hybrids reached 2130000 ha (Basu and Patil 1993). Hybrid cultivars in India as a result had not only yield increase but they also had a great impact on the quality improvement of cotton. Thus, although 40 years ago, 85% of the cotton produced in India was below 22-mm fiber length, today, more than 85% of the Indian cotton belongs to the superior medium, long, and extra long staple categories. This is due to the fact that coarser cotton, represented by *G. arboreum* and *G. herbaceum*, has been replaced firstly by finer cottons of *G. hirsutum* and marginally of *G. barbadense*, and secondly in the 1970s and 1980s, by finer cottons of inter- and intraspecific hybrids. After the release of the first commercial hybrid (H-4) in 1970, the spread of hybrid cultivars in India was spectacular (Fig. 3). Basu and Patil (1993) reported that cotton hybrids are more popular under irrigated as well as under rain-fed conditions because of their higher yield and better stability of performance over the years compared to inbred varieties. In addition, hybrid cottons show a better response to nitrogen and higher plant densities. This is in accordance with the general belief that hybrids often exhibit greater homeostasis by offering alternative genetic pathways, which gives them an advantage under conditions of stress such as crowdedness, adverse weather, disease attacks, etc. (Lewis 1954; MacKey 1970, 1976). It is obvious, therefore, that hybrid cultivars could have a good future in cotton production if the seed cost were not prohibitive. Use of hand emasculation and pollination for the production of hybrid cultivars is out of the question in the developed world, due to the

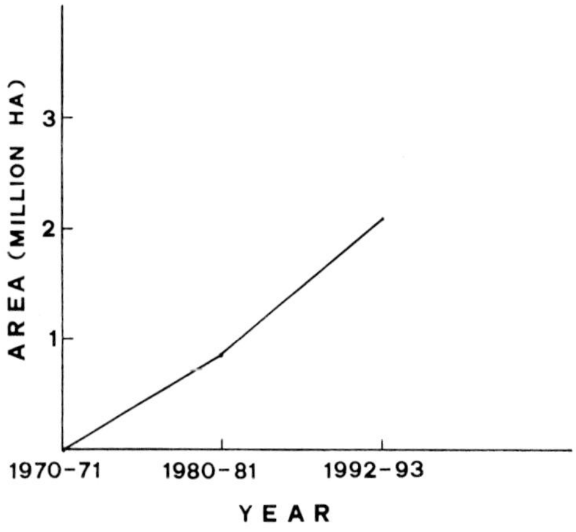

Fig. 3. The spread of hybrid cotton cultivars in India from 1970 to 1993. (Basu and Patil 1993)

expensive labor and the relatively small number of seeds (≈50) produced per flower. Thus, if hybrid cultivars had to reach the farmers, the researchers knew that they had to identify ways for a cheap and effective removal of the male gametophyte.

Eaton (1957) was the first to report that cotton plants sprayed with certain concentrations of 2,3-dichloroisobutyrate (FW-450) produced flowers with little or no functional pollen, but produced fertile seeds when pollinated by pollen from untreated plants. This caused considerable excitement among researchers interested in methods of producing hybrid cotton. However, the results from field experiments conducted to provide additional information on the gametocidal properties of FW-50 were discouraging (Pate and Duncan 1960; Richmond 1962). This disappointment was followed by optimism based on the isolation of the first male sterile systems (Justus and Leinweber 1960). The first report on male sterility was followed by intensive work aiming at the identification and isolation of genetic and cytoplasmic male sterile systems. Indeed, within two decades, a number of such systems had been isolated (for review see Davis 1978). The most effective male sterile systems identified so far are the cytoplasmic male sterile systems developed by Meyer (1971, 1973a,b,c) from *G. harknessii* material. These systems have flowers that are normal, but they have reduced stamens. The seed set with *G. harknessii* steriles has been good (Rosales and Davis 1976); they are environmentally stable and have excellent agronomic properties.

The development of cytoplasmic male sterile cotton flowers and the availability of satisfactory restorer lines could make the production of hybrid cottonseed practical if an insect pollinator could be found to transfer pollen from the donor to the male sterile flowers. McGregor (1959) showed that honeybees are effective pollinators of cotton when present in sufficient numbers. The same conclusion was reached by Moffett and Stith (1972), Moffett et al. (1975), and Rosales and Davis (1976). McGregor conducted experiments in Arizona and was able to obtain excellent cross-pollination of Pima cotton with 10 bees per 100 flowers. In addition, Rosales and Davis (1976) estimated that

2 bees per 100 flowers would be sufficient for near-normal production of the *G. harknessii*-derived cytoplasmic male sterile cottons.

It has to be emphasized, however, that honeybees cross-pollinate cotton cultivars effectively only in areas where there are no other plant species that strongly attract bees (Moffett et al. 1975). It is obvious, therefore, that the area, in combination with bee population levels, may well be the deciding factor in determining whether hybrid seed production is feasible.

With cytoplasmic male sterility at hand and with bees attracted by the *G. harknessii*-derived male sterile lines (Moffett et al. 1975), one could consider that in the middle 1970s, breeders had reached the point where they could produce upland F_1 hybrids that would give a consistent 10 to 20% increase in yield. It was this system that made Davis (1978) state, "the breeder now has in hand all the tools necessary to produce large quantities of bee-pollinated hybrids". If this were applicable today, 20 years later, one would expect part of the cultivated area to be seeded with hybrid cultivars. However, this is not the case. It is becoming clear, therefore, that the early attempt made by researchers to compare the history of maize hybrids with cotton hybrids as a source of optimism was not so successful. Although, before 1930, predominantly open-pollinated maize cultivars were grown in Iowa, the double-cross hybrids increased from nearly zero in the early 1930s to 100% within 10 years (Hallower and Miranda 1981). The same trend was also observed for the use of single-cross hybrids (from nearly zero in 1960 to nearly 100% in the middle 1980s; Lamkey and Smith 1987). In contrast, the production and distribution of cotton hybrids 20 years after the development of cytoplasmic male sterile lines and the use of honeybees as vector for pollen transfer is nearly nil.

The production of successful hybrids via the cytoplasmic-genetic system necessitates three breeding programs in parallel. The first one is to develop successful male sterile female A parents along with their maintainer isogenic B fertile lines; the second one is to develop promising fertility-restoring male parents (R lines); and the third one includes an extensive crossing and testing program to identify the best combinations. The main reason for the limited use of cotton hybrids during the past 20 years is associated with the fact that it has not been an easy task to develop fertility-restorer strains that have both good fiber properties and good specific combining ability for yield (Weaver 1984). In addition, another genetic problem in producing cotton hybrids with the *G. harknessii* cytoplasmic-genetic male sterility system still appears to be the absence of satisfactory restorer lines, especially under high temperatures. Weaver (1984) also reported that the *G. harknessii* cytoplasm caused a reduction in micronaire. This could be due to the fact that the *G. harknessii* cytoplasm hybrids tended to be later in maturing. Another technical factor that could have a negative effect on the production of hybrid cultivars is that pollination is influenced by the local level of pollinating insects, the weather conditions, and the frequency of insecticide use to keep the crop in good condition, which renders the hybrid seed production quite variable from year to year and from field to field. Thus, hybrid seed production is economically not feasible. Another important reason is that, during the time required by researchers to make possible, at least theoretically, the production of hybrid cotton in large quantities (Davis 1978), the breeders succeeded in fixing most of the heterosis observed in cotton. In this way, they produced inbred lines with high yield production, lint quality, and disease resistance (Lee 1983).

Thus, the genetic improvement of upland cotton realized in the United States advanced at the rate of 7 kg/ha/year of fiber during the 1960s and 1970s (Meredith and Bridge 1984). In addition, one has to emphasize the observation that the highest-yielding hybrids have come from crosses among the best cultivars. Therefore, a hybrid cotton program would need to be accompanied by a successful program for conventional breeding. These, together with the potential to select even better genotypes from new cultivars, and the fact that seed companies have no difficulty in keeping the cottonseed production in their hands, gives the impression that upland cotton hybrids are not going to reach the farmer in the near future.

It has to be questioned at this point whether the production of inbred lines in cotton is scientifically more sound than the production of hybrids. Fasoulas (1993), in an attempt to identify the breeding goal (pure lines vs. hybrids) of the various crops, stated that plant breeders have two alternative approaches for the production of monogenotypic cultivars. One is to eliminate deleterious genes by continuous self-pollination and selection, and the other to block their defective action by means of heterozygosity. The first alternative leads to the development of pure-line cultivars that have the lowest load of deleterious genes and breed true when they reproduce by self-pollination. Pure-line cultivars carry a smaller load of deleterious genes, which is revealed in the generation followed selfing of the F_1 hybrids, and is usually estimated indirectly by measuring inbreeding depression. Thus, he suggested that in the self-pollinated crops with a low load of deleterious genes, inbred vigor may become more advantageous than hybrid vigor. This, together with the finding that in almost every study of self-pollinated crops, near-homozygous lines have been developed that equal or surpass the F_1 (Matzinger 1963), and the prediction made by Kearsey and Pooni (1992) that, in the future, inbreds may outyield current hybrids whether the species is normally inbreeding or outbreeding, indicate that in a self-pollinated crop with a low load of deleterious genes, the breeding goal will be pure-line cultivars and not hybrids.

Cotton is a self-pollinated crop with a range of cross-pollination dependent on the level of pollinated insects. Allard (1960) classified cotton as frequently having more than 10% cross-pollination, with ranges from 5 to 50%. Since then, a great many changes in cotton culture have taken place. The most important, which could have a great impact on the level of cross-pollination in cotton, is the large number of insecticide applications. Indeed, in a study of natural crossing in cotton in the Mississippi Delta at 11 locations, Meredith and Bridge (1973) found that the average natural crossing was only 2%. The range from location to location was between 0 and 5.9%. These results indicate that, in the Mississippi Delta, cotton is essentially a self-pollinated crop. This, together with the finding that heterosis in cotton is mainly due to additive and dominance effects (White and Kohel 1964; White 1966; Marani 1968a,b), and the relatively low level of inbreeding depression (Young and Murray 1966;

Marani 1968b; Al-Rawi and Kohel 1969), indicate that the load of deleterious genes is low and heterosis is fixable.

Therefore, at least theoretically, the breeding goal in cotton should be pure lines. Thus, from the scientific point of view, the failure of hybrid cotton to reach the farmer in the developed world in one point was not unexpected. If this is the case, then how could the plant breeder utilize heterosis in cotton? At this point we should consider separately intraspecific and interspecific heterosis.

8.2 Alternative Utilization of Intraspecific Heterosis

Even though heterosis occurs in cotton, it has not been used for the production of hybrid cultivars due to the difficulties mentioned before in producing commercial F_1 hybrids. F_1 heterosis, however, could be utilized in cotton if it could be correlated with the yielding ability of either the F_2 generation in order to produce the less expensive F_2 hybrid seed or the F_2 and advanced generations for production of inbred varieties.

The primary limitation of commercial use of F_1 hybrids of cotton is the cost of producing F_1 seed. This, together with the observation that inbreeding depression was relatively small (Marani 1968b; Al-Rawi and Kohel 1969), led researchers to consider the possibility of utilizing hybrid vigor via the open-pollinated F_2 generation (F_2 hybrids). Thus, Weaver (1984) presented data indicating that some specific F_2 generation seed were capable of producing significant yield increases over their parents. Galanopoulou-Sendouca and Roupakias (1995), in a study of cotton heterosis in the F_1 and F_2 generations, identified 2 out of 13 upland intraspecific hybrids that maintained part of their F_1 useful heterosis in the F_2 generation (Table 7). These F_2 hybrids had lint percentage, micronaire index, and fiber length within the *hirsutum* range. Finally, the percentage of plants infected with *Verticillium* wilt in the F_2 generation was similar to that observed in the F_1 hybrids. In addition, Tang et al. (1993a), in an extensive study of 64 upland F_2 hybrids in four environments, identified six (9%) F_2 hybrids expressing heterosis over the best parent (Table 8). Two of these F_2 hybrids also had a significantly higher number of bolls than the BP. These two hybrids averaged 14.6% higher lint yield and 12.8% higher

Table 7. Mean seed cotton yield (% of best parent Taskend 6) plant height (cm) and percentage of plants infesterd with *Verticillium* wilt in two heterotic upland intraspecific crosses in the F_1 and F_2 generation evaluated in one location over 3 and 4 years, respectively. (After Galanopoulou-Sendouca and Roupakias 1995)

Cross	Yield (%)		Plant height (cm)		*Verticillium* wilt (%)	
	F_1	F_2	F_1	F_2	F_1	F_2
Thalia-2 × Taskend-6	130	109	93	91	6.2	7.3
Pee Dee × Taskend-6	132	115	77	91	6.0	6.6
Taskend-6	295(g/m^2)		95		4.8	

Table 8. Means of nine F_2 hybrids which were significantly higher than best parent or mean of four female cultivars. (After Tang et al. 1993a)

Cross	Lint yield ($kg\,ha^{-1}$)	Percentage (mean of 4 female cultivars)	Percentage (best parent)
DPL.50 × MHR11	1354	7.7	13.8**
DPL.50 × MHR12	1287	2.4	11.4**
DPL.50 × MHR16	1364	8.5	18.1**
ST453 × L.AHG	1383	10.0**	4.5
DES119 × Delcot 344	1384	10.1**	6.7
C315 × MHR11	1385	10.2**	8.8
C315 × ST825	1418	12.8**	11.5*
ST453 × Delcot344	1437	14.2**	10.3*
C315 × Delcot344	1529	21.3**	17.9**
Mean of 4 female cultivars	1257		

*, ** Significantly higher than best parent or mean of female cultivars at the 0.05 and 0.01 levels, respectively.

boll number, when compared with their best parents. The highest-yielding F_2 hybrid, C315 × Delcot 344, expressed a superiority of 18% in yield, 3.6% in boll weight, and 10.9% in boll number, compared to its best parent. In addition, Tang et al. (1993a) observed that six genotypes, including three with nonsignificant F_2 BP heterosis, yielded significantly more lint than the mean of the four female cultivars (Table 8). Thus, they reached the conclusion: if one is to measure F_2 heterosis for lint yield based only upon the best parent, then one could overlook some elite F_2 hybrids that perform as well or better than commercial cultivars.

Furthermore, they recognized that the per se performance of the parents could not always predict F_2 hybrid performance. Therefore, the estimation of F_2 heterosis should be based only on the hybrid performance evaluated in the environments of interest. Fiber traits of most F_2 hybrids studied by Tang et al. (1993b) were relatively stable and similar to the relative heterosis. Genetic control of the fiber properties was largely additive. The same was reported by other workers (Davis 1978; Meredith 1990). Thus commercial F_2 hybrids could be produced only from crosses between pairs of parents having good yielding ability and fiber properties.

Given that inbreeding depression is relatively small and F_2 hybrids should be better than the best cultivars, one has to make a number of crosses between different cultivars with good performance in the area. Then, the F_1 hybrids must be evaluated and those with useful heterosis further evaluated in the F_2 in different environments. The hybrids with the highest useful heterosis and with a relatively small inbreeding depression are then reproduced and distributed to the farmer as F_2 hybrids. It should be noted here, however, that the term F_2 hybrids does not literally describe the genetic make up of the F_2 population because the encountered superiority of F_2 over either parent cannot be attributed entirely to heterosis, but rather to transgressive or inbred vigor (A. Fasoulas, pers. comm.). This is because, in the F_2 50%, on the average, of the heterozygous loci are fixed. Granted that inbred vigor is due to additive genes; it implies that inbred vigor and therefore low inbreeding depression are revealed in crops with a very reduced load of deleterious genes.

It is generally believed that heterotic F_1s with a relatively small inbreeding depression in the F_2 are promising populations to be used as source for further selection. Thus, the crosses with the best combined performance in F_1 and F_2 could be multiplied and utilized as F_2 hybrids, whenever this is economically feasible, and in parallel, they could be selected for further improvement. The new isolated improved inbred varieties (those with better performance than the F_2 hybrids) will then be multiplied and distributed to the farmer. Successively, they could be used in crosses with other cultivars for the production of new F_2 hybrids with useful heterosis, and so on.

The question that has to be raised here is the growing conditions under which the breeder should evaluate the F_1 and F_2 generation. Fasoulas (1973, 1993) and Fasoulas and Fasoula (1995) recognized interplant competition and soil heterogeneity as important factors in reducing the effectiveness of selecting single plants for improved yield, and developed the so-called honeycomb selection of individual plants grown without competition (very low densities). One of the points that Fasoulas (1988) emphasized is that a breeder should evaluate a number of crosses in the F_1 and F_2 generations at low plant densities and in a number of environments, and concentrate his efforts on the population(s) with the best combined performance in the two generations.

Indeed, Roupakias et al. (1996), in an attempt to obtain information on the effectiveness of early-generation selection at two plant densities in faba bean, evaluated eight F_1s and their F_2s under two densities (i.e., 5.7 and 1.4 plants/m^2). It was found that the eight hybrids had a different ranking when they were evaluated under high and low plant density. In addition, the hybrid with the best combined (F_1 + F_2) yield performance from each plant density was selected for two generations (F_2 and F_3) under both low and high plant density. It was found that early-generation selection under low plant density was successful in faba beans. Whether these findings are applicable to cotton is open to be studied.

8.3 Alternative Utilization of Interspecific Heterosis

Gossypium barbadense cottons are grown for their longer, stronger, and finer fibers as compared to the more widely grown *G. hirsutum* cultivars. *G. hirsutum*, however, has the advantage of high yielding potential, early maturity, and good adaptability throughout the world. It was early recognized that the two species cross easily and produce vigorous F_1 hybrids (Loden and Richmond 1951). In general, it was observed in the early 1950s that *G. barbadense* and *G. hirsutum* crosses produce a very large F_1 plant of very little practical value (Stroman 1961). However, Patel and Patel (1952) and Richmond and Lewis (1951) reported some good results in both yield and quality in certain interspecific hybrids. Today, it is widely recognized that economically important heterosis resulting in higher lint yield and improved fiber properties exist in some interspecific hybrids (Davis 1974; Weaver 1979; Call and Weaver 1980; Davis and Palomo 1980; Malm et al. 1983; Davis et al. 1986; Galanopoulou-Sendouca and Roupakias 1995).

In addition, Davis (1974) reported that he obtained shorter-statured, earlier maturing, and more productive hybrids as a result of developing an early semidwarf *G. hirsutum* parent. Based on this report, Percy and Turcotte (1988) thought that they might have even better results if they could develop short-fibered, short-statured, and early-maturing *G. barbadense* strains. Indeed, Percy and Turcotte (1991) found that short-statured and early-maturing *G. barbadense* parent strains improved plant height, earliness, and yield of the interspecific hybrids. These findings, together with the narrow-row high plant population cultural system, could contribute towards the commercial use of interspecific hybrid cotton (Palomo and Davis 1983). It is obvious, therefore, that commercial utilization of interspecific heterosis is currently limited in cotton by our inability to produce F_1 hybrids in an economically feasible way.

Utilization of interspecific heterosis via F_2 hybrids or via selection is not feasible due to the high level of inbreeding depression in F_2 and the "hybrid breakdown" observed in advanced generations (Kohel et al. 1977; Natsiou-Vozike et al. 1988; Galanopoulou-Sendouca and Roupakias 1995).

Galanopoulou-Sendouca and Roupakias (1995) observed that in two interspecific crosses the inbreeding depression in F_2 was 53%. In addition, Natsiou-Vozike et al. (1988) observed that the F_2 population was made up with plants exhibiting the *G. barbadense* or *G. hirsutum* phenotypes and a large range of variation between the two species which could never be selected by a breeder. Kohel et al. (1977) mentioned that the "hybrid breakdown" in advanced generations tends to preserve the identity of the two species, even when they are grown together. This is probably the main reason why the plant breeders have not succeeded, so far, in combining fiber properties of *G. barbadense* with high yielding potential and early maturity of *G. hirsutum* germplasm. This failure has compelled researchers to look for alternative approaches to the whole problem.

The isolation of monosomics (Endrizzi and Ramsay 1979) and, therefore, the possibility of developing interspecific substitution lines in cotton (White et al. 1967) opened the way to identifying individual chromosomes which carry the major genes responsible for the good agronomic characters in *G. hirsutum* and *G. barbadense*. It was suggested under these provisions that a construction could be made, by cytogenetical means, of a new genotype (Fig. 4) carrying the good *hirsutum* chromosomes, together with the good *barbadense* chromosomes in a common nucleus. This genotype was named partial interspecific hybrid (PIH) or measured hybrid (White et al. 1967).

Roupakias and Galanopoulou-Sendouca (1988) attempted unsuccessfully to develop a complete set of substitution lines in two cultivars (Sindos 80, a *hirsutum* cultivar and Pima S4 a *barbadense* cultivar). The failure was due to the lack of plants with the characteristic phenotype of most of the monosomics after the fourth backcross. Thus, the difficulties associated with the development of a complete set of substitution lines in a number of cultivars is the main factor that prohibits the development of the partial interspecific hybrid, at

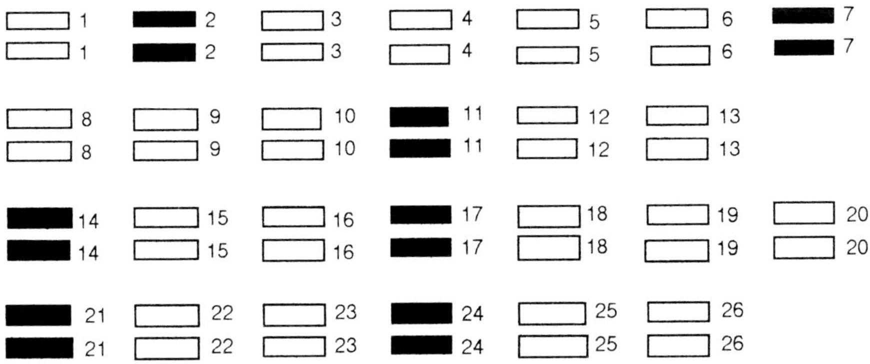

Fig. 4. Partial interspecific hybrid carrying some *G. hirsutum* (*open symbols*) and some *G. barbadense* (*closed symbols*) chromosomes

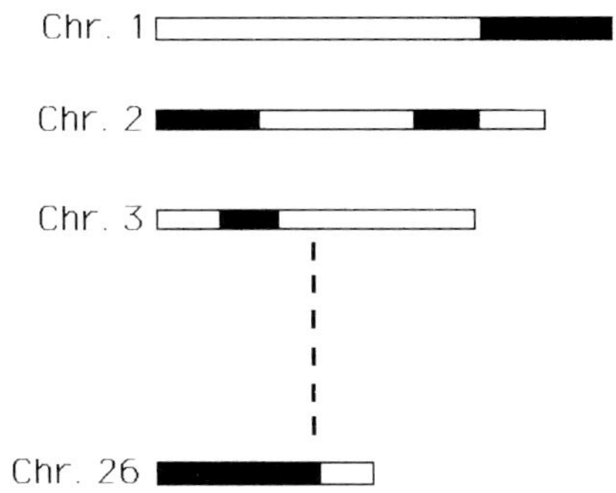

Fig. 5. Chromosome makeup of a pollen grain or an unfertilized egg cell from an F_1 (*G. hirsutum* × *G. barbadense*) plant

least as was originally suggested. It was explained above that a partial interspecific hybrid is a plant that carries in its nucleus some of its non-homologous chromosomes from *G. barbadense* and the rest from *G. hirsutum*. Having this in mind, Mavromatis and Roupakias (1994) extend the definition of PIH as the hybrid between two species that carries, in homozygous condition, part of the chromosomes or chromosome segments from the one species and the rest from the second one. This could become a reality if one could regenerate plants from immature pollen grains or from unfertilized egg cells of F_1 plants (Mavromatis and Roupakias 1994). The chromosome makeup of a pollen grain or unfertilized egg cell from an F_1 plant is given in Fig. 5. This is also the chromosome makeup of the haploid plant regenerated from such a pollen grain. Chromosome dublication of such a haploid will produce a stable line of a partial interspecific hybrid (Fig. 6). The production of such plants in large numbers could very easily mean that one of them could be the one that combines the good qualities of both species.

Several alternative procedures have been suggested for the production of haploids in cotton. Turcotte and Feaster (1963) reported that a doubled haploid, 57-4, produced a high frequency of haploids from single-embryo

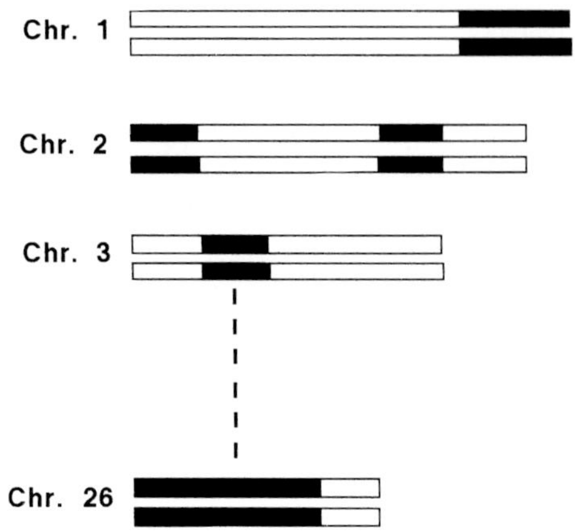

Fig. 6. Chromosome makeup of a doubled haploid or a partial interspecific hybrid in cotton

seeds in Pima cotton, *Gossypium barbadense*. This phenomenon was explained as semigamy, an abnormal type of fertilizàtion in which genetic fusion of the egg with the sperm nucleus is lacking (Turcotte and Feaster 1967, 1969). Since then, several researchers (Barrow and Chaudhari 1976; Mahill et al. 1984) have developed a number of doubled haploids via semigamy. It was found from these studies that agronomic and fiber traits were improved for some doubled haploids and they were as environmentally stable for yield performance as parental lines. The main drawback of this procedure is the relatively small number of haploids produced; thus, anther culture studies have been undertaken. However, most of these studies deal with callus formation and media composition (Barrow et al. 1978; Bajaj 1982; Turaev and Shamina 1986; Mavromatis and Roupakias 1995; Van Overbeke et al. 1995). The subject of in vitro induction of haploidy in cotton has been recently discussed by Bajaj and Gill (1996). The excised anthers of *G. arboreum* in culture formed callus showing a wide range in chromosome number, and the pollen underwent pollen embryogenesis (Bajaj and Gill 1989); however, no haploid plants have been produced to date. Zhou Shi-qui et al. (1991) reported the production of haploid plants via parthenogenesis at a rate of 1.4% when cotton flowers were pollinated with pollen from *Hibiscus cannabinus*. Mavromatis and Roupakias (1995) assumed that a larger number of unfertilized ovules could have been stimulated for embryo development by the alien pollen but they were unable to reach maturity. Thus, they tried to rescue the young embryos via in vitro ovule culture and observed that 9 out of 45 different cross-combinations produced germinated embryos. None of them, however, reached maturity. It is obvious, therefore, that there is a long way to go before the development of a protocol for the commercial production of doubled haploids in cotton is accomplished.

9 Concluding Remarks

The existence of intraspecific and interspecific heterosis in cotton is well documented. Heterosis for yield is mainly due to dominance. Fiber properties, however, are inherited in an additive fashion. In addition, the highest-yielding hybrids obtained so far have come from crosses among the best cultivars. Thus, a hybrid cotton program would need to be accompanied by a successful program for conventional breeding. Hybrid cultivars, however, could have a good future in cotton production only when the system of cytoplasmic male sterility-pollen vector-restoration of pollen fertility in F_1 is improved significantly.

Inbreeding depression of F_1 cotton intraspecific hybrids is relatively small. Thus, F_2 hybrids could be an alternative utilization of intraspecific heterosis. For the time being, however, hybridization and selection within promising F_2 populations seems to be more effective.

Finally, the development of an efficient system for the production of doubled haploids could contribute towards a more effective utilization of intraspecific as well as interspecific heterosis in cotton.

Acknowledgment. The authors thank Professors, A.C. Fasoulas, and C. Goulas for their critical review and valuable contribution.

References

Ali M, Lewis CF (1962) Effects of reciprocal crossing on cytological and morphological features of interspecific hybrids of *Gossypium hirsutum* L. and *G. barbandense* L.. Crop Sci 2:20–22

Allard RW (1960) Principles of plant breeding. John Wiley, New York

Al-Rawi KM, Kohel RJ (1969) Diallel analysis of yield and other agronomic characters in *Gossypium hirsutum* L. Crop Sci 9:779–783

Al-Rawi KM, Kohel RJ (1970) Gene action in the inheritance of fiber properties in intervarietal diallel crosses of upland cotton, *Gossypium hirsutum* L. Crop Sci 10:82–85

Bajaj YPS (1982) Survival of anther- and ovule-derived cotton callus frozen in liquid nitrogen. Curr Sci 51:139–140

Bajaj YPS, Gill MS (1989) Pollen-embryogenesis and chromosomal variation in anther culture of a diploid cotton (*Gossypium arboreum* L.). Sabrao J 21:57–63

Bajaj YPS, Gill MS (1996) In vitro induction of haploidy in cotton. In: Jain SM, Sopory SK, Veilleux RE (eds) In vitro haploid production in higher plants. Kluwer, Dordrecht, pp 165–174

Baker JL, Verhalen LM (1975) Heterosis and combining ability for several agronomic and fibre properties among selected lines of upland cotton. Cotton Grow Rev 52:209–223

Barrow J, Katterman F, Williams D (1978) Haploid and diploid callus from cotton anthers. Crop Sci 18:619–622

Barrow JR, Chaudhari HK (1976) A homozygous interspecific F_2 hybrid of *Gossypium barbadense* × *Gossypium hirsutum* via the semigametic haploid method. Crop Sci 16:441–442

Basu AK, Patil NB (1993) Producing quality cotton from commercial cotton hybrids grown under rainfed conditions in India. Presented at a Tachnical Seminar at the 52nd Plenary meeting of the International Cotton Advisory Committee, New Delhi

Bingham ET, Groose RW, Woodfield DR, Kidwell KK (1994) Review and interpretation: complementary gene interactions in alfalfa are greater in autotetraploids than diploids. Crop Sci 34:823–829

Brewbaker JL (1964) Agricultural genetics. Prentice Hall, Englewood Cliffs, New Jersey
Brown HB (1942) Results from inbreeding upland cotton for a 10-year period. J Am Soc Agron 34:1084–1089
Bruce AB (1910) The Mendelian theory of heredity and the augmentation of vigor. Science 32:627–628
Call TB, Weaver JB (1980) Resistance to Heliothis in intraspecific and interspecific hybrid cotton. Belt Cotton Prod Res Conf, 6–10 Jan, St Louis, pp 95–96
Christidis BG (1955) Hybrid vigour effects with cotton. J Genet 53:224–231
Cook OF (1909) Suppressed and intensified characters in cotton hybrids. USDA Plant Ind Bull 147
Crow JF (1952) Dominance and overdominance. In: Gowen JW (ed) Heterosis. lowa State College Press, Ames IA, pp 282–297
Davenport CB (1908) Degeneration, albinism and inbreeding. Science 28:454–455
Davis DD (1974) Synthesis of commercial F_1 hybrids in cotton. I Genetic control of vegetative and reproductive vigor in *Gossypium hirsutum* L. × *G. barbadense* L. crosses. Crop Sci 14:745–749
Davis DD (1978) Hybrid cotton: specific problems and potentials. Adv Agron 30:129–157
Davis DD (1979) Synthesis of commercial F_1 hybrids in cotton. II Long, strong-fibered *G. hirsutum* L. × *G. barbadense* L. hybrids with superior agronomic properties. Crop Sci 19:115–116
Davis DD, Palomo A (1980) Yield stability of interspecific hybrid NX-1. Belt Cotton Prod Res Conf:81
Davis D, Barrow J, Roberts C, Malm N, Barnes C (1986) Hybrid cotton in New Mexico. A chronology of development, 1967 to 1984. AES Bull 724:1–34
Dobzhansky T (1950) Genetics of natural populations. XIX Origin of heterosis through natural selection in populations of *Drosophila pseudobscura*. Genetics 35:288–302
East EM (1908) Inbreeding in corn. Connecticut Agricultural Experimental Station Rep for years 1907–1908, pp 419–428
Eaton FM (1957) Selective gametocide opens way to hybrid cotton. Science 126:1174–1175
Eberhart SA (1979) Genetics and breeding, In: Hafliger E (ed) Maize. Ciba-Geigy, Basle, Switzerland, pp 6–12
El-Adl AM, Miller PA (1971) Transgressive segregation and the nauture of gene action of yield in an intervarietal cross of upland cotton. Crop Sci 11:381–384
Endrizzi JE, Ramsay G (1979) Monosomes and telosomes for 18 of the 26 chromosomes of *Gossypium hirsutum*. Can J Genet Cytol 21:531–536
Fasoulas A (1973) A new approach to breeding superior yielding varieties. Dept of Genetics and Plant Breeding, Publ No 3, Aristotelian Univ Thessaloniki, Greece
Fasoulas AC (1988) The honeycomb methodology of plant breeding. Thessaloniki, Greece
Fasoulas AC (1993) Principles of crop breeding. Thessaloniki, Greece
Fasoulas AC, Fasoula VA (1995) Honeycomb selection designs. Plant Breed Rev 13:87–139
Fehr WR (1987) Principles of cultivar development, vol 1. Macmillan, New York
Galal HE, Miller PA, Lee JA (1966) Heterosis in relation to development in upland cotton, *G. hirsutum* L. Crop Sci 6:555–559
Galanopoulou-Sendouca S (1987) Performance of cotton hybrids in Greece. Agric Rec 11:325–342
Galanopoulou-Sendouca S (1988) Heterosis in cotton. Proc 2nd meeting of the Greek Plant Breeding Soc, Thessaloniki, 20–21 October 1988, pp 236–244
Galanopoulou-Sendouca S, Roupakias DG (1995) Heterosis of cotton F_1 hybrids and its relationship to the mean yield of advanced unselected generations. J Genet Breed (submitted)
Hageman RH, Leng ER, Dudley JW (1967) A biochemical approach to corn breeding. Adv Agron 19:45–86
Hallower AR, Miranda JBFO (1981) Quantitative genetics in maize breeding. lowa State Univ Press, Ames
Harris HB, Loden HD (1954) The relative growth rates of F_1 hybrid of *Gossypium hirsutum* L. and its two parents. Agron J 46:492–495

Hawkins BS, Peacock HA, Ballard WW (1965) Heterosis and combining ability in upland cotton – effect on yield. Crop Sci 5:543–546

Jones JE, Loden HD (1951) Heterosis and combining ability in upland cotton. Agron J 43:514–516

Justus N, Leinweber CL (1960) A heritable partially male sterile character in cotton. J Hered 51:191–192

Kearsey MJ, Pooni HS (1992) The potential of inbred lines in the presence of heterosis, In: Dattee Y, Dumas C, Gallais A (eds) Reproductive biology and plant breeding. Springer, Berlin Heidelberg New York, pp 371–385

Kohel RJ, Endrizzi JE, White TG (1977) An evaluation of *Gossypium barbadense* L. chromosomes 6 and 17 in the *G. hirsutum* L. genome. Crop Sci 17:404–406

Krishnaswami R, Kothandaraman R (1977) Heterosis in interspecific hybrids of *Gossypium*. Indian J Genet Plant Breed 37:40–45

Lamkey KR, Smith OS (1987) Performance and inbreeding depression of populations representing seven eras of maize breeding. Crop Sci 27:695–699

Lee JA (1983) Is there much genetical room left for hybrid cotton? In: Brown JM (ed) Proc Beltwide Cotton Prod Res Conf, San Antonio, Tex Jan 2–6, Natl Cotton Council, Memphis, TN, pp 92–94

Lee JA (1987) Cotton. In: Fehr WR (ed) Principles of cultivar development, vol 2. Macmillan, New York, pp 126–160

Lee JA, Miller PA, Rawlings JO (1967) Interaction of combining ability effects with environments in diallel crosses of upland cotton (*Gossypium hirsutum* L.). Crop Sci 7:477–481

Lewis D (1954) Gene-environment interaction: a relationship between dominance, heterosis, phenotypic stability and variability. Heredity 8:333–356

Liang GH, Reddy CR, Dayton AD (1972) Heterosis, inbreeding depression and heritability estimates in a systematic series, of grain sorghum genotypes. Crop Sci 12:409–411

Loden HD, Richmond TR (1915) Hybrid vigor in cotton-cytogenetic aspects and practical applications. Econ Bot 5:387–408

Luckwill LC (1937) Studies in the inheritance of physiological characters. IV Hybrid vigour in the tomato, part 2. Manifestation of hybrid vigour during the flowering perior. Ann Bot (Lond) 1:379–408

Mac Key J (1970) Significance of mating systems for chromosomes and gametes in polyploids. Hereditas 66:165–176

Mac Key J (1976) Genetic and evolutionary principles of heterosis, In: Janossy A, Lupton FGH (eds) Heterosis in plant breeding. Proc 7th Congr Eucarpia Budapest, 24–29 June 1974, pp 17–57

Mahill JF, Jenkins JN, McCarty JC Jr, Parrott WL (1984) Performance and stability of doubled haploid lines of upland cotton derived via semigamy. Crop Sci 24:271–277

Malm N, Davis DD, Fowler J, Barrow J, Smith F (1983) Cotton research, breeding, genetics and production. N M Agric Exp Stn Spec Rep 57

Maner BA, Worley S, Harrell DC, Culp TW (1971) A geometrical approach to yield models in upland cotton (*Gossypium hirsutum* L.). Crop Sci 11:904–908

Marani A (1963) Heterosis and combining ability for yield and components of yield in a diallel cross of two species of cotton. Crop Sci 3:552–555

Marani A (1967) Heterosis and combining ability in intraspecific and interspecific crosses of cotton. Crop Sci 7:519–522

Marani A (1968a) Inheritance of lint quality characteristics in intraspecific crosses among varieties of *Gossypium hirsutum* and of *G. barbadense* L. Crop Sci 8:36–38

Marani A (1968b) Heterosis and F_2 performance in intraspecific crosses of *Gossypium hirsutum* L. and of *G. barbadense* L. Crop Sci 8:111–113

Marani A (1968c) Heterosis and inheritance of quantitative characters in interspecific crosses of cotton. Crop Sci 8:299–303

Marani A (1968d) Inheritance of lint quality characteristics in interspecific crosses of cotton. Crop Sci 8:653–657

Marani A, Avieli E (1973) Heterosis during the early phases of growth in intraspecific and interspecific crosses of cotton. Crop Sci 13:15–17

Matzinger DF (1963) Experimental estimates of genetic parameters and their applications in self-fertilizing plants. In: Hanson WD, Robinson HF (eds) Statistical genetics and plant breeding. Natl Acad Sci, Natl Res Coun Publ 982, pp 352–379

Mavromatis AG, Roupakias DG (1994) Biotechnology: a hope for a partial interspecific hybrid in cotton (*Gossypium* spp.) Proc meeting of the Working Group Cotton Biotechnology of the FAO – International Cooperative Research Network on Cotton, 22–23 October 1993, Leuven, Belgium, pp 29–36

Mavromatis AG, Roupakias DG (1995) Haploid production via intergeneric pollination and in-ovule culture in cotton (*Gossypium* spp.). Proc 2nd meeting of the WG Cotton biotechnology of the FAO – Int Cooperative Res Network on Cotton, 31st March and 1st April 1995, Thessaloniki, Greece, pp 179–184

McGregor SE (1959) Cotton-flower visitation and pollen distribution by honey bees. Science 129:97–98

Meredith WR Jr (1979) Inbreeding depression of selected F_3 cotton progenies. Crop Sci 19:86–88

Meredith WR Jr (1984) Quantitative genetics. In: Kohel RJ, Lewis CF (eds) Cotton. ASA, CSSA, SSSA, Publishers. Madison, Wisconsin, USA, pp 131–150

Meredith WR Jr (1990) Yield and fiber-quality potential for second-generation cotton hybrids. Crop Sci 30:1045–1048

Meredith WR Jr, Bridge RR (1972) Heterosis and gene action in cotton, *Gossypium hirsutum* L. Crop Sci 12:304–310

Meredith WR Jr, Bridge RR (1973) Natural crossing in cotton (*Gossypium hirsutum* L.) in the delta of Missisipi. Crop Sci 13:551–552

Meredith WR Jr, Bridge RR (1984) Genetic contribution to yield changes in upland cotton. In: Fehr WR (ed) Genetic contributions to yield change in five major crop plants. Crop Sci Soc Am, Madison, Wisconsin, pp 75–87

Meyer VG (1971) Cytoplasmic effects on anther numbers in interspecific hybrids of cotton. II *Gossypium herbaceum* and *G. harknessii*, J Hered 62:77–78

Meyer VG (1973a) A study of reciprocal hybrids between upland cotton (*Gossypium hirsutum* L.) and experimental lines with cytoplasms from seven other species. Crop Sci 13:439–444

Meyer VG (1973b) Registration of 16 germplasm lines of upland cotton. Crop Sci 13:778–779

Meyer VG (1973c) Fertility-restorer genes for cytoplasmic male sterility from *Gossypium harknessii*. Belt Cotton Prod Res Conf Proc, 9–10 Jan, Phoenix, p 65

Miller PA, Lee JA (1964) Heterosis and combining ability in varietal top crosses of upland cotton, *Gossypium hirsutum* L. Crop Sci 4:646–649

Miller PA, Marani A (1963) Heterosis and combining ability in diallel crosses of upland cotton, *Gossypium hirsutum* L. Crop Sci 3:441–444

Moffett JO, Stith LS (1972) Pollination by honeybees of male-sterile cotton in cages. Crop Sci 12:476–478

Moffett JO, Stith LS, Burkhardt CC, Shipman CW (1975) Influence of cotton genotypes on floral visits of honey bees. Crop Sci 15:782–784

Natsiou-Vozike T, Roupakias DG, Galanopoulou-Sendouca S, Tsaftaris A (1988) Interspecific hybrids between *Gossypium hirsutum* L. and *Gossypium barbadense* L. in the F_1 and the following generations. Proc second meeting of the Greek Plant Breed Soc, 20–21 October 1988, Thessaloniki, pp 267–273

Palomo A, Davis DD (1983) Response of F_1 interspecific (*Gossypium hirsutum* L. × *G. barbadense* L.) cotton hybrid to plant density in narrow rows. Crop Sci 23:1053–1056

Pate JB, Duncan EN (1960) Evaluation of sodium 2,3-dichloroisobutyrate as a selective male gametocide in cotton. Agron J 52:506–508

Patel GB, Patel CT (1952) Long staple cotton from hybrid seed of crosses between *hirsutum* and *barbadense* species. Indian Cotton Grow Rev 6:206–210

Pathak RS, Kumar P (1975) Combining ability studies in uplant cotton (*Gossypium hirsutum* L.). Z Pflanzenzücht 75:297–310

Percy RG, Turcotte EL (1988) Development of short and coarse-fibered American Pima cotton for use as parents of interspecific hybrids. Crop Sci 28:913–916

Percy RG, Turcotte EL (1991) Early-maturing, short-statured American Pima cotton parents improve agronomic traits of interspecific hybrids. Crop Sci 31:709–712

Powers L (1944) An expansion of Jones' theory for the explanation of heterosis. Am Nat 78:275–280

Richmond TR (1962) Effects of sodium 2,3-dichloroisobutyrate on six characteristics of American upland cotton. Crop Sci 2:58–60

Richmond TR, Lewis CS (1951) Evaluation of varietal mixture of cotton. Agron J 43:66–70

Rosales FE, Davis DD (1976) Performance of cytoplasmic male sterile cotton under natural crossing in New Mexico. Crop Sci 16:99–102

Roupakias D, Galanopoulou-Sendouca S (1988) Partial interspecific hybrids in cotton: prospects and limitations. Proc 2nd Meet of the Greek Plant Breeding Soc, Thessaloniki, 20–21 Oct 1988, pp 260–266

Roupakias DG, Zesopoulou A, Kazolea S, Dlakalitses G, Mavromatis A, Lazaridou T (1996) Effectiveness of early generation selection under two plant densities in faba bean (*Vicia faba* L.). Euphytica 93:63–70

Shepherd RL (1974) Transgressive segregation for root-knot nematode resistance in cotton. Crop Sci 14:872–875

Shrivastava HK (1972) Mitochondrial complementation and hybrid vigor. Indian J Genet Plant Breed 32:215–228

Shull GH (1908) The composition of a field of maize. Am Breed Assoc 4:296–301

Shull GH (1914) Duplicate genes for capsule form in *Bursa bursa bastoris*. J Ind Abst Vererb 12:97–149

Shull GH (1952) Heterosis. Iowa State College Press, Ames, Iowa, p 48

Simpson DM, Duncan EN (1953) Effect of selecting within selfed lines on the yield and other characters of cotton. Agron J 45:275–279

Sinha SK, Khanna R (1975) Physiological, biochemical, and genetic basis of heterosis. Adv Agron 27:123–174

Sprague GF (1983) Heterosis in maize: Theory and practice. In: Frankel R (ed) Heterosis: reappraisal of theory and practice. Springer, Berlin Heidelberg New York, pp 47–70

Srinivasan K, Santhanam V, Rajasekaran S (1972) Development of hybrid cotton utilising male-sterile line. Cotton Dev 2:37–39

Stewart JMcD, Hsu CL (1978) Hybridization of diploid and tetraploid cottons through in ovule embryo culture. J Hered 69:404–408

Stroman GN (1961) An approach to hybrid cotton as shown by intra- and interspecific crosses. Crop Sci 1:363–366

Stuber CW (1995) Mapping and manipulating quantitative traits in maize. Trends Genet 11:477–481

Tang B, Jenkins JN, McCarty JC, Watson CE (1993a) F_2 hybrids of host plant germplasm and cotton cultivars. I Heterosis and combining ability for lint yield and yield components. Crop Sci 33:700–705

Tang B, Jenkins JN, McCarty JC, Watson CE (1993b) F_2 hybrids of host plant germplasm and cotton cultivars. II Heterosis and combining ability for fiber properties. Crop Sci 33:706–710

Thomson NJ (1971) Heterosis and combining ability of American and African cotton cultivars in a low latitude under high-yield conditions. Aust J Agric Res 22:759–770

Thomson NJ, Luckett DJ (1988a) Heterosis and combining ability effects on cotton. I Combining ability. Aust J Agric Res 39:973–990

Thomson NJ, Luckett DJ (1988b) Heterosis and combining ability effects in cotton. II Heterosis. Aust J Agric Res 39:991–1002

Tsaftaris AS (1990) Biochemical analyses of inbreds and their heterotic hybrids in maize. In: Ogita A, Markert C (eds) Isozymes: structure, function and use in biology and medicine. Wiley-Liss, New York, pp 639–664

Tsaftaris AS (1995) Molecular aspects of heterosis in plants. Physiol Plant 94:362–370

Turaev AM, Shamina ZB (1986) Optimization of the medium for cotton anther culture. Sov Plant Physiol 33:439–444

Turcotte EL, Feaster CV (1963) Haploids: high-frequency production from single-embryo seeds in a line of Pima cotton. Science 140:1407–1408

Turcotte EL, Feaster CV (1967) Semigamy in Pima cotton. J Hered 58:55–57

Turcotte EL, Feaster CV (1969) Semigametic production of haploids in Pima cotton. Crop Sci 9:653–655

Turner JH Jr (1953a) A study of heterosis in upland cotton. I Yield of hybrids compared with varieties. Agron J 45:484–486

Turner JH Jr (1953b) A study of heterosis in upland cotton. II Combining ability and inbreeding effects. Agron J 45:487–490

Van Overbeke G, Mavromatis A, Roupakias D, Swennen R, Peeters C (1995) Initiation of cotton anther cultures. Proc 2nd Meet of the WG Cotton biotechnology of the FAO Int Cooperative Res Network on Cotton, 31 March and 1 April 1995, Thessaloniki, Greece, pp 185–189

Verhalen IM, Murray JC (1967) A diallel analysis of several fiber property traits in upland cotton (*Gossypium hirsutum* L.) Crop Sci 7:501–505

Verhalen LM, Murray JC (1969) A diallel analysis of several fiber property traits in upland cotton (*Gossypium hirsutum* L.) II Crop Sci 9:311–315

Walton PO (1971) Heterosis in spring wheat. Crop Sci 11:422–424

Weaver JB Jr (1979) Production and performance of interspecific hybrid cotton. Belt Cotton Prod Res Conf Proc, Phoenix, pp 72–74

Weaver JB Jr (1984) Agronomic properties of F_1 hybrids and open-pollinated F_2s among 12 cultivars of cotton. Belt Cotton Prod Res Conf, 7–11 Jan, Phoenix, pp 103–104

Wells R, Meredith WR Jr (1986) Heterosis in upland cotton. I Growth and leaf area partitioning. Crop Sci 26:1119–1123

White TG (1966) Diallel analyses of quantitatively inherited characters in *Gossypium hirsutum* L. Crop Sci 6:253–255

White TG, Kohel RJ (1964) A diallel analysis of agronomic characters in selected lines of cotton, *Gossypium hirsutum* L. Crop Sci 4:254–257

White TG, Richmond TR (1963) Heterosis and combining ability in top and diallel crosses among primitive, foreign and cultivated American upland cottons. Crop Sci 3:58–63

White TG, Richmond TR, Lewis CF (1967) Use of cotton monosomes in developing interspecific substitution lines. Crop Res, ARS, 34–91, pp 3–15

Woodfield DR, Bingham ET (1995) Improvement in two-allele autotetraploid populations of alfalfa explained by accumulation of favorable alleles. Crop Sci 35:988–994

Xiao J, Li J, Yuan L, Tanksley SD (1995) Dominance is the major genetic basis of heterosis in rice as revealed by QTL analysis using molecular markers. Genetics 140:745–754

Young EF, Murray JC (1966) Heterosis and inbreeding depression in diploid and tetraploid cottons. Crop Sci 6:436–438

Zhou Shi-Qui, Qian De-Qi, Cao Xiu-Yun (1991) Induction of parthenogenesis and chromosome behaviour in plants of parthenogenetic origin in cotton (*Gossypium hirsutum*). Genome 34:255–250

Section III
Anther/Pollen Culture; Parthenogenesis and Haploid Breeding

III.1 Anther Culture Studies and Pollen Embryogenesis in Cotton

Y.P.S. BAJAJ[1] and M.S. GILL[2]

1 Introduction

The potential use of haploids in genetics and breeding of cotton has been emphasized (Harland 1936; Roux 1958; De Garcia 1962; Endrizzi 1966). Their production in cotton is valuable for obtaining homozygous lines in one generation, and these homozygous lines can be evaluated in the field. Because doubled haploids possess additive variance only, early-generation testing of segregating populations is possible. Haploids can also be used to increase the efficiency of existing breeding methods through improved reliability of selection.

Natural occurrence of cotton haploids has been occasionally reported in various species, i.e., *G. barbadense* (Harland 1920), *G. davidsonii* (Skovsted 1935), *G. hirsutum* (Harland 1936), *G. arboreum*, and *G. herbaceum* (Dergach 1971). Cotton haploids have also been induced by conventional methods, i.e., polyembryony and semigamy. Polyembryony (Blank and Allison 1963; Lee 1970), in which one or both members of a pair of twin seedlings may be haploid, is a rare phenomenon. Semigamy (Turcotte and Feaster 1974; Chaudhari 1978) is restricted to the two tetraploid species (*G. hirsutum* and *G. barbadense*), and does not work for the diploid species. Agronomic evaluations of semigametically derived doubled haploids of *G. hirsutum* in *G. barbadense* cytoplasm revealed superiority for some traits but inferiority for others, compared to the cultivars from which they were derived (Mahill et al. 1984). Another method of haploid formation is by induced parthenogenesis (Zhou et al. 1991). In general, the frequency of haploid induction by conventional methods is rather low, and for any cotton breeding/improvement program they are needed in large numbers. Thus, anther culture studies have been undertaken.

The in vitro culture of anthers has been successfully employed to produce haploids, under defined conducive conditions, in a number of agricultural crops (see Bajaj 1983a, 1990). However, in cotton, early attempts to induce androgenesis resulted in only callus formation (Barrow et al. 1978). Later, Bajaj and Gill (1989) working with *G. arboreum*, reported the induction of

[1] Biotechnology in Agriculture and Forestry, A-137 New Friends Colony, New Delhi 110065, India
[2] Punjab Agricultural University, Ludhiana (Punjab), India

Biotechnology in Agriculture and Forestry, Vol. 42
Cotton (ed. by Y.P.S. Bajaj)

pollen embryos and callus with haploid cells as well as a wide range of variation in chromosome number (Bajaj and Gill 1985). The present chapter is primarily based on our published work (Bajaj 1982; Bajaj and Gill 1985, 1989, 1996).

2 In Vitro Culture of Anthers

Various studies conducted on in vitro culture of excised anthers of *Gossypium hirsutum*, *G. barbadense* and *G. arboreum* are summarized in Table 1, and discussed in the following.

2.1 Materials and Method

In our work (Bajaj 1982; Bajaj and Gill 1989), flower buds from field-grown plants of a diploid *Gossypium arboreum* cv. G27 and a tetraploid *G. hirsutum* L. cv. F414 were collected, and those containing anthers at the uninucleate pollen stage were used for culture. The buds were sterilized with freshly prepared chlorine water for 8–10 min, and washed twice with sterilized distilled water. Anthers were excised aseptically, and 20–25 of them were cultured per test tube on SH (Schenk and Hildebrandt 1972) and MS medium (Murashige and Skoog 1962) containing various concentrations and combinations of indoleacetic acid (IAA), kinetin (kin), benzyladenine (BA), naphthalene acetic acid (NAA), 2,4-dichlorophenoxy acetic acid (2,4-D), casein hydrolysate (CH), and coconut milk (CM). Twenty replicates were made for each treatment, ten of which were kept at 4–5 °C in a refrigerator for 48 h in the dark prior to transfer to the culture room. For androgenic studies, the anthers were periodically fixed in acetic alcohol (1:3), and stained with acetocarmine (1%). For chromosome counts, fast-growing callus was subcultured on a medium containing 3% glucose instead of sucrose, and was fixed in acetic alcohol

Table 1. Anther culture studies on cotton

Species	Response/remarks	Reference
G. hirsutum	Haploid and diploid callus	Barrow et al. (1978)
G. hirsutum	Induction of callus	Baluch (1979)
G. hirsutum	Callus formation	Hsi and Wu (1981)
G. arboreum, and *G. hirsutum*	Survival of anther callus at –196 °C	Bajaj (1982)
G. arboreum	Chromosomal variation in anther callus – haploid to hexaploid	Bajaj and Gill (1985)
G. hirsutum	Callus formation	Turaev and Shamina (1986); Shamina et al. (1986)
G. arboreum	Multinulleate/multicellular pollen, pollen embryos, callus	Bajaj and Gill (1989, 1996)

for 1 h. Staining was done by subjecting the callus to 60–70 °C in 2% acetic orcein for 2–3 min. After this, the callus was kept in the stain for 3–10 days. Squash preparations were made in 45% acetic acid. The slide was heated to 60–70 °C with two changes in acetic acid, and studied under the microscope.

2.2 Induction and Establishment of Anther Callus

Excised anthers of *G. barbadense* and *G. hirsutum* (Barrow et al. 1978), when plated on Linsmaier and Skoog (1965) and DBM-II (Gresshoff and Doy 1972) media supplemented with NAA (5 mg/l) and BA (1 mg/l), formed callus in about 3 weeks. The response was similar in both the species. The callus emerged from within the anthers or from the anther wall. Bajaj (1982) induced callus from anthers of *G. hirsutum* and *G. arboreum* on SH medium containing 5 mg/l of 2,4-D. The callus obtained by bursting of the anther, obviously of pollen origin, was compact, whereas that originating from the parts of the anther other than pollen was relatively loose. The callus of *G. arboreum* grew faster than that of *G. hirsutum*.

In a later study (Bajaj and Gill 1989), the cultured anthers of *G. arboreum* increased in size within 7 days (Fig. 1A). Callus formation was observed during the second week (Fig. 1C), and a mass of callus was formed in 4 weeks (Fig. 1B). The percentage of anthers that formed callus on various media is shown in Table 2. Callus induction was greatest on MS + 2 mg/l NAA + 1 mg/l BAP and MS + 2 mg/l 2,4-D + 0.2 mg/l kin + 7% CM, with 45 and 37% of cultured

Table 2. Effect of various media and cold treatment (4 °C for 48 h) in the dark on callus induction in cultured anthers of *G. arboreum*. (Bajaj and Gill 1989)

Medium	Without cold treatment			Cold-treated		
	No. of anthers cultured	No. of anthers callusing	Callusing (%)	No. of anthers cultured	No. of anthers callusing	Callusing (%)
1. MS + 1AA (2 mg/l) + kin (0.5 mg/l)	369	29	7.86	340	19	5.59
2. MS + IAA (2 mg/l) + kin (0.05 mg/l)	290	14	4.83	250	11	4.40
3. MS + IAA (4 mg/l) + kin (2 mg/l)	315	79	25.06	275	48	17.45
4. MS + IAA (1 mg/l) + kin (0.2 mg/l) + CH (500 mg/l)	336	91	27.08	270	52	19.26
5. MS + 2,4-D (2 mg/l) kin(0.2 mg/l) + CM (70 ml/l)	285	105	36.84	270	59	21.85
6. MS + NAA (2 mg/l) + BAP (1 mg/l)	415	185	44.58	294	81	27.55

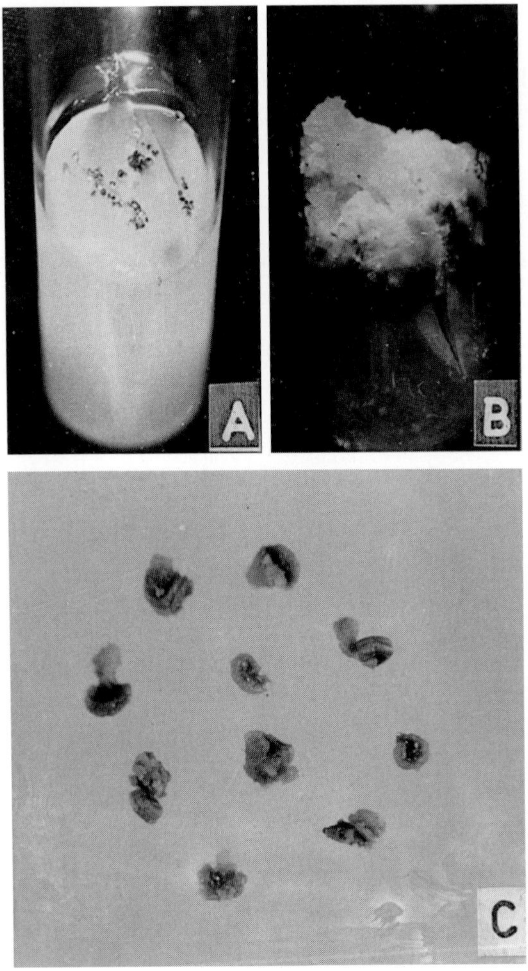

Fig. 1A–C. Callus formation from excised anthers of *Gossypium arboreum* cultured at the uninucleate pollen stage. **A** Proliferating anthers 1 week after culture on MS + NAA (2 mg/l) + BAP (1 mg/l) + sucrose 3%. **B** Anther-derived callus after 4 weeks of subculture on Smith et al.'s medium (1977) containing NAA (2 mg/l) + BAP (0.5 mg/l). (Bajaj and Gill 1989). **C** A 2-week-old culture of proliferating anthers. (Bajaj 1982)

anthers forming callus, respectively. Two other media, MS + 4 mg/l IAA + 2 mg/l kin and MS + 1 mg/l IAA + 0.2 mg/l kin + 500 mg/l CH, also produced good callus growth. Callus that formed on medium containing NAA and BA was compact and pinkish; however, callus on 2,4-D medium was friable and off-white. Callus subcultured on media containing glucose (3%) grew rapidly.

2.3 Chromosomal Variation in Anther-Derived Callus

Anther-derived callus of *G. hirsutum* (Barrow et al. 1978) and *G. arboreum* (Bajaj and Gill 1989) showed a wide range in chromosome number. The cell suspensions of *G. hirsutum* were primarily diploid; however, 13 cultures, that grew more than four times as fast as 527 diploid cultures, had 83% haploid cells. Therefore, haploid tissue was produced in 2–3% of the anthers only.

The chromosome numbers in callus of *G. arboreum* varied from haploid (2n = x = 13) to hexaploid (2n = 6x = 78; Figs. 2, 3). The haploid cells occurred at a low frequency (3.5%). Most of the cells were diploid (38%) or tetraploid (35%); however, triploid (15%), pentaploid (7%), and hexaploid (1.5%) cells were also present (Fig. 3). Aneuploid cells that were observed were grouped with the nearest euploid category. Some of the cells at anaphase showed laggards.

The occurrence of only a low frequency of haploid cells may indicate that most of the anther-derived callus was derived from somatic tissue, although instability of haploid callus may also be responsible for this phenomenon. Endomitosis and other factors may have resulted in the gain or loss of chromosomes in anther-derived callus of wheat (Hu 1986). Thus, in addition to the normal formation of haploids through pollen embryogenesis, the induction of gametoclonal variation through pollen-derived callus would be useful for induction of genetic variability in cotton (also see Chap. IV.1, this Vol.).

2.4 Cryopreservation of Anther-Derived Callus

Haploid cell cultures are known to be genetically unstable, and have a tendency to revert to their diploid form by endomitosis. In view of the importance of cotton haploid cell cultures in mutation and biochemical genetics (Barrow et al. 1978), it is highly desirable to develop methods for their preservation. To overcome the problem of instability in cultures, cryopreservation (see Bajaj 1995) has been experimented successfully with pollen embryos and excised anthers of a number of plant spect (Bajaj 1987). Herein are summarized results on the freeze preservation of anther-derived cotton callus in liquid nitrogen (−196 °C), based on earlier work (Bajaj 1982).

Flower buds obtained from the field-grown plants of two species (*G. hirsutum* L. cv. F414, and *G. arboreum* cv. G 27) were cut open under sterile conditions. Twenty anthers were cultured per test tube containing MS and SH media supplemented with 5 mg/l 2,4-D. After 10–15 days, the proliferating anthers (Fig. 1C) were again transferred to SH medium having 5 mg/l 2,4-D + 3% DMSO and incubated further for 4 days. A total of 100 proliferating anthers were pooled together in a Petri dish on a sterile filter paper, moistened with cryoprotectants (mixture of 5% each of sucrose, glycerol, and DMSO), and were maintained for 2–3 h. During this period, Petri dishes were placed in ice. The specimens were wrapped in sterile aluminum foil and subjected to freezing for 5 min in liquid nitrogen. After freezing treatment, the aluminum-wrapped material was thawed in warm water (35–38 °C), and retrieved anthers were recultured on SH medium containing 5 mg/l 2,4-D at 23–25 °C under diffused light. The resumption of growth in retrieved anthers was considered as a selection survival parameter for anthers. Cell suspension obtained from the retrieved callus resumed active growth, and contained highly cytoplasmic aggregates of cells.

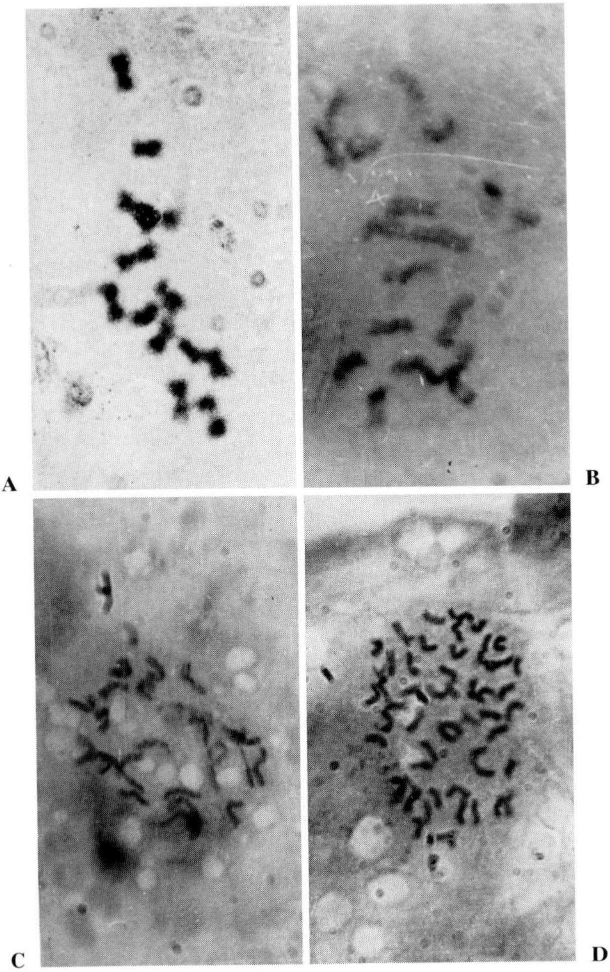

Fig. 2A–D. Range of chromosome numbers in anther-derived callus of *Gossypium arboreum*. **A** Haploid number of chromosomes (n = 13). **B** 16 or 17 chromosomes. **C** Diploid (2n = 26). **D** Tetraploid (52 chromosomes)

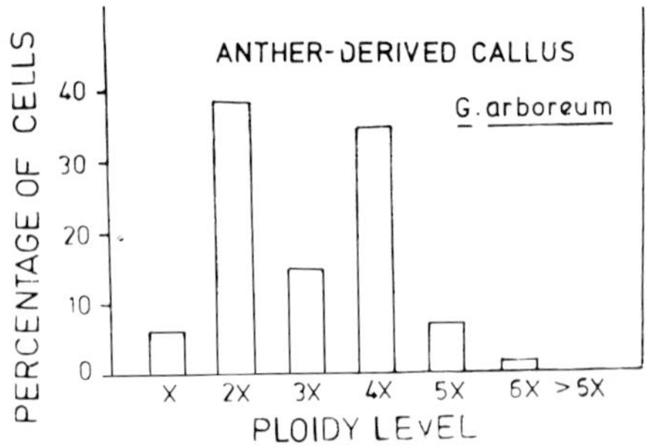

Fig. 3. Histogram showing the extent of genetic variability in anther-derived callus; the chromosome number varied from haloid to hexaploid. (Bajaj and Gill 1985)

Table 3. Effect of various cryoprotectants on survival of cotton anther callus frozen in liquid nitrogen. Data based on the resumption of growth. (Bajaj 1982)

Cryoprotectant	Survival of control cultures (%)
Sucrose 8%	7
Glycerol 10%	15
DMSO 10%	29
Sucrose + Glycerol + DMSO (5% each)	34

The effect of various cryoprotectants on the survival of cells is shown in Table 3. A mixture containing 5% of each sucrose, glycerol, and DMSO increased viability of cells as compared to chemicals used individually. Anther-derived callus was more sensitive than that obtained from ovules (Bajaj 1982).

These cryopreservation studies are in conformity with the earlier work on anther-derived haploid callus of tobacco (Bajaj 1976), androgenic anthers of *Atropa* (Bajaj 1978), rice (Bajaj 1980), *Primula* (Bajaj 1981), and pollen embryos of various species of *Arachis*, *Brassica*, and *Triticum* (Bajaj 1983b).

3 Pollen Embryogenesis in Cultured Anthers (Fig. 4)

Pollen at the uninucleate stage were observed to undergo both equal and unequal divisions. They exhibited three modes of development (1) repeated nuclear division resulting in multinucleate pollen, (2) after a few divisions start of proliferation to form callus, (3) repeated cell division resulting in the formation of pollen embryos.

Most of the multinucleate pollen did not undergo further development and remained at the same stage, or aborted; however, some proliferated to form small irregular masses of callus. In another mode of development, after the second haploid mitosis/cell division there was repeated segmentation to form round proembryos, which eventually developed into pollen embryos showing various stages of development observed in zygotic embryogenesis.

Cold treatment (4 °C for 48 h in the dark) influenced the anther culture in two ways. The cold-treated anthers produced less callus than the untreated ones (Table 2) but, on the other hand, formed more multinucleate and multicellular pollen (Table 4).

On MS + 2,4-D (2 mg/l) + kinetin (0.2 mg/l) + coconut milk (7%), the microspores showed various modes of development (Fig. 4). The frequency of pollen embryos and multinucleate pollen was higher in the cold-treated anthers. The percentage of embryo formation was low (0.9%). Most of the pollen were still uninucleate; however, binucleate (10.7%), tetranucleate (1.2%), and multinucleate pollen (1.5%) were observed.

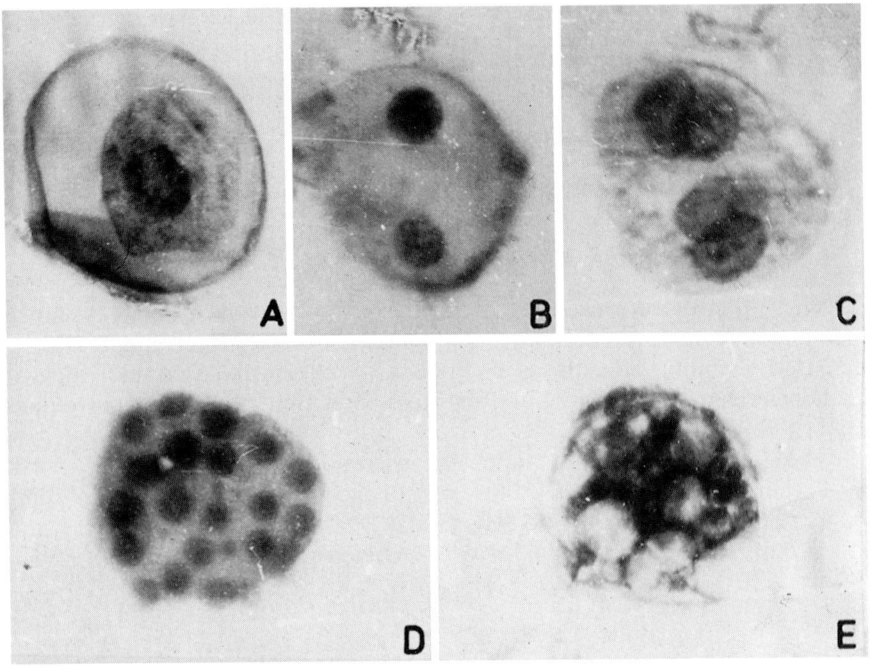

Fig. 4A–E. Pollen embryogenesis in excised anthers of *Gossypium arboreum* cultured on MS + 2,4-D (2 mg/l) + kinetin (0.2 mg/l) + CM (7%) + sucrose 3%. **A–C** Uni-, bi-, and tetra-nucleate pollen from anthers cultured for 1, 2, and 3 weeks, respectively. **D** A multinucleate pollen from 5-week-old culture. **E** A young pollen-embryo from a 5-week-old cultured anther (Bajaj and Gill 1989). Such pollen embryos later resulted in the formation of heart-shaped embryos

Table 4. Effect of cold treatment (4 °C for 48 h) on pollen embryogenesis in excised anthers of *G. arboreum* cultured on MS + 2,4-D (2 mg/l) + kin (0.2 mg/l) + CM 7% + sucrose 3%. (Bajaj and Gill 1989)

	No. of pollen studied	Uninucleate	Binucleate	Tetranucleate	Multi-nucleate	Pollen embryos
Control	880	839 (95.34%)	41 (4.66%)	–	–	–
Cold-treated	895	766 (85.25%)	96 (10.73%)	11 (1.21%)	14 (1.53%)	8 (0.89%)

4 Summary and Conclusion

The frequency of haploid induction by conventional methods in cotton is very low. To overcome this problem, studies have been conducted on culture of excised anthers, and the literature reviewed. In our work, the excised anthers (at the uninucleate microspore stage) of a diploid Asiatic cotton (*Gossypium arboreum*) and a tetraploid (*G. hirsutum*) in culture underwent profuse proliferation (Bajaj 1982). Highest callus formation was observed on MS medium supplemented with NAA (2 mg/l) and BAP (1 mg/l). The callus showed a wide range of chromosome numbers, varying from haploid to hexaploid and aneuploids (Bajaj and Gill 1985).

Cold treatment (4 °C) to anthers promoted nuclear division in pollen, resulting in the formation of multinucleate pollen and pollen embryos (Bajaj and Gill 1989).

The low frequency of haploid cells may be due to the instability of haploid callus. To overcome this, experiments were performed to cryopreserve anther-derived callus in liquid nitrogen (Bajaj 1982).

Since the success of a crop improvement program depends on the extent of genetic diversity in the base population, some of the in vitro-derived genetic variability (Bajaj and Gill 1985) or gametoclonal variation could be utilized in cotton breeding. Furthermore, methods of micropropagation of cotton have been developed (Bajaj and Gill 1992) that could be utilized in multiplication of a few cotton haploid plants.

References

Bajaj YPS (1976) Regeneration of plants from cell suspensions frozen at −20, −70 and −196 °C. Physiol Plant 37:263–268

Bajaj YPS (1978) Effect of super-low temperature on excised anthers and pollen embryos of *Atropa*, *Nicotiana* and *Petunia*. Phytomorphology 28:171–176

Bajaj YPS (1980) Induction of androgenesis in rice anthers frozen at −196 °C. Cereal Res Commun 8:365–369

Bajaj YPS (1981) Regeneration of plants from ultra-low frozen anthers of *Primula obconica*. Sci Hortic 14:93–95

Bajaj YPS (1982) Survival of anther-, and ovule-derived cotton callus frozen in liquid nitrogen. Curr Sci 51:139–140

Bajaj YPS (1983a) In vitro production of haploids. In: Evans DA, Sharp WR, Ammirato PV, Yamada Y (eds) Handbook of plant cell culture, vol 1. Macmillan, New York, pp 228–287

Bajaj YPS (1983b) Regeneration of plants from pollen embryos of *Arachis*, *Brassica* and *Triticum* spp. cryopreserved for 1 year. Curr Sci 52:484–486

Bajaj YPS (1987) Cryopreservation of pollen and pollen embryos, and the estabolishment of pollen banks. In: Giles KL, Prakash J (eds) Pollen development and cytology. Academic Press, New York, pp 397–420

Bajaj YPS (1990) In vitro production of haploids and their use in cell genetics and plant breeding. In: Bajaj YPS (ed) Biotechnology in agriculture and forestry, vol 12. Haploids in crop improvement I. Springer, Berlin Heidelberg New York, pp 3–44

Bajaj YPS (1995) Cryopreservation of plant germplasm I. Biotechnology in agriculture and forestry, vol 32. Springer, Berlin Heidelberg New York

Bajaj YPS, Gill MS (1985) In vitro induction of genetic variability in cotton (*Gossypium* spp.). Theor Appl Genet 70:363–368

Bajaj YPS, Gill MS (1989) Pollen-embryogenesis and chromosomal variation in anther culture of a diploid cotton (*Gossypium arboreum* L.). SABRAO J 21:57–63

Bajaj YPS, Gill MS (1992) Micropropagation of cotton (*Gossypium* species). In: Bajaj YPS (ed) Biotechnology in agriculture and forestry, vol 19. High-Tech and micropropagation III. Springer, Berlin Heidelberg New York pp 483–504

Bajaj YPS, Gill MS (1996) In vitro induction of haploidy in cotton. In: Jain SM, Sopory SK, Veilleux RE (eds) In vitro haploid production in higher plants, vol 5. Kluwer, Dordrecht, pp 165–174

Baluch ZAM (1979) In-vitro anther culture of *Gossypium* species. Pak Cottons 2:161–163
Barrow J, Katterman F, Williams D (1978) Haploid and diploid callus from cotton anthers. Crop Sci 18:619–622
Blank ML, Allison DC (1963) Frequency of polyembryony in certain strains of *Gossypium hirsutum* L. Crop Sci 3:97–98
Chaudhari HK (1978) The use of semigamy in the production of cotton haploids. Bull Torrey Bot Club 105:98–103
De Garcia OB (1962) Utilization of haploids in the improvement of cotton. Turrialba 12:101 (in Spanish)
Bergach GV (1971) Haploidy in the cotton plant. Genetich issled. Klopvhatnika, Tashkent, Uzbek SSR, pp 270–275 (in Russian)
Endrizzi JE (1966) Use of haploids in *Gossypium barbadense* L. as a source of aneuploids. Curr Sci 35:34–35
Gresshoff PM, Doy CH (1972) Development and differentiation of haploid *Lycopersicum esculentum* (tomato). Planta 107:161–170
Harland HC (1920) A note on a peculiar type of rogue in Sea Island cotton. Agric News, Barbados 19:29
Harland HC (1936) Haploids in polyembryonic seeds of Sea Island cotton. J Hered 27:229–231
Hsi YL, Wu KI (1981) Callus formation induced from the anther culture of upland cotton (*Gossypium hirsultum* L.). Abstr Proc Symp Plant Tissue Culture, Pitman Publ, London, pp 239–240
Hu H (1986) Wheat: Improvement through anther culture. In: Bajaj YPS (ed) Biotechnology in agriculture and forestry, vol 2. Crops I. Springer Berlin Heidelberg New York, pp 55–72
Lee JA (1970) On the origin of haploid/diploid twinning in cotton. Crop Sci 10:453–454
Linsmaier EM, Skoog F (1965) Organic growth factor requirements of tobacco tissue cultures. Physiol Plant 18:100–127
Mahill JF, Jenkins JN, McCarty JC Jr, Parrott WL (1984) Performance and stability of doubled haploid lines of upland cotton derived via semigamy. Crop Sci 24:271–277
Murashige T, Skoog F (1962) A revised medium for rapid growth and bio-assays with tobacco tissue cultures. Physiol Plant 15:473–497
Roux JB (1958) Haploidy in the cotton. Cotton Fibre Trop 13:289–292
Schenk RU, Hildebrandt AC (1972) Medium and techniques for induction and growth of monocotyledonous and dicotyledonous plant cell cultures. Can J Bot 50:199–204
Shamina ZB, Turaev AM, Musaev DA (1986) Capacity for induction of callus formation in a cotton anther culture. Cytol Genet 20:40–44
Skovsted A (1935) Some new interspectific hybrids in the genus *Gossypium* L. J Genet 30:447–463
Smith RH, Price HJ, Thaxton JB (1977) Defined conditions for the initiation and growth of cotton callus in vitro I. *Gossypium arboreum*. In Vitro 13:329–334
Turaev AM, Shamina ZB (1986) Optimization of the medium for cotton anther culture. Sov Plant Physiol 33:439–444
Turcotte EL, Feaster CV (1974) Methods of producing haploids: semigametic production of cotton haploids. In: Kasha KJ (ed) Haploids in higher plants – advances and potential, Univ Guelph Press, Guelph, Ontario, pp 53–64
Zhou SQ, Qian DQ, Cao XY (1991) Induction of parthenogenesis, and chromosome behaviour in plants of parthenogenetic origin in cotton (*Gossypium hirsutum*). Genome 34:255–260

III.2 In Vitro Culture of Isolated Pollen in Cotton

A.M. Touraev and Z.B. Shamina[1]

1 Introduction

The culture of isolated immature pollen in vitro has some advantages over anther culture. The possibility of plant regeneration from anther somatic tissues is avoided, the effect of the anther wall is eliminated, and embryogenesis can be studied starting from a single cell (Reinert et al. 1975; Bajaj 1978). It may also have wider applications, especially in in vitro selection and plant transformation (Huang 1992; Jäne et al. 1994; Stöger et al. 1995).

Considerable efforts have been made to obtain and use doubled haploids in cotton. Several methods including semigamy (Turcotte and Feaster 1974; Chaudhari 1978) and parthenogenesis (Zhou et al. 1991) were described for obtaining relatively large numbers of doubled haploids, but these methods are restricted to some genotypes or species (Bajaj and Gill 1989). Therefore, anther/microspore culture would be a rather attractive method for cotton haploid production (Bajaj and Gill 1996).

Our laboratory has put considerable effort into cotton anther/microspore culture in vitro (Touraev and Shamina 1986, 1993; Touraev 1990; Touraev et al. 1990). One of the greatest problems of culturing cotton anthers in vitro proved to be the formation of callus from the anther wall tissues and from filaments. Histological experiments have shown that callus tissues originate from the endothecium of cotton anthers (Touraev 1990). This can be inhibited by cold pretreatment of flower buds for several days in the dark. However, even after the inhibition of somatic callus formation, sporophytic divisions from cold-treated microspores could not be induced efficiently. Thus, isolated microspore culture was initiated in order to better control the factors responsible for the induction of embryogenesis from microspores. The present chapter is primarily based on our published work (Touraev et al. 1990; Touraev and Shamina 1993).

[1] Timiriazev Institute of Plant Physiology, Russian Academy of Sciences, Botanicheskaya ul. 35, Moscow 127276, Russia

Biotechnology in Agriculture and Forestry, Vol. 42
Cotton (ed. by Y.P.S. Bajaj)

2 Isolation and Culture Studies

2.1 Plant Material and Growth Conditions

Three cotton species, tetraploid *G. hirsutum L.*, *G. barbadense L.*, and the diploid *G. herbaceum L.*, as well as several inbred lines from the collection of Prof. D. Musaev which were established at the Tashkent State University (Musaev 1979) were grown under both field and greenhouse conditions. Plants could be used for a long time by removing old or flowering buds.

2.2 Stages of Pollen Development and Staining Techniques

Cotton flower bud contains approximately 60 anthers, which develop synchronously inside the bud (Rumy 1969). There is good correlation between bud length and the stage of pollen development (A.M. Touraev et al., unpubl.). Of course, the length of buds and the corresponding pollen development stages differ between species and even cultivars, and depend on growth conditions. We would suggest that this correlation is established de novo when working with new donor plants.

Correct staging of pollen development is usually achieved by staining of anthers/pollen with dyes and analysis under a light or fluorescence microscope. Success in staining the nucleus of microspores or the nuclei of developing pollen largely depends on the structure of the pollen and the pollen wall. Cotton pollen forms a thick wall and spinules relatively early in development, already at the unicellular stage (Rumy 1969), and it is not easy to stain its nucleus. We have tried a number of stains, including acetocarmine, orcein, hemotoxiline, DAPI, and others according to several protocols.

Staining with Acetocarmine. Fixed anthers or microspores were incubated in acetocarmine solution, prepared in acetic acid (Pausheva 1973), at 65 °C for several hours, cleaned with a chloralgidrate and glycerol solution, and preparations were observed under the light microscope. This procedure gives good visualization of the nucleus in tetrads and microspores until the formation of pollen wall spinules (Fig. 1A–D).

Staining with Hemotoxiline. There are a number of different procedures for the preparation of hemotoxiline solutions (Pausheva 1973; Prozina 1960). We used hemotoxiline prepared in glacial acetic acid (Prozina 1960). Depending on the duration of incubation, one can stain all stages of cotton pollen development. The problem with the use of hemotoxiline is that cells sometimes contain pieces/crystals of dye, and it is hard to distinguish the nucleus from other unrelated substances.

Feulgen Staining. This is not reliable for the staining of cotton pollen. We could stain almost all stages with Feulgen only after using the sophisticated

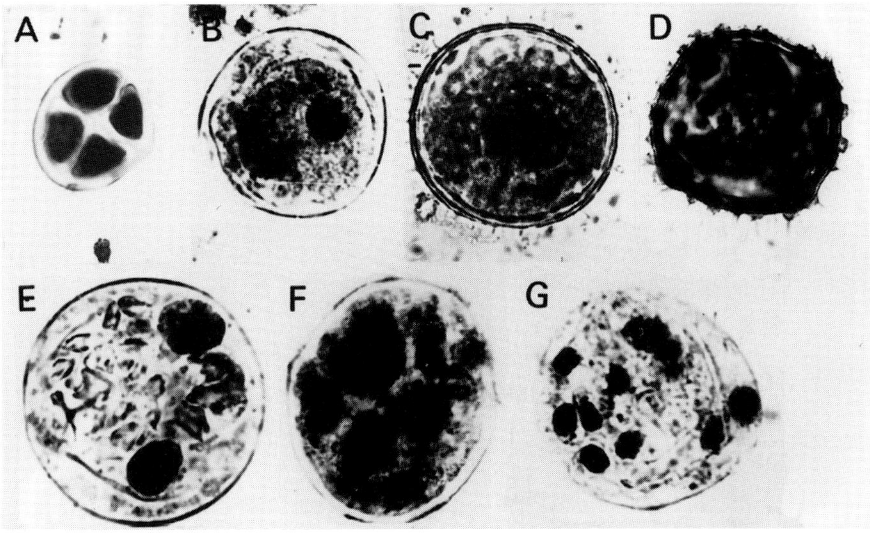

Fig. 1A–G. Cotton pollen developmental stages after staining with acetocarmine in vivo and sporophytic divisions of the microspores after culturing in vitro. **A** Tetrads. **B** Early-uninucleate microspore. **C** Mid-uninucleate microspores. **D** Late microspores with well-developed exine and spinules. **E** Microspore with two nuclei after symmetrical division. **F,G** Multinuclear structures, observed in isolated cotton microspore culture after 10 days

modification of the staining procedure developed by Prof. I.P. Ermakoff (pers. comm.). This stain is not recommended for staging due to its complexity and the duration of the procedure, but it gives excellent results when combined with other dyes in histological analysis (Touraev 1990).

FDA Staining for Viability. The young stages of pollen development can be stained easily with FDA, and this assay can be used routinely to estimate the viability of young cotton microspores with a thin exine and without spinules. However, because of the intensive autofluorescence of the pollen wall at later stages, FDA is not recommended for these.

DAPI (4.6-Diamidino – 2-Phenilindole-2HCl) Staining. DAPI staining of cotton pollen at different developmental stages was successful using the protocol described by Vergne et al. (1987); but, again, at later stages, autofluorescence is so intense that it is hard to visualize the nuclei. Additionally, DAPI staining requires a microscope with fluorescence lamps and equipment. When this is available, DAPI staining is the fastest and most reliable method to stage cotton pollen at different developmental stages (Fig. 2a–h).

The number and size of cotton pollen differ among different species and at different developmental steps. Approximately 400–1000 pollen grains per anther were observed. Cotton pollen is rather large and ranges from 30–60 μm in diameter (Touraev 1990).

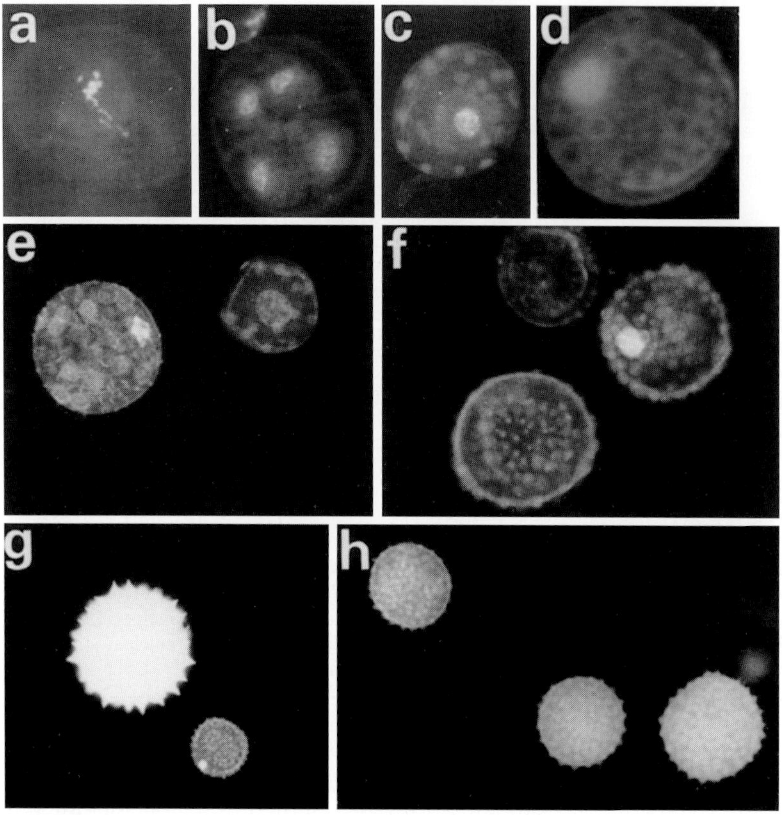

Fig. 2a–h. Cotton pollen stages visualized after DAPI staining. **a** Meiotic cell. **b** Tetrads. **c** Early microspores. **d,e,f** Late microspores. **g,h** Pollen grains

2.3 Isolation Procedures

Sterilization. Cotton flower buds of different sizes were excised and surface sterilized in a sterilizing agent (Na- or Ca-hypochlorate, 70% ethanol, or others) described by Sigma (USA) from seconds to several minutes, depending on the agent. Then sterile buds were washed three times with sterile water.

Anther and Microspore Isolation. Cotton anthers are very small in size and difficult to isolate without using a binocular microscope. In the case of in vitro anther culture, one should carefully take individual anthers, using a fine forceps. For isolated microspores, whole buds can be blended in a Waring blender (after removing the sepal and petal layers), stirred in small beakers with a magnetic stirrer, or vortexed at low speed or crushed with a glass rod. All of these methods lead to the effective release of microspores from anthers, but with different levels of viability, which can be monitored for each case individually. We used the crushing method, but would like to suggest blending or stirring as more effective and reliable.

As an isolation medium, a 0.4 M mannitol solution was used. The viability of microspores was very high (up to 70–80%) in this solution. The homogenized anthers can be filtered through a nylon mesh (pore size 80–90 μm) for the removal of small anther debris. Then the suspension of microspores in 0.4 M mannitol is centrifuged twice at 100 *g* for 5 min, and the final pellet can be resuspended in various induction media at the appropriate density.

2.4 Fractionation of Pollen at Different Developmental Stages

The separation of pollen at different developmental stages allows the identification of the stage of pollen development which is optimal for the induction of embryogenesis. To obtain a homogeneous population of microspores at certain defined stages (in order to find an optimal stage for the induction of embryogenesis), we tried to establish the method of fractionation of cotton microspores at different developmental stages by using discontinuous Percoll gradient centrifugation. There are several reports now on the use of Percoll for the separation of microspores, pollen, or young microspore-derived embryos (Kyo and Harada 1986; Olsen 1991). Cotton microspores differ quite significantly in size during their development, and this was the basis for fractionation experiments. We prepared different concentrations of Percoll by mixing appropriate volumes of Percoll and 0.4 M mannitol to obtain several concentrations ranging from 10 to 70% Percoll. One ml of each Percoll concentration was layered one on top of the other with a fine Pasteur pipet to produce a discontinuous gradient. The mixture of pollen at different developmental stages was resuspended in 1 ml of 0.4 M mannitol and layered onto the surface of the gradient in a sterile glass 12-cm centrifuge tube. After centrifugation at 350 *g* for 15 min, a number of fractions were observed in the gradient. Each visible band was carefully collected with a Pasteur pipet in a separate centrifuge tube, resuspended in 0.4 M mannitol, and washed three times with the same solution by centrifugation at 100 *g* for 5 min. As shown in Table 1 several fractions were obtained in this first preliminary step of separation. In the next step, four distinct fractions of pollen were purified by using particular concentrations of Percoll (Fig. 3). The stages of pollen were confirmed by DAPI staining:

Table 1. The main fractions of cotton pollen at different developmental stages, obtained by discontinious Percoll gradient centrifugation

Main fractions	Percoll concentrations in 0.4 M mannitol (%)							
	15	20	30	40	50	60	70	
Small anther debris	+	–	–	–	–	–	–	–
Tetrads	–	–	–	+	++	–	–	–
Early microspores	+	++	+	–	–	–	–	–
Mid-microspores	–	–	+	++	++	–	–	–
Late microspores	–	–	–	–	+	++	++	+
Pollen grains	–	–	–	–	–	+	++	+
Large anther debris	–	–	–	–	–	–	–	++

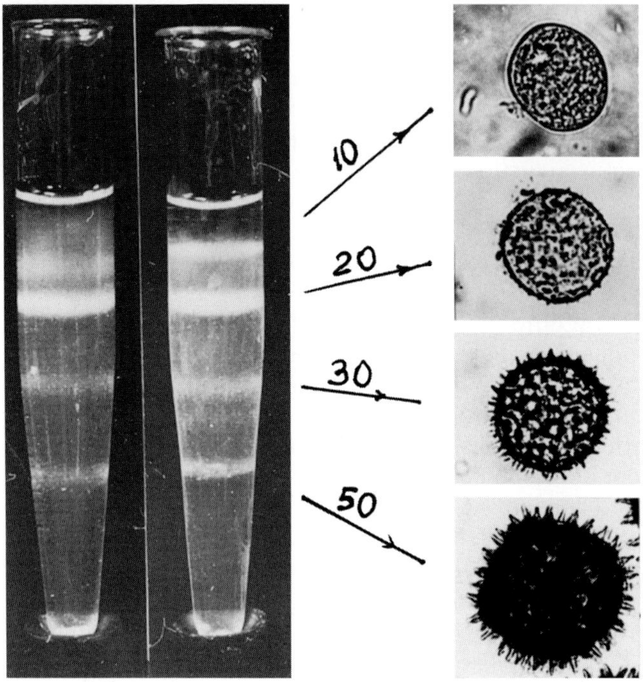

Fig. 3. Fractionation of cotton microspores in discontinuous Percoll gradient. Explanation in text

10/20 – early-uninucleate microspores with thin exines and lacking spinules;
20/30 – mid-uninucleate microspores with thin exines, incomplete spinules and pores;
30/50 – late microspores with well-developed exine, spinules, and pores. Most of the population of tetrads is present in this fraction;
50/70 – pollen at late developmental stages. In this fraction visualization of nuclei and the determination of pollen stage was impossible due to the thick exine and well-developed spinules.

High cell purity (90–95%) was observed in fractions 10 and 20, while the remaining fractions required a second separation. Tetrads seem to have a bouyant density, similar to the late microspores.

Each pollen fraction was resuspended in N6 medium with 5% glucose, 300 mg/l glutamine, 0.1 mg/l kinetin, and 0.1 mg/l 2,4-D, and plated in 30-mm Petri dishes at a density of 4×10^5 cells/ml. Usually, 10–15 flower buds were required to obtain the appropriate density. After incubation for 30 days in N6 media containing 5% glucose, fractions 10 and 20 exhibited high viability.

2.5 Immobilization of Cotton Microspores in Agarose and Culture

Several authors reported that immobilization of cells in agarose or alginate can stimulate better cell division and plant regeneration (Hayashi and Shimamoto 1988). Since the solid/liquid culture conditions of cotton microspores are similar to those published for other species, we describe here the protocol for culturing cotton microspores in agarose blocks. A scheme of the steps involved in this method is shown in Fig. 4.

Double-concentrated (3%) low melting agarose (Sigma, type VII, USA) was dissolved in deionized water and autoclaved before use. The isolated microspores were resuspended in double-concentrated liquid N6 medium with various additions, mixed with molten and cooled agarose (1.5 ml suspension + 1.5 ml agarose), and immediately poured into small 30-mm glass Petri dishes for solidification. The solidified agarose was cut into four pieces and transferred to a new 30-mm Petri dish containing liquid N6 induction medium, which was sometimes conditioned by anther or flower extracts. Plates were sealed with Parafilm and incubated in the dark at 27 °C for 25–30 days, and then used for cytological analysis.

2.6 Culture Media and Culture Conditions

Numerous variations in media components and culture conditions were examined in order to induce embryogenesis from cotton microspores. The results of some experiments have been published (Touraev 1990; Touraev et al. 1990;

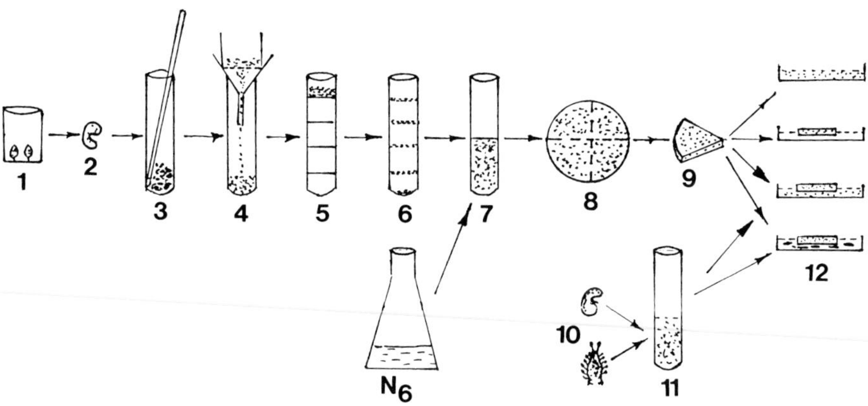

Fig. 4. Scheme of the isolation, fractionation, and culture of cotton microspores in agarose. *1* Sterilization of flower buds; *2* isolation of anthers; *3* homogenization of anthers with a glass rod; *4* filtration through a nylon mesh (80–90 μm); *5* layering of the pollen suspension onto the Percoll gradient; *6* fractionation of microspores by centrifugation at 300 *g* for 15 min; *7* mixing of the microspore suspension with melted 3% agarose; *8* immobilization of the microspores in agarose; *9* cut agarose block with microspores; *10* anthers and flower buds without sepals and petals, as conditioning factors; *11* extracts from anthers and flower buds; *12* various culture conditions

Touraev and Shamina 1993). Unfortunately, none of the media or culture conditions can be recommended as optimal for cotton microspore embryogenesis. In several basic media (MS, Nitsch, N6, Miller) cotton microspores are viable for 3–4 weeks and even longer. However, the frequency of sporophytic divisions is very low in all media and culture conditions tested so far by us and others (Barrow et al. 1978; Bajaj and Gill 1989). In some cases, multicellular structures were observed (Bajaj and Gill 1989; Touraev and Shamina 1993; Fig. 1E–G), which failed to develop into plants.

3 Summary

The growth conditions of donor plants, procedures for the staging of pollen development, isolation, fractionation, and culture of cotton microspores at various culture conditions are described.

To the best of our knowledge, doubled haploid production using cotton anthers or by microspore culture has not been reported so far. The few reports on this topic describe the induction of sporophytic divisions and the formation of multicellular structures in anthers (Bajaj and Gill 1989) or microspore cultures (Touraev and Shamina 1993). Further studies are necessary to find out optimal conditions for the induction of embryogenesis from isolated cotton microspores.

Acknowledgments. The authors would like to thank Prof. R. Butenko and Prof. D. Musaev for their continuous support of this work and for much helpful advice, and Dr. Cathal Wilson for critical reading of the manuscript.

Referenecs

Bajaj YPS (1978) Regeneration of haploid tobacco plants from isolated pollen grown in drop culture. Indian J Exp Biol 16:407–409

Bajaj YPS, Gill MS (1989) Pollen embryogenesis and chromosomal variation in anther culture of a diploid cotton (*Gossypium arboreum L.*). SABRAO J 21:57–63

Bajaj YPS, Gill MS (1996) In vitro induction of haploidy in cotton. In: Jain SM, Sopory SK, Veilleux RE (eds) In vitro haploid production in higher plants. Kluwer Dordrecht, pp 165–174

Barrow J, Kattermann F, Williams D (1978) Haploid and diploid callus from cotton anthers. Crop Sci 18:619–622

Chaudhari H (1978) The use of semigamy in the production of cotton haploids. Bull Torrey Bot Club 105:98–103

Hayashi Y, Shimamoto K (1988) Wheat protoplast culture: embryogenic colony formation from protoplasts. Plant Cell Rep 7:414–417

Huang B (1992) Genetic manipulation of microspores and microspore-derived embryos. In Vitro Cell Dev Biol 28P:53–58

Jähne A, Becker D, Brettschneider R, Lörz H (1994) Regeneration of transgenic, microspore-derived fertile barley. Theor Appl Genet 89:525–533

Kyo M, Harada H (1986) Control of the developmental pathway of tobacco pollen in vitro. Planta 168:427–432
Musaev DA (1979) The cotton genetic collection. FAN, Tashkent, 163 pp (in Russian)
Olsen FL (1991) Isolation and cultivation of embryogenic microspores from barley (*Hordeum vulgare* L.). Hereditas 115:255–266
Pausheva ZP (1973) Practical approaches to plant cytology. Kolos, Moscow, 255 pp (in Russian)
Prozina EH (1960) Botanical microtechniques. Nauka, Moscow, 270 pp (in Russian)
Reinert J, Bajaj YPS, Heberle E (1975) Induction of haploid tobacco plants from isolated pollen. Protoplasma 84:191–196
Rumy WA (1969) Cotton embryology. FAN, Tashkent, 189 pp (in Russian)
Stöger E, Fink C, Pfosser M, Heberle-Bors E (1995) Plant transformation by particle bombardment of embryogenic pollen. Plant Cell Rep 14:273–278
Touraev AM (1990) Cotton anther and microspore culture in vitro. PhD Thesis, Moscow, 260 pp
Touraev AM, Shamina ZB (1986) The optimization of media components for the in vitro cotton anther culture. Sov Plant Physiol 33:565–571
Touraev AM, Shamina ZB (1993) Cotton microspore culture in agarose. Sov Plant Physiol 40:319–322
Touraev AM, Moon AN, Shamina ZB, Butenko RG (1990) The isolation, fractionation and culture of cotton microspores in vitro. Dokl Acad Nauk SSSR 311:371–373
Turcotte E, Feaster C (1974) Semigametic production of cotton haploids. In: Kasha K (ed) Haploids in higher plants – advances and potential. Univ Guelph Press, Guelph, pp 53–64
Vergne P, Delvallée I, Dumas C (1987) Rapid assessment of microspore and pollen development stage in wheat and maize using DAPI and membrane permeabilization. Stain Technol 62:299–304
Zhou SQ, Qian DQ, Cao XY (1991) Induction of parthenogenesis, and chromosome behaviour in plants of parthenogenetic origin in cotton (*Gossypium hirsutum L*). Genome 34:255–260

III.3 Parthenogenesis and Haploid Breeding in *Gossypium hirsutum*

S.Q. ZHOU and D.Q. QIAN[1]

1 Introduction

Doubled haploids for achieving stable homozygosity rapidly are of considerable interest for plant breeders. The functions of haploids produced, both through induced parthenogenesis and anther culture are the same. The priority of parthenogenesis lies in that it does not require a well-equipped laboratory, it can be manipulated in the field without culture medium, and the procedure is comparatively simple.

Many researchers have tried to induce apomixis by using different agents and methods such as high or low temperature, chemicals, X-rays, delayed pollination, hybridization culture in vitro, etc. (van Overbeek et al. 1941; Smith 1946; Deanon 1957; Kimber and Riley 1963; Hermsen 1969; Al-Yasiri and Rogers 1971; Barrow and Chaudhari 1974; Turcotte and Feaster 1974; Byrne and Jelenkovic 1976; Luo et al. 1976; Zhou 1980, 1985; Bajaj 1982; Zhao and Gu 1984; Bajaj and Gill 1985, 1989; Zhou et al. 1991, 1992).

In cotton, Turcotte and Feaster (1974) reported semigamy in Pima cotton (*Gossypium barbadence*) and obtained chimeric plants. Barrow and Chaudhari (1974) obtained upland cotton (*G. hirsutum*) haploids which were produced by semigamy. The doubled haploids are being compared with the parental material for various agronomic and morphological traits; so far there has been no report of success in producing artificial parthenogenesis in upland cotton.

The purpose of the present series of experiments was to develop methods for the induction of haploid parthenogenesis in *G. hirsutum*, to study the possibility of purifying cotton hybrids by using a single step of parthenogenesis, and to study the chromosome behavior and genetics of parthenogenetically derived lines.

[1] Institute of Genetics, Shenyang Agricultural University, Shenyang 110161, People's Republic of China

Biotechnology in Agriculture and Forestry, Vol. 42
Cotton (ed. by Y.P.S. Bajaj)

2 Induced Parthenogenesis

2.1 Artificial Induction of Parthenogenesis

Flower buds of F_1 plants from 15 cross-combinations of *G. hirsutum* (2n = 4x = 52) cultivars with various traits were used as test materials. Chemicals and physical and biological factors were used as inducing agents. For chemical treatment, flower buds of different ages were employed. The results showed that 1 to 4 days before flowering was the best time for inducing parthenogenesis. Therefore, for physical methods, flower buds at that age were used. For biological methods, flower buds were taken 1 day before flowering (Zhou 1980, 1985).

Chemical treatments included different combinations of various chemicals, such as aqueous solution of 0.2% dimethyl sulfoxide (DMSO) with 0.2% colchicine, 0.3% DMSO with 0.3% colchicine, 0.4% DMSO and 0.4% colchicine, 2 mg/l kinetin, 2 mg/l triacontanol with 4 mg/l kinetin, 4 mg/l triacontanol with 4 mg/l adenine, etc. A petroleum-derived growth promoter was added in a few instances. The solution was injected into a wheat straw segment 3–4 cm long, sealed with paraffin at one end. The straw was placed over the stigma of an emasculated flower. The control was treated in the same way but with distilled water only (Zhou 1980).

For physical treatments, the stigmas of emasculated flowers were covered with wheat straw segments, and then the whole plant was treated with high or low temperature. For high-temperature treatments in the field, the whole plants were covered with pieces of transparent plastic sheet. Sunlight was used to increase the temperature of the covered space. The temperature was recorded, and regulated by ventilation. After 4 h the plastic was removed, and the treated buds were marked. The treatments were repeated for 3 consecutive days (4 days in one experiment). Low temperatures were induced by sprinkling water at 15 °C over flower buds with wheat straws for 5 h. Untreated but emasculated flower buds were used as controls.

For biological treatments, flower buds were emasculated and isolated with cellophane bags on the day before anthesis. The next day, the stigmas were pollinated with pollen of *Hibiscus cannabinus* and covered with wheat straws. The cellophane bags were replaced. Tomato juice, containing complexes of sugars, organic acids, amino acids, hormones, etc., and extracts of maize pollen and *G. hirsutum* pollen, were injected into wheat straws, as with the chemical treatments, and applied to the stigmas of emasculated flowers.

In order to provide an environment for good development, a mixture of 40 mg/l gibberellic acid plus 40 mg/l NAA (1:1) was dropped between the calyx and bract at emasculation. This enables the flower buds and other parts of the plant to increase the auxin gradient, which improves transfer of nutrients to the flowers and can delay shedding.

2.2 Frequency of Induced Parthenogenesis

From 1978 to 1982, a total of 1673 flower buds were treated with 43 different treatments, of which 11 treatments involving 553 flower buds were successful. In the remaining 32 treatments involving 1120 buds, no parthenogenetic seeds were formed. In Tables 1, 2, and 3 only the results of the 11 successful treatments are given.

In the chemical treatments, only the combinations of dimethyl sulfoxide and colchicine were successful, and of the 21 combinations tested only 4 produced parthenogenetic seeds. The most effective was the combination of 0.2% DMSO with 0.2% colchicine (Table 1).

Among the physical treatments, a temperature of 36–42 °C was the most effective (Table 2). Above 43 and below 33 °C, no parthenogenetic seeds were formed, but a few were obtained under low-temperature conditions (Zhou 1985).

Pollination with *H. cannabinus* pollen was the most effective of the biological treatments (Table 3). Other treatments and the controls did not result in parthenogenetic seeds.

2.3 Characters of Parthenogenetic Bolls, Seeds, and Plants

The authors treated 1673 flower buds (about 50, 190 ovules) either chemically, physically, or biologically to induce the ovules to produce seed through artificial parthenogenesis on the maternal plant. A large quantity of parthenogenetic bolls were produced, all of which were less plump than the controls. Styles and stigmas grew along with some bolls, and usually were not shed until 30–40 days after anthesis (Fig. 1). A total of 92 parthenogenetically derived seeds from treated flowers was obtained. The seeds were more roundish than the controls (Fig. 2) and did not germinate readily. Sometimes, removal of the seed hull was required for germination. However, after one generation of

Table 1. Effect of various chemicals on induction of parthenogenesis in cotton

Treatment	Year	No. of treated flower buds	No. of treated ovules	No. of partheno-genetic seeds	Partheno-genetic seeds ovules (%)
0.2% DMSO + 0.2% Col + 0.04% PGPS	1978	46	1380	2	0.15
1.0% DMSO + 0.2% Col + 0.04% PGPS	1978	51	1530	3	0.20
0.2% DMSO + 0.2% Col	1978	41	1230	15	1.22
0.4% DMSO + 0.4% Col	1979	71	2130	3	0.14
Control	1978	28	840	0	0.00

Col = colchicine; PGPS = petroleum growth-promoting substance; Control = emasculated and treated with distilled water. DMSO = dimethyl sulfoxide.

Table 2. Effect of physical factors on inducing parthenogenesis

Treatment[a]	Year	No. of treated flower buds	No. of treated ovules	No. of partheno-genetic seeds	Partheno-genetic seeds/ ovules (%)
Low temp. (15 °C)	1980	57	1710	15	0.88
High temp. 1	1980	10	300	8	2.66
High temp. 2	1981	9	270	4	1.48
High temp. 3	1982	37	1110	18	1.62
High temp. 4	1982	16	480	2	0.42
High temp. 5	1982	24	720	0	0
High temp. 6	1982	19	570	0	0
High temp. 7	1982	45	1350	0	0
High temp. 8	1982	50	1500	0	0
Control	1980	12	360	0	0
Control	1981	22	660	0	0
Control	1982	45	1350	0	0

[a] Low temp. = Sprinkled with water at 15 °C for 5 h (as rain), totaling 70 mm.
High temp. 1 = 42 °C, 4 h/day, 3 days.
High temp. 2 = 41.5 °C, 4 h/day, 4 days.
High temp. 3 = 36.2 °C, 4 h/day, 3 days.
High temp. 4 = 40.5 °C, 4 h/day, 3 days.
High temp. 5 = 32.8 °C, 4 h/day, 3 days.
High temp. 6 = 30.8 °C, 4 h/day, 3 days.
High temp. 7 = 43.5 °C, 4 h/day, 3 days.
High temp. 8 = 43.3 °C, 4 h/day, 3 days.
Control = emasculated, but untreated.

Table 3. Effect of biological factors on induction of parthenogenesis

Treatment	Year	No. of treated flower buds	No. of treated ovules	No. of partheno-genetic seeds	Parthno-genetic seeds/ ovules (%)
Tomato juice	1980	28	840	1	0.12
Hibiscus cannabinus pollen	1981	10	300	7	2.33
Hibiscus cannabinus pollen	1982	41	1230	14	1.14
Control[a]	1980	12	360	0	0.00
Control	1981	22	660	0	0.00
Control	1982	45	1350	0	0.00

[a] Control = emasculated, but untreated.

selfing, dehulling was no longer necessary. Eleven seeds with incompletely developed embryos did not germinate or died after germination, and 12 seeds did not germinate as a result of other causes. Plants (Pa_1) were obtained from 69 seeds and, from these, Pa_2 progeny populations were derived. All plants (Fig. 3) of both generation flowered normally, and set viable seed. Table 4 indicates that there is no significant difference in seed set between the control F_2s and the Pa_1 plants: seed formation of Pa_2 was normal.

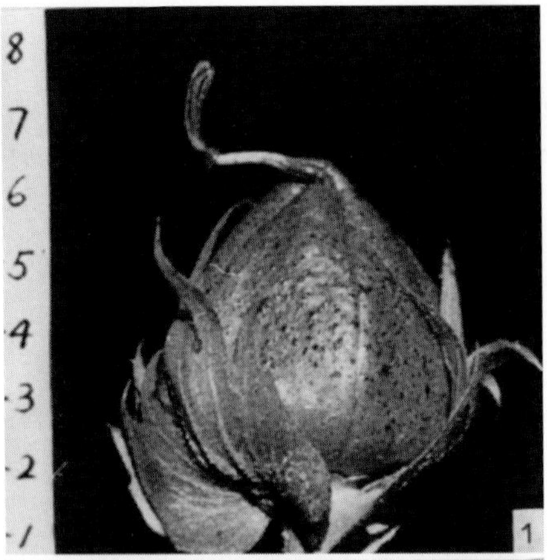

Fig. 1. Parthenogenetic boll, the stigma is still vital

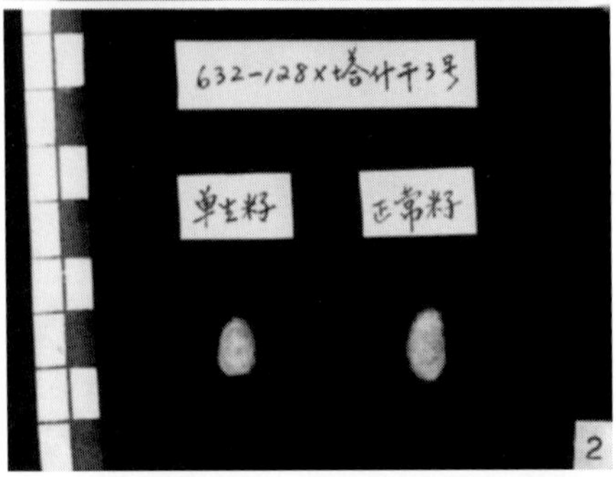

Fig. 2. *Left* Parthenogenetic seed; *right* normal seed

Fig. 3. Plant of parthenogenesis; normal blossom, set viable seeds

Table 4. The average seed number/boll/plant in Pa_1 and its control (F_2)

Plant no.	Seed number per boll (average)		
	Pa_1	F_2	t-test
1	19.8 (10)[a]	26.8 (7)	NS[b]
2	19.0 (10)	14.0 (6)	NS
3	20.4 (11)	21.4 (10)	NS
4	33.2 (15)	30.0 (9)	NS
5	16.9 (11)	14.0 (8)	NS
6	22.9 (9)	26.2 (8)	NS
7	24.1 (16)	21.5 (9)	NS
8	29.0 (6)	20.6 (7)	NS
9	15.0 (6)	11.2 (8)	NS
10	18.7 (11)	28.8 (8)	NS

[a] The numerals in parentheses show the number of bolls per plant.
[b] NS = difference not significant.

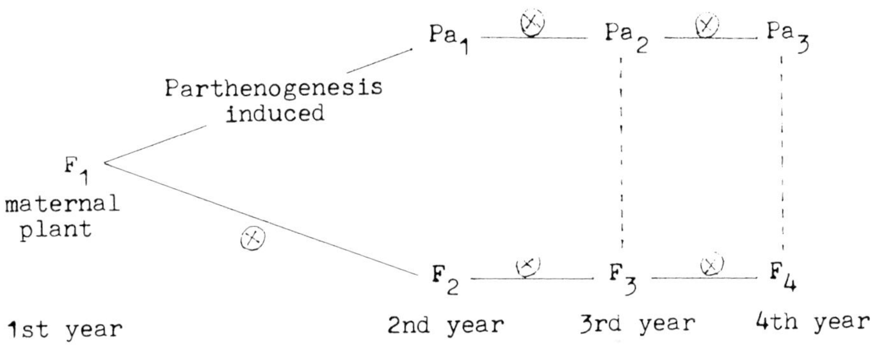

Fig. 4. An outline of the pedigrees of a parthenogenetic line and an inbred line (control)

3 Morphological Characters of the Parthenogenetic Lines

3.1 Procedures

The procedures employed are shown in Fig. 4. In the first year, one flower for each maternal plant was self-pollinated as control. All flowers of every parthenogenetic plant and control (i.e., F_1, F_2, and F_3) were self-pollinated with cellophane bags and/or with collodion plus acetone (3:1) which was dropped onto the top of the flower buds under strictly isolated conditions, to obtain inbred parthenogenetic lines and their corresponding control.

The seeds of Pa_1 (initial generation) obtained from parthenogenetic development were germinated, the seedlings were planted in pots, and the flowers were selfed to produce the Pa_2 (second generation from parthenogenetic origin). The seeds originating from Pa_1 self-pollination were harvested separately from individual plants in the order of the fruit branches and the position of the bolls, and bagged on the basis of one bag for one boll. In the following year, half the seeds collected from one boll of each individual plant (one parthenogenetic line) were germinated and subsequently planted in separate field plots for character evaluation. The remaining half were used for cytological analysis and/or as reserve. Since the number of Pa_2 seeds harvested from each plant varied, the sizes of the plots were different. The parthenogenetic lines and the controls were planted near each other in the field. Internode length of the main stem, internode length of fruit branch, boll shape, number of bolls per plant, maturity date, and fiber length were determined. The flowers of Pa_2 were selfed to produce Pa_3. The selfed seeds were harvested separately from each plot and from each plant. Two seeds were taken from each plant. The seeds from one line were mixed and planted in a separate plot in the field. All lines were treated in the same way. The plants thus produced were the Pa_3 generation. The same characters as studied in the Pa_2 were analyzed statistically. Except for the uniformity of boll shape, which was expressed as a percentage, the variation of all characters was expressed as the coefficient of variation. The variation in Pa_2 and Pa_3 was compared and a t-test was applied to determine whether the difference was significant, for the purpose of establishing if these characters were stably heritable.

For internode length of the main stem, the third, fourth, and fifth internodes of all plants were measured in each plot.

For internode length of the fruit branch, the internodes between the first and second node of the third fruit branch were measured, and again all plants in a plot were measured.

For boll shape, the number of plants with one or more abnormal bolls was determined, including plants in which all the bolls were different from those of other plants in a plot. All plants of each plot were counted.

The number of bolls per plant is the number of bolls per plant on plants 3 to 12 in the second row in the plot. Maturity date is the ratio of the number of fully mature bolls per plant before frost to the total number of bolls per plant for plants 3 to 12 in the second row of the plot; this value was then used as a variable to calculate the coefficient of variation of maturity date.

Fiber length is the average fiber length of all first bolls on fruit branches 3 to 5. All plants in the plot were measured.

3.2 Comparison of Uniformity of Parthenogenetic Lines and Its Control

The uniformity of some characters among individuals in the parthenogenetic lines (Pa_2) and control was investigated. The internode length of the main stem and the fruit branches of each plant in all plots for the parthenogenetic lines and the control were measured, respectively, and the significance of differences in internode length among individuals was examined by analysis of variance. The number of plots with good uniformity in both characters was compared between the parthenogenetic lines and the control. Table 5 shows that the uniformity in the internode length of the fruit branch and the main

Table 5. Test of significance of difference in internode length between plants within a plot of Pa_2 and the control

	Total no. of plots tested	Not significant		$P < 0.05$		$P < 0.01$	
		No. of plots	% of total	No. of plots	% of total	No. of plots	% of total
				Fruit branch			
Pa_2	60	33	55.00	16	26.70	11	18.30
Control	60	26	43.33	19	31.60	15	25.00
				Main stem			
Pa_2	47	42	89.36	2	4.25	3	6.38
Control	47	34	72.34	6	12.77	7	14.89

Note: 1 The 1st and 2nd internode length of the 3rd, 4th, and 5th fruit branches.
2 The 3rd, 4th, and 5th internode length of main stem.

stem in the parthenogenetic lines was higher than that in the control. The purity of the parthenogenetic lines is higher than that of the control.

The abnormal boll plants (a plant which carried one abnormal boll among others was considered an abnormal boll plant) in the parthenogenetic lines averaged 28.53%, against 36.21% in the control.

3.3 Comparison of Purity Between the Parthenogenetic Lines and Inbred Lines

The purification efficiency between the progeny of the parthenogenetic lines and inbred lines was compared, and it was found that 40.9% of the partherogenetic lines (Pa_2) reached stable standard stock variatal purity through a single step of parthenogenesis. Although the vigor in the Pa_1 and Pa_2 of the parthenogenetic lines was less than in the F_2 and F_3 of the inbred lines, this became less apparent in the third and fourth generations. Most importantly, the parthenogenetic method reduced by half the time required to purify the lines compared with the conventional selection method.

3.4 Genetic Stability of the Parthenogenetic Lines

To investigate the genetic stability of the main characters in parthenogenetic lines, the coefficient of variation (CV) for three characters in Pa_3 plants obtained from the mixed seeds of Pa_2 (taking two seeds from each plant) was compared with that of the control F_4 which were obtained from the mixed seeds of F_3 in the same way; 33 samples were measured. Table 6 shows that the purity of Pa_3 was more or less superior to Pa_2, as was F_4 to F_3. This confirmed that the purity of the parthenogenetic lines can be maintained through many generations.

Table 6. Comparison of uniformity between Pa_2 and F_3, Pa_3 and F_4

Item	Pa_2	F_3	Pa_3	F_4
CV of internode length of fruit branches	20.54	26.03	21.60	22.98
CV of internode length of main stem	19.48	19.53	16.89	17.93
Plants with abnormel boll (%)	27.30	46.95	24.63	46.91

Table 7. Uniformity of characters between Pa_2 and Pa_3

Character	Generation	Coefficient of variation of parthenogenetic lines					Mean	t-test
		0195	212	283	0038	051		
Internode length of fruiting branch	Pa_2	26.92	17.89	10.13	19.85	18.47	18.652	ns[a]
	Pa_3	23.43	19.99	12.63	20.57	23.56	20.036	
Internode length of main stem	Pa_2		16.85	14.79	20.93	15.09	16.915	ns
	Pa_3	25.98	18.78	14.53	28.15	19.18	21.324	
Maturity date	Pa_2	54.85	56.14	26.31	36.64	29.23	40.634	ns
	Pa_3	91.49		61.43	73.00		75.307	
No. of boll/plant	Pa_2	20.2	20.07	42.38	22.37	24.57	25.918	ns
	Pa_3	32.37	37.5	37.54	35.34	37.09	35.968	
Fiber length	Pa_2	4.6689	4.4299	4.0896	4.6255	4.5267	4.4681	ns
	Pa_3	4.6207	3.1508	2.7162	4.6471	4.7795	3.9829	

[a] ns = not significant.

Table 8. Percentage of plants with one or more abnormal bolls between Pa_2 and Pa_3

Generation	Parthenogenetic line		Mean	t-test
	0195	283		
Pa_2	45.8	4.16	24.98	ns[a]
Pa_3	34.37	10.25	22.31	

[a] ns = not significant.

Six plant characters were identified (Tables 7, 8). There was no significant difference between Pa_2 and Pa_3 in uniformity of internode length of the fruit branch, internode length of the main stem, maturity date, number of bolls per plant, fiber length, and rate of plants with abnormal shape of bolls. These characters were inherited stably from parents to offspring. There were considerable differences between lines as a result of their varied origin.

3.5 Growth Vigor in Parthenogenetic Lines

Owing to the common belief that these parthenogenetic lines lack hybrid vigor, they have rarely been put into production. According to our investiga-

Table 9. Comparison of the growth status between Pa_2 and F_3, Pa_3 and F_4

		Pa_2	F_3	Pa_3	F_4
Disease boll	No. of plots investigated	53	53	13	13
	Disease bolls/plant[a]	0.692	0.483	0.5077	0.6538
	t-test[b]	t = 2.085,	DS	t = 1.0411,	NS
No. of bolls	No. of plots investigated	65	65	13	13
	Bolls/plant[a]	7.680	7.4969	7.415	7.615
	t-test[b]	t = 0.745	NS	t = 0.404,	NS
Early-stage wilting	No. of plots investigated	65	65		
	No. of plots with wilting plants[c]	11	10		
	% of total plot	16.92	15.38		

DS = difference significance; NS = not significant.
[a] The mean of all plants of parthenogenetic line or its control.
[b] With average/plant of each parthenogenetic line or its control as a variable.
[c] A plot which bore one early-stage wilting plant among others was considered as a plot with early-stage wilted plant.

tion, the germinating ability of seeds of Pa_1 was very weak. Some of them did not germinate normally even if the hastening method was adopted before planting; but all the seeds of Pa_2 germinated normally.

Early-stage wilting and number of diseased and normal bolls in all plants in each plot were compared between the parthenogenetic lines and control for 2 years. The significance of difference was examined by t-test. No significant difference was observed in the growth condition between Pa_3 and F_4 (Table 9).

4 Cytology of Parthenogenetic Lines

4.1 Chromosome Counts

The number of chromosomes in root-tip cells and pollen mother cells in progeny of parthenogenetic origin was examined. To facilitate chromosome observation, the root tips of the seedlings were fixed in Carnoy's fluid. After fixation, the material was placed in a refrigerator at −1 to −2 °C for 4 h and then stained with Heidenhain's hematoxylin. After this, the stained root tips were treated with aqueous solution of 2% cellulase and 2% pectinase for the cells to spread, and then transferred to a slide onto which 45% acetic acid was dropped until the chromosomes became dark blue and the cytoplasm nearly colorless. Flame drying of the slides made the chromosomes spread well. Slides on which the cytoplasm was intact and chromosomes spread well were selected for chromosome counting to minimize error due to chromosome loss.

For counting the number of chromosomes in pollen mother cells, the flower buds, after removal of the calyx and petal, were treated like the root tips, except that the pollen mother cells were treated with an aqueous solution of 2% snail-stomach enzyme and 2% cellulase for softening.

4.2 Variation in Somatic Chromosome Number

The number of chromosomes was determined for eight lines. The average number of chromosomes in root-tip cells in Pa_2 for six lines was approximately 2n = 26, and two lines had approximately 2n = 33 (Table 10). It is noteworthy that all of these plants showed variation in chromosome numbers (Fig. 5), but they were normal in flowering and seed setting and not deformed in appearance. A comparison between chromosome numbers in Pa_2 and Pa_3 showed that in lines with an average of 26 chromosomes, the chromosome number in Pa_3 was similar to that in Pa_2. Statistical analysis indicated that there were no significant differences in this regard, suggesting stability in chromosome number. On the other hand, in lines with 33 chromosomes, there was a significant difference in the number of chromosomes between Pa_2 and Pa_3 (the difference was especially significant, $P < 0.01$, in line 051.4.2), suggesting instability in chromosome number. Nevertheless, whether the average chromosome number was 26 or 33, except in line 212.12.1, the coefficient of variation in chromosome number in all the other lines was invariably greater in Pa_2 than in Pa_3, suggesting that in aneusomatic plants, with the advance in

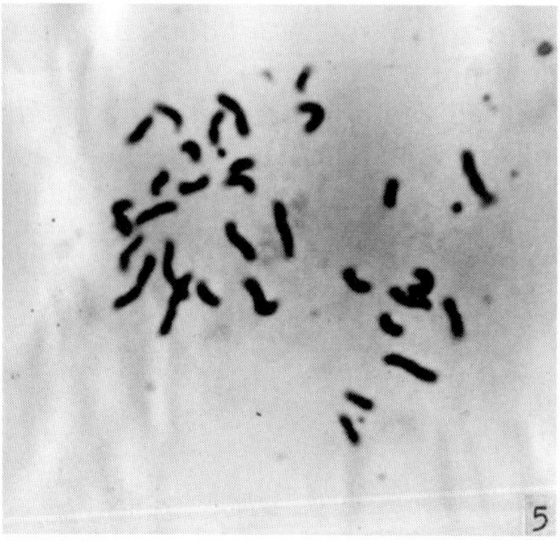

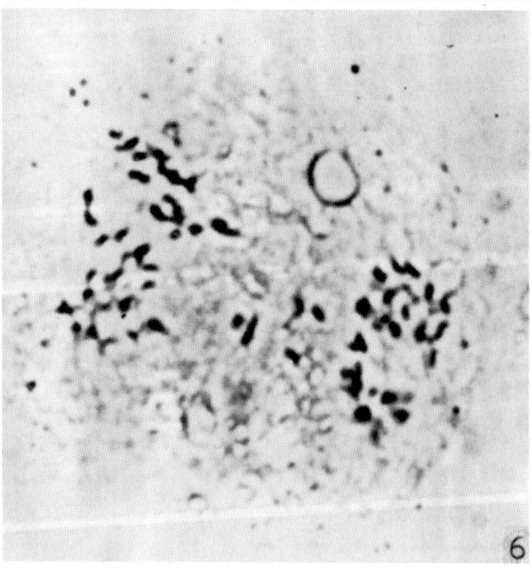

Fig. 5. Root-tip cell with 31 chromosomes

Fig. 6. Abnormal division of root-tip cell, lagging chromosomes

Table 10. Comparison of chromosome numbers in root tip cells between parents and offspring

Parthenogenetic lines no.		No. of cells examined	Chromosome numbers										$\overline{X} \pm S$	t-test	Coefficient of variation
			10–15	16–20	21–25	26	27–31	32–36	37–41	42–46	47–51	52<			
0195.13	Pa_2	100	2	21	16	23	16	12	6	2	2	0	26.48 ± 7.29	t = 1.25	27.53
	Pa_3	100	0	15	20	24	16	12	7	2	3	1	27.8 ± 7.6	ns[a]	27.33
212.12.1	Pa_2	100	3	15	22	22	17	15	4	2	0	0	26.3 ± 6.4	t = 0.82	24.33
	Pa_3	100	3	11	28	26	19	10	1	2	0	0	25.61 ± 5.45	ns	21.28
212.12.1	Pa_2	100	3	15	22	22	17	15	4	2	0	0	26.3 ± 6.4	t = 0.68	24.33
	Pa_3	100	5	9	22	28	15	10	4	7	0	0	26.95 ± 7.07	ns	26.32
0038	Pa_2	100	8	12	20	14	10	16	10	9	1	0	27.93 ± 9.07	t = 1.6145	32.47
	Pa_3	100	0	19	26	16	23	9	7	0	0	0	26.2 ± 5.71	ns	21.79
283.4.2	Pa_2	100	7	18	26	9	10	15	8	6	0	1	26.76 ± 8.65	t = 1.58	32.32
	Pa_3	200	11	34	54	33	42	12	11	2	1	0	25.26 ± 6.32	ns	25.02
051.4.2	Pa_2	200	11	17	31	20	24	36	14	21	6	20	33.59 ± 14.93	t = 2.813	44.45
	Pa_3	200	7	20	29	33	25	46	18	13	4	5	30.22 ± 9.87	[b]	32.66
64	Pa_2	109	3	34	29	13	16	5	6	2	1	0	24.49 ± 6.84		
69	Pa_2	96	1	7	15	10	13	23	7	12	3	5	32.86 ± 10.49		
77	Pa_2	200	12	40	50	22	43	14	7	5	2	5	26.08 ± 9.77		

[a] ns = difference not significant.
[b] Significantly different at the 1% level.

generation, the chromosome number of cells tended to funnel into dihaploid (2n = 26) and become stabilized.

4.3 Variation in PMC Chromosome Number

Six lines were recorded for the number of chromosomes in pollen mother cells. The plants developed from these aneusomatic seedlings exhibited also a variation in the chromosome number in pollen mother cells. For example, in the plants of line 0038 the chromosome number in PMCs of Pa_2 ranged from 20 to 35, in the plants of line 283 from 20 to 47, and in the plants of line 051 from 23 to 39 (Table 11).

It is noteworthy that the average chromosome number in PMCs and in the root-tip cells were similar. In PMCs of lines 0038 and 051, the average chromosome number was lower than that in the root-tip cells; in line 283 the average chromosome number was similar to that in the root-tip cells.

4.4 Abnormality at Mitosis and Meiosis

In root-tip cells, lagging chromosomes (Fig. 6), multipolar spindles (Fig. 7) and cells with two or three nuclei (Fig. 8) were usually observed (Zhou 1980).

In PMCs, chromosomes were able to pair at metaphase I (Fig. 9). Configuration from univalents to nonavalents were observed. Because the chromosomes of the A genome are almost twice the size of the chromosomes of the D genome, the genome of the chromosomes in the different configurations could be determined (Figs. 10, 11). About 30% of the chromosomes were involved in multivalents with a large number of A-D associations, mostly in A-D-A and D-A-D sequences. A-D chiasmata accounted for 68.4% of all chiasmata, indicating that a large number of chiasmata or associations resembling chisamata must have formed between chromosomes within the A and D genomes. In the 25 PMCs analyzed, the number of A chromosomes was larger than the number of D chromosomes: the ratio was almost 1.2:1 (Table 12).

The chromosomes segregated into unequal groups in 89.4% of the anaphase I cells, with a difference in number varying from 0 to 8, and an average of 2, 3 (Fig. 12).

Table 11. Chromosome number at metaphase I in pollen mother cells of Pa_2

Parthenogenetic line no.	No. of PMCs examined	Range	Mean chromosome no.
283	9	20–47	30.22 ± 8.51
0038	11	20–35	25.82 ± 4.42
051	5	23–39	29.80 ± 6.61
64	15	13–57	27.00 ± 12.30
69	45	20–48	29.69 ± 6.55
77	60	13–50	28.13 ± 7.49

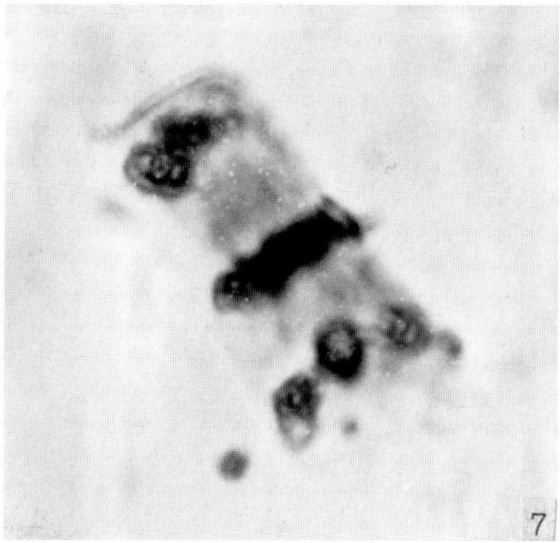

Fig. 7. Abnormal division of root-tip cell, multipolar spindles

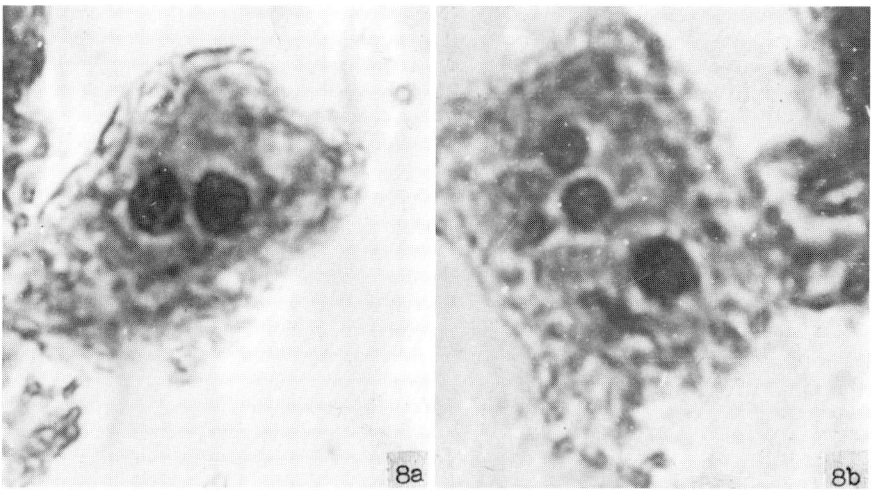

Fig. 8a,b. A root-tip cell with two nuclei (**a**) and three nuclei (**b**)

5 The Problem of Further Study of the Progeny of Parthenogenetic Plants

In this experiment, all the plants of progeny originating from parthenogenesis were aneuploidy aneusomatics. These aneusomatic plants can grow and develop, displayed normal phenotypical characters, and set seeds normally. So far, these characteristics still remained stable from parents to their selfed offspring. What causes these plants to remain aneuploidy aneusomatics stably from generation to generation? One possible explanation might be that the aneusomatics were produced through continual parthenogenesis from generation to generation. In 1993 and 1994, the authors examined embryogenesis of

Table 12. Frequency of chromosome pairing type at metaphase I in PMCs

Parthenogenetic plant no.	Univalents		Bivalents			Multivalents	No. of chromosomes		
	A	D	AA	DD	AD		Total	A	D
0038	3	1	4	2	4	1IV (DDDA)	28	16	12
0038	1	1	2	4	4	1III + 2V (ADA, DADDA, AAADD)	35	16	19
0038	1	1	2	2	3	1IV (AAAA)	20	12	8
0038	3	2	1		7		21	12	9
0038	1	1	8	1	2		24	19	5
0038	2		4		2	2III + 1V (ADA, ADA, ADADD)	25	18	7
0038	1	1	1	1		2III + 1IV + 1IX (ADA, DAD, AAAA, DDDAAADDA)	25	14	11
0038	3		1	2	2	1VIII (DDAAADDA)	21	11	10
0038	2	2	4		1	2III + 1IV + 1V (ADA, DAD, AADD, ADAAA)	29	20	9
0038	2		1	2	4	2III + 1V (DAD, DAD, AAADA)	27	14	13
0038	2	1	1	5	2	2III + 1IV (ADA, DAD, ADDA)	29	11	18
283	1		1	1	3	2V + 1VI (DADAA, DADDA, DDDAAD)	27	13	14
283	3	1	7	4	2	2III (ADA, AAD)	36	23	13
283	1		2	3	4	1III (ADA)	22	11	11

283			4	2	4	2III (ADA, DDA)	26	15	11
283	4	2	5		2		20	16	4
283	1	2	4	2	2	2III (DAD, DAA)	25	14	11
283	2	4	1	8		2III + 1IV (ADA, ADA, ADAD)	34	10	24
283		2	7	7	4	1III + 1VI (DAD, AAADDD)	47	22	25
283		1	7	5	2	2III (ADA, DAD)	35	19	16
051	3	3	2	1	1	1III + 2IV + 1V (AAD, ADAD, ADAD, AAAAD)	30	18	12
051	4	3	1	2	4	1III (ADA)	24	12	12
051	1		1	1	3	6III + 1IV (2ADA, 3DAD, DDA, DAAA)	33	17	16
051			1	4	2	3III (DAD, DAD, ADA)	23	8	15
051	2	1	10	7	1		39	23	16
Total	43I	29I	82II (23.3%)	66II (18.7%)	65II (18.4%)	34III, 91V, 8V, 2VI, 1VIII, 1IX	705	384	321
Total (%)	10.2%		60.4%			29.4%			

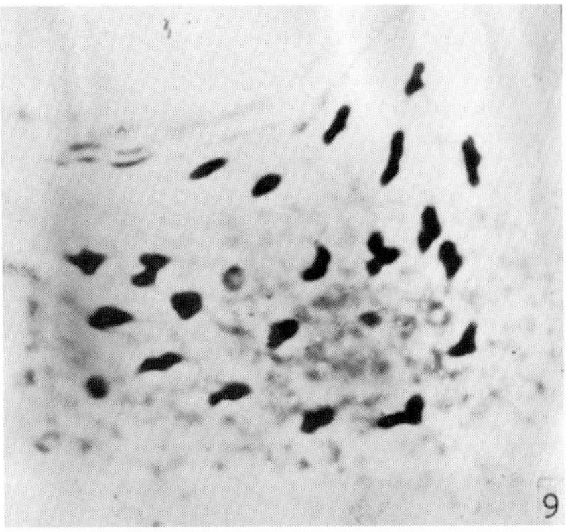

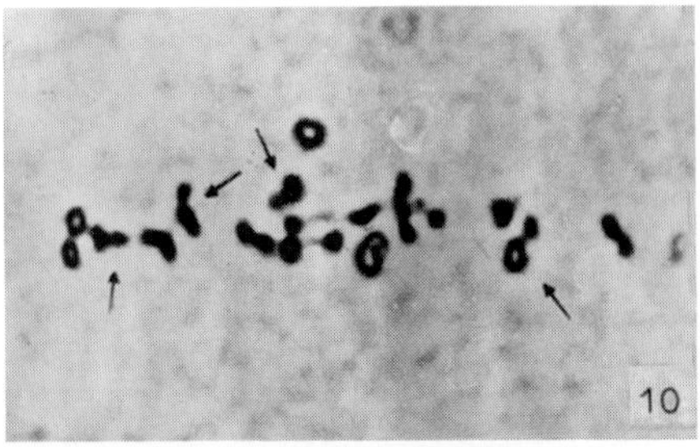

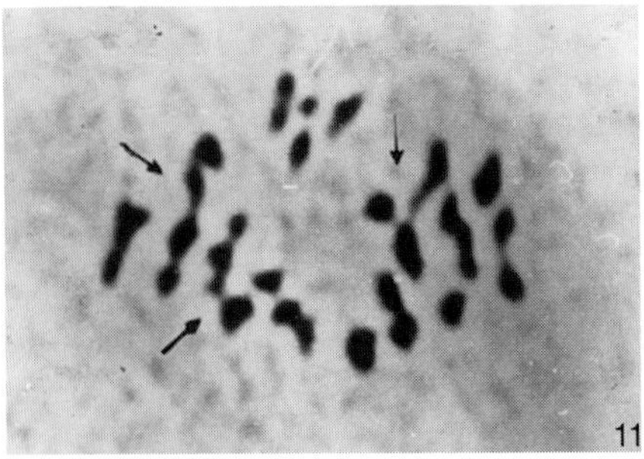

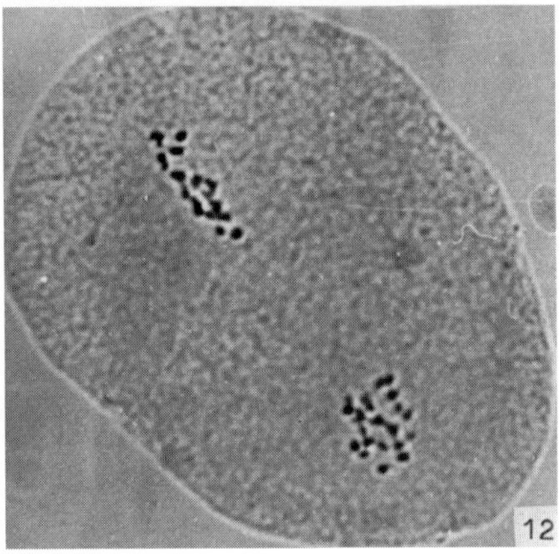

Fig. 9. Metaphase I: chromosomes pairing with each other

Fig. 10. Metaphase I: chromosome in A genome and chromosome in D genome pairing with each other

Fig. 11. Metaphase I. *Arrow* Trivalent or multivalent

Fig. 12. Anaphase I. *Upper* 13; *lower* 19

these aneusomatic plants. A total of 70 ovules was collected at different times 25–72 h after selfing and examined continuously in paraffin section. In all cases, it was seen that the pollen tube penetrated the embryo sac and discharged some material into the degenerating synergid. In some cases, two nucleoli could be seen in an egg nucleus: the large one was female, and the smaller male. These findings showed clearly that the aneusomatic plants were produced by a sexual process.

Why do the aneusomatics remain stable? Based on the authors' investigation, the following points might be a possible cause to explain these phenomena.

1. *Tolerance to Chromosome Deficiency.* In general, compared with diploids, polyploids were more tolerant of chromosome deficiency because of the buffering effect of the presence of more than two genomes. Myles and Endrizzi (1989) reported that *G. hirsutum* plants can tolerate the loss or absence of one or more chromosomes or arms of chromosomes and transmit the deficiency to some of the progeny, and suggest that egg cells with 17 chromosomes of the n = 26 haploid complement are functional.

2. *Compensating Effect of Chromosomes.* Jauhar and Joppa (1996) indicated that, in wheat, the complete absence of a specific chromosome was compensated for by four doses of one of its homoeologs in either of the other two genomes. The corresponding chromosomes of the A, B, and D genomes of bread wheat are genetically equivalent. Although in *G. hirsutum* similar reports have not yet been seen, in upland cotton, there are two genomes. Chromosomes of the A genome have been numbered from 1 to 13, and those of the D genome from 14 to 26. It was known that the chromosomes 1 and 15, 7 and 16, and 12 and 26 are homoeologs (quoted from Stelly et al. 1995). Thus, we have reason to deduce that the homoeologous or homologous chromosomes might have a compensating effect.

3. *Cytomixis of PMCs.* Cytomixis was observed in PMCs in this experiment (Fig. 13). The frequency ranged from 15.09 to 87.95% (when the flower buds developed under a mean temperature maximum in the day of 25.3 °C and

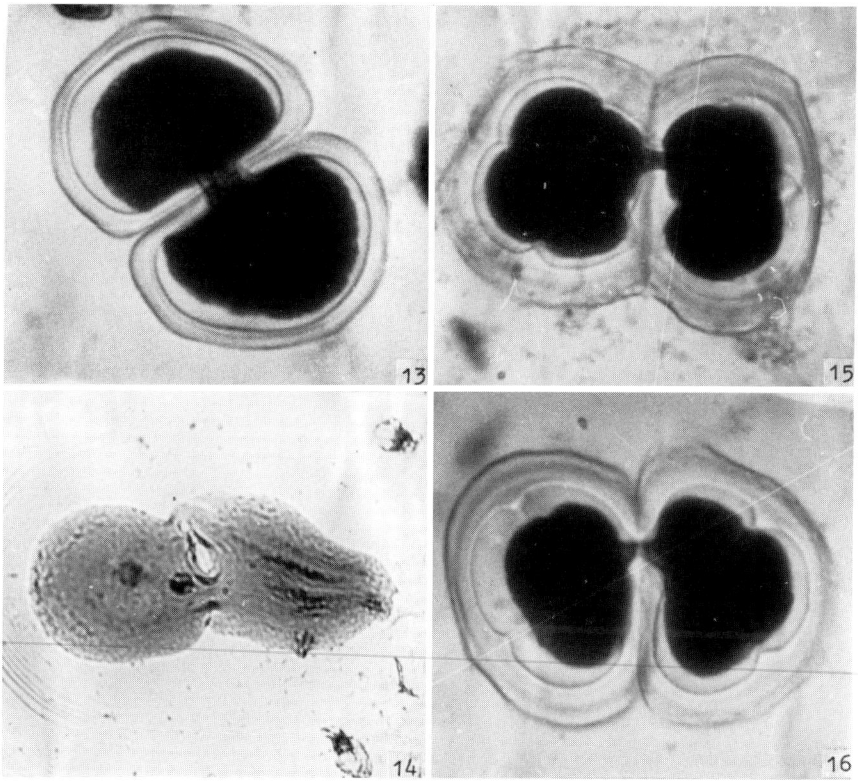

Fig. 13. Cytomixis of PMCs

Fig. 14. Nucleus migrating between neighboring cells

Figs. 15, 16. Direction of migration of cell contents could be in one way, reciprocal, or cyclic among PMCs minimum at night of 11.15 °C). Cytomixis occurred in every stage of meiosis. Chromosomes, nucleus, and cytoplasm could migrate between neighboring cells (Fig. 14). The direction of migration could be in one way, reciprocal, or cyclic among PMCs (Figs. 15, 16, 17). The flower buds set seeds normally. Chromosome numbers in somatic cells and the traits of plant derived from cytomixis seed were normal. It is believed that cytomixis of PMCs was a normal process, occurring under stress in unfavorable circumstances. This process was a self-adjustment to adapt the plant to external changes for maintaining normal reproduction and hereditary stability.

4. *Selection in Fertilization*. This might help hereditary stability. In these aneusomatic plants there were 1634–3080 pollen grains per stigma, and about

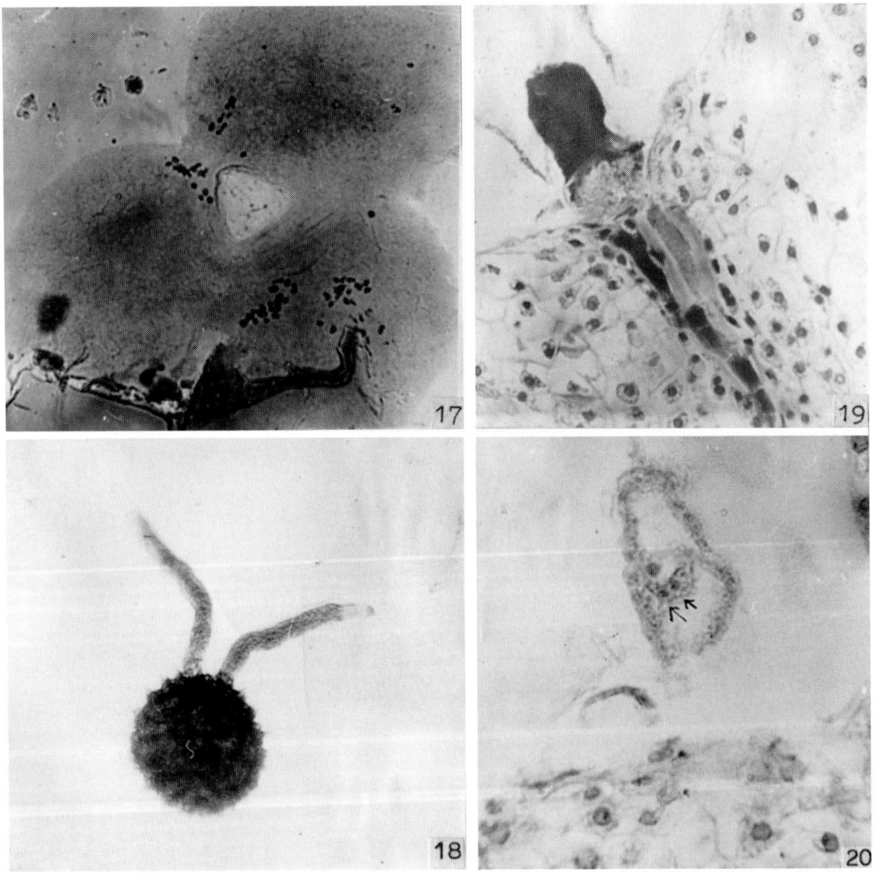

Fig. 17. Direction of migration of cell contents could be in one way, reciprocal, or cyclic among PMCs

Fig. 18. A pollen grain with two pollen tubes

Fig. 19. More than one pollen tube entered one embryo sac

Fig. 20. One large and two small nucleoli in an egg nucleus

777–1827 pollen grains germinated on the stigma; but there were only 32–40 ovules in a flower. Therefore, the scope for selecting the most favorable pollen grains to fertilize to maintain normal development of progeny was ample. Sometimes two or more tubes from a single pollen grain penetrated simultaneously into a stigma (Fig. 18) and such pollen grains accounted for 5.07–18.23% of the total germinated pollen grains. Sometimes, the pollen tubes from several different pollen grains entered one embryo sac (Fig. 19), and found one large and two small nucleoli in an egg nucleus (Fig. 20). These phenomena played more or less a role of adjusting and balancing fertilization.

There are still many points which need further investigation of evidence to demonstrate why aneusomatic plants could maintain hereditary stability. The authors will make use of genome analysis by cytogenetic and molecular cytogenetic methods to explore the inheritance and generative mechanism of aneuploidy aneusomatics in upland cotton.

6 Summary

Different methods were used to induce artificial parthenogenesis in *G. hirsutum*. The frequency of induced parthenogenesis was 0–2.66%. All the progeny plants were aneuploidy aneusomatics, and could grow, and set seeds normally. Chromosome numbers ranged from 12 to 56 in somatic cells. At meiotic metaphase I in PMCs the chromosomes showed uni-, bi-, tri-, and multivalents, and showed AA, DD, and AD pairing. Anaphase I segregation was often unequal. The morphological characters in Pa_2 of these plants in the same parthenogenetic line were uniform. The mean chromosome numbers and phenotypical characters remained stable from parents to their selfed offspring. Embryological investigation showed that all the aneusomatic plants were produced through a sexual process.

Acknowledgments. The authors thank Ms. Cao Xun-Yun for participating in this work. The project was supported by the National Natural Science Foundation of China. Part of illustration and Tables 1–3 and 12 and Tables 7, 8, and 10 have been reproduced with permission of Genome, Vols. 34 and 35, respectively.

References

Al-Yasiri S, Rogers OM (1971) Attempting chemical induction of haploidy using toluidine blue. J Am Soc Hortic Sci 96(1):126–127

Bajaj YPS (1982) Survival of anther-, and ovule-derived cotton callus frozen in liquid nitrogen. Curr Sci 51:139–140

Bajaj YPS, Gill MS (1985) In vitro induction of genetic variability in cotton (*Gossypium* spp.). Theor Appl Genet 70:363–368

Bajaj YPS, Gill MS (1989) Pollen embryogenesis and chromosomal variation in anther culture of a diploid cotton (*Gossypium arboreum* L.). SABRAO J 21:57–63

Barrow J, Chaudhari H (1974) Cotton haploids produced by semigamy from specific breeding lines. In: Kasha KJ (ed) Haploid in higher plants. University of Guelph, Guelph, Ontario, p 274

Byrne D, Jelenkovic G (1976) Cytological diploidization in the cultivated octoploid strawberry *Fragaria* × *ananassa*. Can J Genet Cytol 18:653–659

Deanon JR Jr (1957) Treatment of sweet corn silks with maleic hydrazide and colchicine as a means of increasing the frequency of monoploids. Philipp Agric 41:364–377

Hermsen JGTh (1969) Induction of haploids and aneuhaploids in colchicine-induced tetraploid *Solanum chacoense* Bitt. Euphytica 18:183–189

Jauhar PP, Joppa LR (1996) Chromosome pairing as a tool in genome analysis: merits and limitation. In: Jauhar PP (ed) Methods of genome analysis in plants. CRC Press, Boca Raton, p 21

Kimber G, Riley R (1963) Haploid angiosperms. Bot Rev 29:480–531

Luo P, Liao YH, Lan ZJ, Liu TL (1976) The study of the artificial induction of gynogenetic haploids of rape. Acta Genet Sin 3(4):293–298

Myles EL, Endrizzi JE (1989) Aneuploids induced by deficiencies of chromosome 9 and analysis of the time of nondisjunction in cotton. Genome 32:12–18

Riley R (1960) The diploidisation of polyploid wheat. Heredity 15:407–429

Smith FH (1946) Haploidy in einkorn. J Agric Res 73:291–301

Stelly DM, Price HJ, McKnight TD (1995) Molecular-meiotic cytogenetic analysis in cotton. In: Raupp WJ, Gill BS (eds) Kansas agricultural experiment station report 95-352-d. Kansas State Univ, Manhattan, pp 148–156

Turcotte EL, Feaster CV (1974) Methods of producing haploids: semigametic production of cotton haploids. In: Kasha KJ (ed) Haploidy in higher plants. University of Guelph, Guelph, Ontario, pp 53–64

Van Overbeek J, Conklin ME, Blakeslee AF (1941) Chemical stimulation of ovule development and its possible relation to parthenogenesis. Am J Bot 28:647–656

Wang JJ, Chu ZG, Sun JS (1980) Studies on the induction of pollen sporophyte of *Coix lacryma* Jobic *L.* Acta Bot Sin 22(4):316–322

Zhao ZY, Gu MG (1984) Production of diploid pure lines of maize through parthenogenesis induced by chemicals. Acta Genet Sin 11(1):36–46

Zhou SQ (1980) A preliminary study on the partheongenesis in cotton (*Gossypium hirsutum*). Acta Genet Sin 7 (3):247–258

Zhou SQ (1985) Physically induced parthenogenesis in cotton (*Gossypium hirsutum*) and the variation of chromosome number in parthenotes. Acta Gossypii Sin 1:34–40

Zhou SQ, Qian DQ, Cao XY (1991) Induction of parthenogenesis, and chromosome behavior in plants of parthenogenetic origin in cotton (*Gossypium hirsutum*). Genome 34:255–260

Zhou SQ, Qian DQ, Cao XY (1992) Inheritance of characters and chromosome number in aneusomatic progeny of upland cotton (*Gossyium hirsutum*) of parthenogenetic origin. Genome 35:35–38

Section IV
Somaclonal Variation/ Genetic Variability; Plants Tolerant to Salt and Water Stress, High Temperature and Nematodes (for plants resistant to insects and herbicides see Chaps V.2,3, this Vol.)

IV.1 Somaclonal Variation in Cotton

Y.P.S. BAJAJ[1]

1 Introduction

The extent of genetic variability in the base population determines the success of any crop improvement program. During the past few years, various in vitro methods have been employed to induce genetic variability/somaclonal variation for resistance to pests, diseases, salt, high temperature, etc. in a number of agricultural crops (see Bajaj 1996). The cell cultures on prolonged storage at ordinary temperatures or on periodical subculturing undergo genetic erosions, such as endomitotis, chromosome loss, polyploidy, translocations, gene amplification, mutations, etc. (D'Amato 1985; Bajaj 1990) Such changes can be augmented further by various manipulations. Moreover, it has been speculated (Larkin and Scowcroft 1981) "that tissue culture may generate environments for enhancing chromosome breakage and reunion events, and thus a tissue culture cycle of the hybrid material may provide the means for obtaining the genetic exchange needed between two genomes in the interspecific hybrid. The hybrid callus may enhance the frequency of requisite exchange".

In cotton cell cultures, various changes in chromosomes have been reported (Bajaj and Gill 1985; Stelly et al. 1989), and plants showing tolerance to high temperature, and salt-tolerant cell lines have been obtained. Moreover, variability in cotton through embryo rescue/culture has been induced and hybrid callus obtained (Gill and Bajaj 1987). Recently, through genetic transformation, cotton plants resistant to insects, herbicides, and in vitro induction of male sterility have been reported (Reynaerts and De Sonville 1994). Some of these developments are discussed in details in various chapters of Section IV and V of this Volume, and are summarized here.

2 In Vitro-Induced Genetic Variability

2.1 Chromosomal Variations in Callus Culture

Various chromosomal changes in cell/callus culture of *Gossypium* species are summarized in Table 1. In our work (Bajaj and Gill 1985), the callus cultures

[1] Biotechnology in Agriculture and Forestry, A-135 New Friends Colony, New Delhi 110065, India

Biotechnology in Agriculture and Forestry, Vol. 42
Cotton (ed. by Y.P.S. Bajaj)

Table 1. Somaclonal variation and chromosomal changes studies on callus cultures of cotton

Species	Origin of callus/ cell suspension	Variation/Remarks	Reference
G. hirsutum (2n = 52)	Anther callus	Haploid and diploid callus	Barrow et al. (1978)
G. arboreum (2n = 26)	Ovule callus	Haploid to hexaploids	Bajaj and Gill (1985)
G. arboreum × *G. anomalum* *G. hirsutum* × *G. arboreum*	Hybrid embryo callus	26 to about 80 chromosomes, less than 26 to more than 45	Bajaj and Gill (1985)
G. herbaceum (2n = 26)	Hypocotyl callus	Diploid to hexaploids	Bajaj et al. (1986)
G. arboreum	Anther callus	Haploid (3.5%) to hexaploid (1.5%), mostly diploid (38.3%) and tetraploid (34.8%)	Bajaj and Gill (1989)
G. hirsutum	Different explants	Regenerated plants showed 49–53, mostly aberrant in phenotype, 84% plants failed to set bolls	Stelly et al. (1989) Li et al. (1989)
G. hirsutum	Hypocotyl callus	Cell lines/plants resistant to high temperature	Trolinder and Shang (1991)
G. hirsutum	Hypocotyl callus	Several mitotic abnormalities, 16 to 78 chromosomes in aneuploids	Zheng Sijun (1992)

raised from the embryos, ovules, anthers, and hypocotyl segments of *G. arboreum* and *G. herbaceum* were fast-growing. In both species the ovule-derived callus grew faster than that from the anthers and the hypocotyl. *G. herbaceum* callus was whitish, whereas that of *G. arboreum* was reddish – the color of the donor variety, G 27. The extent of genetic variability in terms of chromosome numbers is shown in Figs. 1 and 2. The basic chromosome number in both species under study is 26 (2n); however, in the callus, the number ranged from haploid (13) to a highly polyploid (>104).

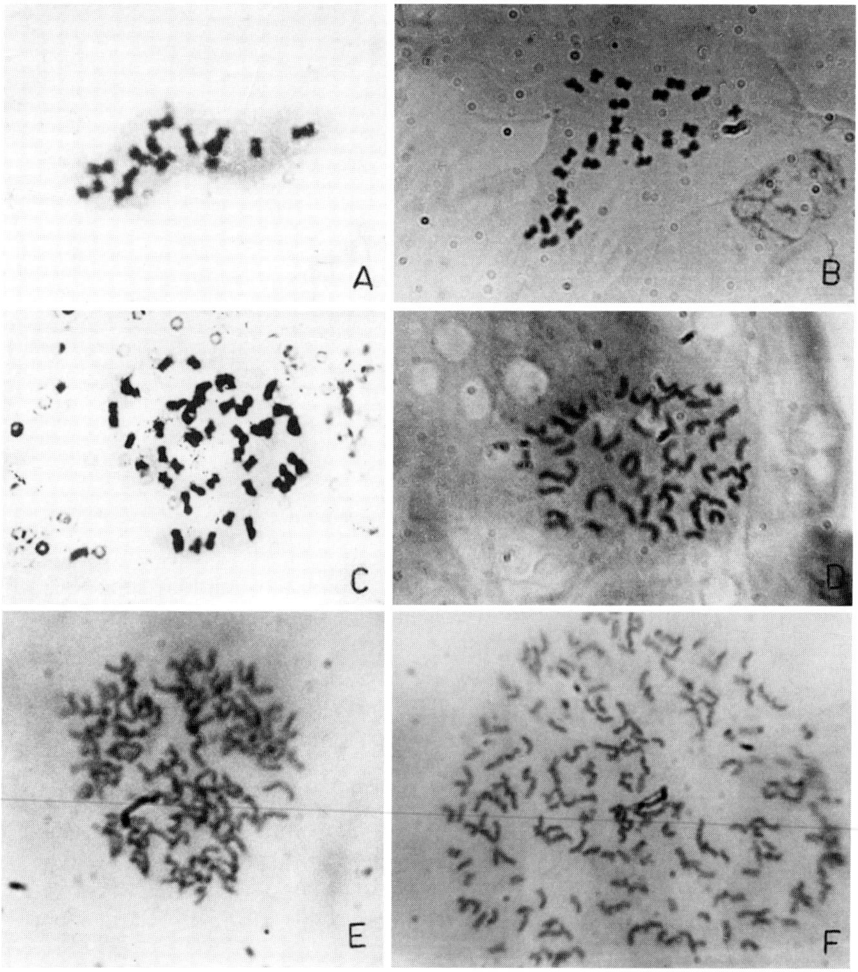

Fig. 1A–F. Genetic variability in ovule- and embryo-derived callus, showing chromosome numbers varying from haploid to highly polyploid. **A** Haploid (n = 13). **B** Diploid (2n = 26). **C** Triploid (3n = 39). **D** Tetraploid (4n = 52). **E** Polyploid (6n = 78). **F** Highly polyploid. (Bajaj and Gill 1985)

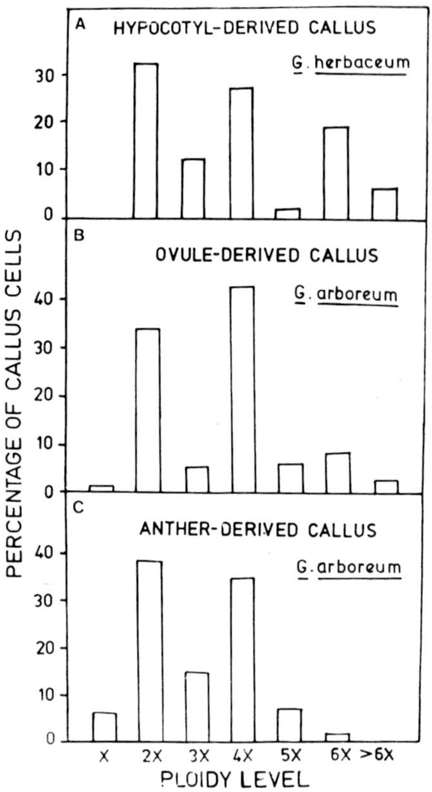

Fig. 2A–C. Histograms showing the extent of genetic variability of callus cells obtained from excised hypocotyl, ovules, and anthers of various species of *Gossypium*. Data based on 500 dividing cells in each case. (Bajaj and Gill 1985)

Hypocotyl-Derived Callus. The ploidy level of callus cells varied from haploid to more than hexaploid (6n = 78). The frequency of haploid cells was quite low (0.74%). Most of the cells had a diploid (45.2%) or tetraploid (40.5%) number. Triploid (4.6%), pentaploid (2.6%), hexaploid (4.4%), or even a higher ploidy than hexaploid (2.9%) cells were also observed (Fig. 2A).

Variability in chromosome number and structures of hypocotyl-derived cell cultures of *G. hirsutum* was also observed by Zheng Sijun (1992). There was a wide range in chromosome number varying from diploid to hexaploid in euploids, and 16 to more than 78 in aneuploid lines. The normal tetraploid cells were found in only 21%. Several mitotic abnormalities, such as multinuclei and micronucleus in interphase and chromosome rings and telophase bridges in the mitotic stage, etc. were observed; 85% of the cells in suspension culture were normal. The variation in cultures tended to increase with time in culture.

Ovule-Derived Callus. The range of genetic variability was greater in ovule-derived callus of *G. herbaceum* than in that of *G. arboreum*. In *G. herbaceum* callus, the haploid, triploid, pentaploid, hexaploid, and higher than hexaploid

Table 2. Variability in chromosome numbers in hybrid callus cells obtained from embryos of a cross *Gossypium hirsutum* (2n = 52) × *G. arboreum* (2n = 26). Data based on 477 dividing cells. (Bajaj and Gill 1985)

Chromosome no.	No. of cells studied	Age (%)
25	7	1.47
26–29	70	14.67
30–33	105	22.01
34–37	161	33.75
38–41	123	25.79
42–45	7	1.47
45	4	0.84

were present in higher frequency, whereas *G. arboreum* callus had a higher number of diploid and tetraploid cells. In general, ovule-derived callus of both species contained cells with a more variable number of chromosomes than the hypocotyl-, or anther-derived callus (Fig. 2B).

Anther-Derived Callus. The callus cells obtained from young anthers (Bajaj 1982) showed a wide range of chromosome numbers, varying from haploid (n = 13) to hexaploid (6x = 78). In most of the cells (73.1%), the number was close to the diploid (38.3%) or tetraploid (34.8%) level. Cells with a haploid (3.5%), triploid (15.1%), pentaploid (6.8%), and hexaploid (1.5%) number of chromosomes were present at relatively low frequencies (Fig. 2C), whereas cells with more than 78 chromosomes were not present (Bajaj and Gill 1989).

Hybrid Embryo-Derived Callus. The embryos of a cross, *G. hirsutum* × *G. arboreum*, (Gill and Bajaj 1987) proliferated to form callus. This was then subcultured periodically on MS + NAA (2 mg/l) + BAP (0.5 mg/l). The callus was creamy-white, soft, and friable. Cytological examination of 1-year-old callus showed a wide range of chromosome number, varying from less than 26 to more than 45 (Table 2, Fig. 3). However, the majority of the cells belonged to the category 26–41. Cells with less than 26, and more than 45 chromosomes were rare, i.e., 1.4 and 0.84%, respectively (Bajaj and Gill 1985).

2.2 Variation in Regenerated Somaclonal Plants

Stelly et al. (1989) studied somaclonal variation in *G. hirsutum* plants regenerated from callus cultures. Thirty-five of the 117 somaclones regenerated from 18-month-old callus cultures were analyzed for meiotic abnormalities. The populations were extremely varied in phenotype, most plants being strikingly aberrant in phenotype. Chromosome numbers in 32 plants studied ranged from 49 to 53, and only one plant was hyperaneuploid. No plant was polyploid. The high frequencies of aneuploidy and tertiary monosomy indicate that cytogenetic anomalies are a major source of somaclonal variation in cotton.

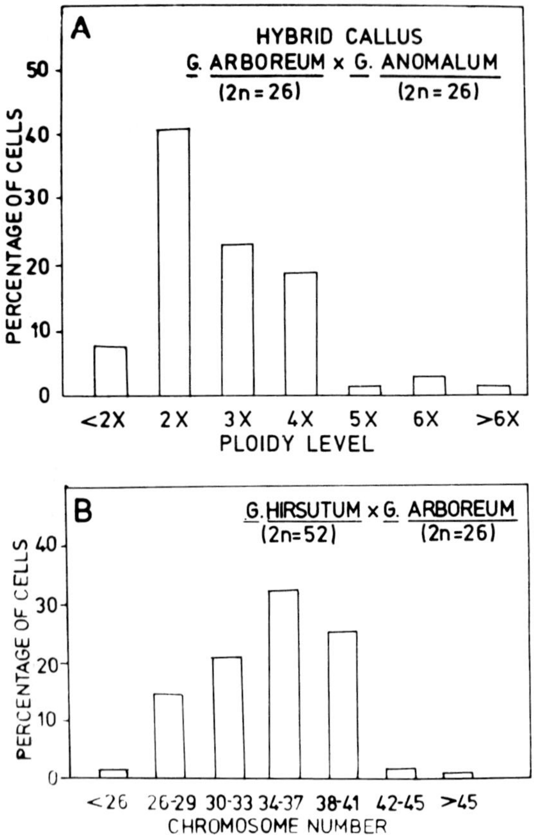

Fig. 3A,B. Chromosomal variability in interspecific embryo culture-derived hybrid callus of *G. arboreum* × *G. anomalum*, and *G. hirsutum* × *G. arboreum*. (Bajaj et al. 1986)

Many somaclones (34%) formed strikingly abnormal flowers that lacked a corolla and an androecium, and sometimes a gynoecium, but included numerous abnormal small bract-like floral parts. Fertility was generally poor. Although ample numbers of insects assured cross-pollination of the plants in the field, 84% of the somaclonal plants failed to set any bolls. Trolinder and Goodin (1988) also observed a similar rate of infertility (84.6%) among cotton somaclones.

2.3 High Temperature-Resistant Somaclones

Cotton is generally grown in areas of the world that are subject to recurring periods of high temperature and drought stress. Pollen fertility is often reduced by excessive temperatures, and drought stress affects both quantity and quality of the fiber produced. In this regard, selection of high temperature-resistant somaclonal cells and their subsequent regeneration is of great significance. Trolinder and Shang (1991) regenerated high temperature-resistant plants from selected somaclones of *G. hirsutum* cv. Coker. The procedure involved exposing cell suspensions to various temperature: time treatments in order to select cell lines resistant to high-temperature stress. Cells were exposed to 45 °C for 3 h each day until the total accumulated hours of stress were 0, 10, 75, 100, or 105 h. After stress treatment, the cells were plated onto embryo development medium, and plants were regenerated. The embryogenic calli that were recovered were subcultured monthly for 6 months and tested for increased resistance to the temperature: time treatments previously determined to be lethal and to water stress as imposed by PEG. All of the selected cell lines were more resistant to both types of stress than the control cell lines. At least ten plants per cell line, including the nonselected control line, were regenerated and tested for resistance of leaf and petiole tissue to high-temperature stress. Leaf tissue of stress-selected plants formed and maintained callus growth when incubated at 38 °C; whereas tissue excised from nonselted controls rarely formed callus or quickly became necrotic.

2.4 Salt-Tolerant Cell Lines

Salinity induces nutritional imbalance (Leidi et al. 1991) and affects cotton yield and fiber quality (Razzouk and Whittington 1991). Thus, induction of genetic variability for salt tolerance in cotton is an important aspect. Although extensive studies have been conducted on this aspect (see Chap. IV.2, this Vol.), in vitro induction of genetic variability has its own advantages. Somaclonal variation for salt tolerance has been induced in a number of agricultural crops (Tal 1990, 1994), and completely, salt-tolerant plants regenerated (Bajaj and Gupta 1987). The work on cotton has recently been initiated. Trolinder and Allen (1994) observed that transgenic cotton cells overexpressing Mn-superoxide dismutase grew better than control cells when cultured in polyethylene glycol-containing medium. Gossett et al. (1994) studied the effect of NaCl on antioxidant enzyme activites in callus tissue of salt-tolerant and salt-sensitive cotton cultivars, and linked antioxidant enzyme activites to salt tolerance. The results from their study also showed that varietal differences in salt tolerance observed in whole plants were also apparent at the cellular level.

2.5 Nematode Resistance in Cotton

Nematodes cause tremendous losses to cotton yield and quality. In 1992, in the USA alone, nematodes accounted for yield loss of 528 000 bales valued at US$ 184 951 200 (Goodell 1993). Later, studies (Creech et al. 1995; Jenkins et al. 1995) were conducted on penetration, reproduction, and development of *Meloidogyne incognita* in cotton. However, gnotobiotic studies through cell and tissue culture offer a number of advantages (Tanda et al. 1980); especially the somaclones selected from these cultures can be regenerated into nematode-resistant plants. Studies conducted on potato and eggplant have resulted in lines showing resistance to nematodes (Wenzel and Uhrig 1981; Gleddie, et al. 1985; Fassuliotis 1990).Such studies in cotton, in future, may lead to the regeneration of nematode-resistant plants.

3 Summary and Conclusion

The work on various chromosomal changes in callus cultures and the regeneration of somaclonal plants in cotton is summarized. The callus derived from various explants, such as hypocotyls, anthers, ovules, hybrid embryos, etc. of both diploid and tetraploid *Gossypium* species showed a wide range of variation in chromosome number and structure. The number varied from haploid to more than hexaploid, highly polyploids and aneuploids. Several mitotic abnormalities, such as multinucleate and micronuclei in interphase, chromosome, rings, and telophase bridges, etc. were observed (Zheng Sijun 1992).

Hybrid (*G. hirsutum* × *G. arboreum*) embryo-derived callus is a rich source of variability, and may augment genetic exchange in inter-specific hybrids (Bajaj and Gill 1985).

The plants regenerated from somaclones varied extremely in phenotype. Many somaclonal plants formed abnormal flowers, fertility was generally poor, and 84% plants failed to set bolls (Stelly et al. 1989).

High temperature-resistant cotton plants were regenerated from selected embryogenic somaclones (Trolinder and Shang 1991). Leaf tissue from stress-selected plants formed and maintained callus growth when incubated at 38 °C, whereas tissue excised from nonselected controls rarely formed callus or quickly became necrotic. In addition, salt-resistant cell lines have been obtained.

To conclude, the in vitro induction of genetic variability in callus cultures, the regeneration of plants with various types of morphology, showing resistance to salt, high temperature, insects, and herbicides, and male sterile lines would help to enrich the existing gene pool of cotton for exploitation in crop improvement.

References

Bajaj YPS (1982) Survival of anther- and ovule-derived cotton callus frozen in liquid nitrogen. Curr Sci 51:139–140

Bajaj YPS (1990) Somaclonal variation – origin, induction, cryopreservation, and implications in plant breeding. In: Bajaj YPS (ed) Biotechnology in agriculture and forestry, vol 11. Somaclonal in crop improvement I. Springer, Berlin Heidelberg New York, pp 3–48

Bajaj YPS (ed) (1996) Biotechnology in agriculture and forestry, vol 36. Somaclonal variation in crop improvement II. Springer, Berlin Heidelberg New York

Bajaj YPS,Gill MS (1985) In vitro induction of genetic variability in cotton (*Gossypium* spp.). Theor Appl Genet 70:363–368

Bajaj YPS, Gill MS (1989) Pollen-embryogenesis and chromosomal variation in anther culture of a diploid cotton (*Gossypium arboreum* L.). SABRAO J 21:57–63

Bajaj YPS, Gupta RK (1987) Plants from salt-tolerant cell lines of napier grass *Pennisetum purpureum* Schum. Indian J Exp Biol 25:58–60

Bajaj YPS, Gill MS, Mahapatra D (1986) Somaclonal and gametoclonal variation in wheat, cotton and *Brassica*. In: Semal J (ed) Somaclonal variations and crop improvement. Martinus Nijhoff, Dordrecht, pp 160–169

Barrow J, Ketterman F, William D (1978) Haploid and diploid callus from cotton anthers. Crop Sci 18:619–622

Bayley C, Trolinder N, Ray C, Morgan M, Quisenberry JE, Ow DW (1992) Engineering 2,4-D resistance into cotton. Theor Appl Genet 83:645–649

Creech RG, Jenkins JN, Tang B, Lawrence GW, McCarty JC (1995) Cotton resistance to root-knot nematode: I. Penetration and reproduction. Crop Sci 35:365–368

D'Amato F (1985) Cytogenetics of plant cell and tissue cultures and their regenerates. CRC Crit Rev Plant Sci 3:73–112

Fassuliotis G (1990) Somaclonal variation for nematode resistance. In: Bajaj YPS (ed) Biotechnology in agriculture and forestry, vol. II. Somaclonal variation in crop improvement I. Springer, Berlin Heidelberg New York, pp 258–268

Gill MS, Bajaj YPS (1987) Hybridization between diploid (*Gossypium arboreum*) and tetraploid (*Gossypium hirsutum*) cotton through ovule culture. Euphytica 36:625–630

Gleddie S, Fassuliotis G, Keller WA, Setterfield G (1985) Somatic hybridization as a potential method for transferring nematode and mite resistance into eggplant. Z Pflanzenzucht 94:348–351

Goodell PB (1993) Cotton nematodes. In: Cotton nematodes – our hidden enemies. Beltwide Cotton Nematode Survey and Education Committee. The Cotton Foundation, Memphis, TN, pp 1–3

Gossett DR, Millhollon EP, Lucas MC, Banks SW, Marney MM (1994) The effects of NaCl on antioxidant enzyme activities in callus tissue of salt-tolerant and salt-sensitive cotton cultivars (*Gossypium hirsutum* L.). Plant Cell Rep 13:498–503

Jenkins JN, Creech RG, Tang B, Lawrence GW, McCarty JC (1995) Cotton resistant to root-knot nematode: II. Post-penetration development. Crop Sci 35:369–373

Larkin PJ, Scowcroft WR (1981) Somaclonal variation – a novel source of variability from cell culture for plant improvement. Theor Appl Genet 60:197–214

Leidi EO, Nogales R, Lips SH (1991) Effect of salinity on cotton plants grown under nitrate or ammonium nutrition at different calcium levels. Field Crops Res 26:35–44

Li R, Stelly D, Trolinder N (1989) Cytogenetic abnormalities in cotton (*Gossypium hirsutum* L.) cell cultures. Genome 32:1128–1134

Perlak FJ, Deaton R, Armstrong T, Fuchs R, Sims S, Greenplate J, Fischoff D (1990) Insect-resistant cotton plants. Biotechnology 8:939–943

Razzouk S, Whittington WJ (1991) Effects of salinity on cotton yield and quality. Field Crops Res 26:305–314

Reynaerts A, De Sonville A (1994) Potential application of engineered male sterility in production of cotton hybrids. In Cotton biotechnology, REUR Tech Ser 32. FAO, Rome, pp 37–42

Stelly DM, Altman DW, Kohel RJ, Rangan TS, Commiskey E (1989) Cytogenetic abnormalities of cotton somaclones from callus cultures. Genome 32:762–770

Tal M (1990) Somaclonal variation for salt resistance. In: Bajaj YPS (ed) Biotechnology in agriculture and forestry, vol II. Somaclonal variation in crop improvement I. Springer, Berlin Heidelberg New York, pp 236–257

Tal M (1994) In vitro selection for salt tolerance in crop plants. Theoretical and practical considerations. In Vitro Cell Dev Biol Plant 30:175–180

Tanda AS, Atwal AS, Bajaj YPS (1980) Reproduction and maintenace of root-knot nematode (*Meloidogyne incognita*) in excised roots and callus cultures of okra (*Abelmoschus esculentus*). Indian J Exp Biol 18:1340–1342

Trolinder NL, Allen RD (1994) Molecular approaches to stress tolerance in cotton. Intern Cotton Res Conf. Brisbane, Australia, p 57

Trolinder NL, Goodin JR (1988) Somatic embryogenesis in cotton (*Gossypium*). II. Requirements for embryo development and plant regeneration. Plant Cell Tissue Organ Cult 12:43–54
Trolinder NL, Shang Xaiomin (1991) In vitro selection and regeneration of cotton resistant to high temperature stress. Plant Cell Rep 10:448–452
Wenzel G, Uhrig H (1981) Breeding for nematode and virus resistance in potato via anther culture. Theor Appl Genet 59:333–340
Zheng Sijun (1992) Chromosome variation in callus culture of *Gossypium hirsutum* L. ••

IV.2 Salt and Water Stress-Tolerant Cotton

E.O. LEIDI and J. GORHAM[1]

1 Introduction

Salinity and drought are the main constraints to crop productivity in many regions of the world with semiarid climates. The response to water or salt stress and the impact on productivity varies depending on the intensity of the stress, the time of appearance, climatic factors, etc.

Adaptation to stress is a complex phenomenon which integrates different responses. When considering tolerance from an agronomic point of view, the complexity of the phenomenon is even greater. Both types of stress frequently appear associated and share some common aspects. In this chapter are summarized some physiological attributes related to stress adaptation, plant characters associated with biological and agronomic tolerance to saline and water stress, as well as the prospects for a biotechnological approach.

2 Salt Stress

Cotton is considered rather tolerant to salinity (Mass 1986), and yield may even be increased by low salinities (Joham 1955). It is grown on saline and reclaimed saline soils, where the presence of soluble salts affects yield of more sensitive crops. Salinity induces nutritional imbalances (Kent and Läuchli 1985; Leidi et al. 1991, 1992; Martinez and Läuchli 1994) and affects cotton growth, yield (Thomas 1980; Nawar et al. 1994), and fiber quality (Razzouk and Whittington 1991). Tolerance to salinity varies from germination to vegetative stage: cultivars that appear fairly tolerant during germination can be sensitive during the vegetative stage (Läuchli et al. 1981) and there is a considerable genotypic variation for tolerance at both growth stages (Abul-Naas and Omran 1974; Leidi 1994). The response is also sensitive to environmental conditions such as relative humidity, nutrient (particularly calcium) supply, and temperature (Neimann and Poulsen 1967; Gorham and Bridges 1995). This differential behavior indicates the importance of choosing suitable selection criteria and screening methods for salt tolerance.

[1] Instituto de Recoursos Naturales y Agrobiologia de Sevilla, Reina Mercedes 10, Apartado 1052, 41080 Sevilla, Spain

Biotechnology in Agriculture and Forestry, Vol. 42
Cotton (ed. by Y.P.S. Bajaj)

2.1 Ion Exclusion or Inclusion?

In wild cotton species (Gorham and Young 1996) and in *Gossypium barbadense* (Rathert 1982), a negative relation between Na uptake and tolerance has been found. However, for some *G. hirsutum* genotypes, positive correlations between Na accumulation and salt tolerance were observed (Läuchli and Stelter 1982; Saiz and Leidi 1994; Leidi and Saiz 1997). This differential behavior to salinity between *Gossypium* species reflects the diversity of mechanisms involved in plant adaptation to stress (Leidi and Saiz 1997; Leidi and de Castro 1997). Using vacuolar membranes from roots of a salt-sensitive cotton line, Lin et al. (1997) have shown that the proton ATPase was unaffected by Na, suggesting that salt-sensitivity might result, in part, from a lack of the driving force required for Na compartmentation. Uptake of Na and Cl from the saline medium and accumulation in stems and leaves (with adequate ion compartmentation) could provide an inexpensive system for osmotic adjustment in comparison with the accumulation of organic solutes. Two cotton genotypes ranked as sensitive and tolerant at seedling stage under controlled conditions differed in Na and K uptake and distribution (Saiz and Leidi 1994). Salt tolerance was positively correlated with Na uptake, but other physiological differences, such as higher water contents and K redistribution in the tolerant genotype, were observed (Leidi and Saiz 1997). Differences in water uptake, Na levels, and water potential could be partially responsible for differential sensitivity to salinity (Leidi and de Castro 1997).

A further complication is compartmentation of ions between different tissues. This is shown in Fig. 1 for the effect of leaf age on ion concentrations in leaves, and in Fig. 2 for partitioning between leaves, petioles, and stems. X-ray microanalysis of bulk frozen tissues demonstrates considerable differences in distribution of elements between different cell types, and even between different cells (J. Gorham, in prep.). Since there is little information on subcellular or apoplastic compartmentation in cotton, interpretations of the effects of gross ion accumulation in shoots on growth may be misleading. Kerimov et al. (1993) found that of two cultivars, the one with the greatest sensitivity to salt at the whole plant level was the most resistant as callus tissue. Thus, tissue tolerance of salinity may not be apparent at the level of the whole plant.

2.2 Organic Osmolytes and Protective Compounds

Organic compounds may have a function in adaptation to stress (Wyn Jones and Gorham 1983). High levels of sugars and amino acids were detected in salt-stressed cotton genotypes, but their levels did not correlate with salt tolerance (Saiz and Leidi 1994). Compatible solutes like quaternary ammonium compounds and polyols may have a protective role in plants subjected to salt stress. At least under greenhouse conditions, proline and polyol accumulation are insignificant in salt-stressed cotton. In contrast, glycinebetaine accumulates to high concentrations (sometimes in excess of $100\,mol/m^3$), both in the laboratory and in the field (Gorham 1995, 1996). Glycinebetaine accumu-

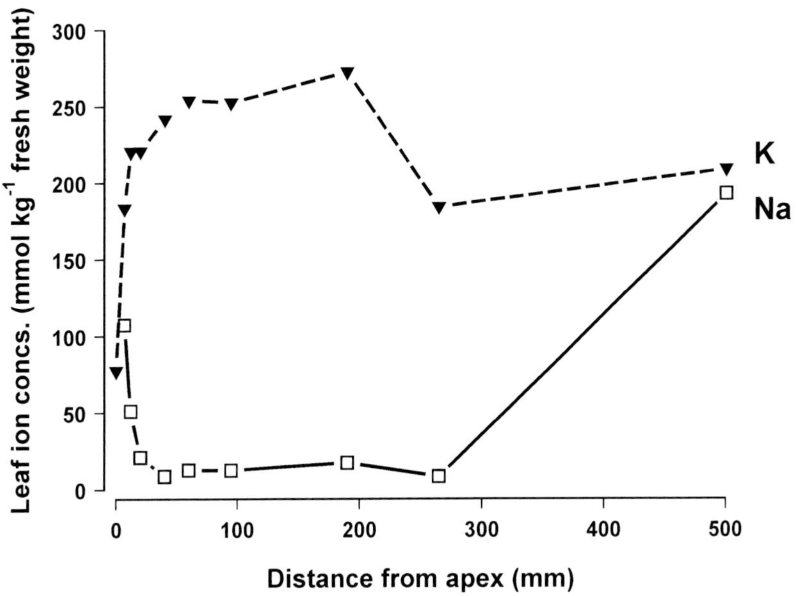

Fig. 1. Changes in leaf ion concentrations with distance from the shoot apex (i.e., with leaf age) in *Gossypium longicalyx* grown at 200 mol/m^3 NaCl

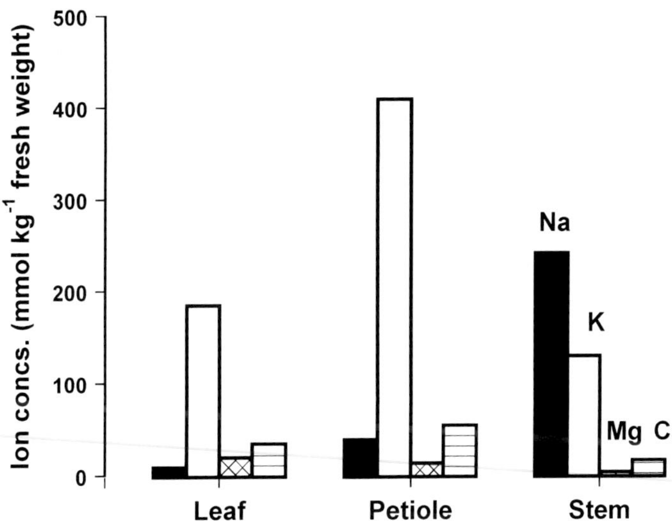

Fig. 2. Ion concentrations in leaves, stems, and petioles 270 mm behind the shoot apex of *Gossypium longicalyx* grown at 200 mol/m^3 NaCl

lation is a characteristic of most members of the Malvaceae, even under non-stressed conditions. The quantities involved suggest that most of the glycinebetaine is contained in vacuoles, i.e., it is not only a compatible cytoplasmic solute but also contributes to overall osmotic adjustment.

2.3 Mechanism of Growth Reduction

As in other species, it appears that the most immediate effect of increased salinity is osmotic, but that osmotic adjustment is quite rapid. In the medium term, reduction of leaf area through inhibition of cell extensibility is more important than reduced stomatal conductance, carbon dioxide fixation, or nutrient acquisition. Chlorophyll fluorescence analysis at this stage shows little effect of salinity on electron transport associated with photosystem II. Toxic effects of salt, evidenced by loss of chlorophyll, are only apparent with prolonged exposure and high salinities.

3 Drought Stress

Much of the research on the effects of drought in cotton has dealt with the effects of water deficit on photosynthesis (Berlin et al. 1982; Hutmacher and Krieg 1983; Radin et al. 1988; Ephrath et al. 1990, 1993), stomatal conductance (Petersen et al. 1991; Radin 1992), and osmotic adjustment (Ackerson and Hebert 1981; Oosterhuis and Wullschleger 1987a). However, there is little information on mechanisms of drought tolerance and genetic variability of traits related to drought resistance.

Under water stress, cotton plants reduce shoot and root growth differentially, increasing the root/shoot ratio (Malik et al. 1979; Ball et al. 1994). This general phenomenon is considered to be triggered in response to drought sensing by roots and their export of ABA to shoots (Davies et al. 1994). The adaptation to drought brings about a change in plant architecture which improves water economy by increasing water uptake capacity and reducing transpiration. As a result of the reduction in leaf area and stomatal conductance, photosynthetic activity is decreased, reducing in turn fruiting-site development, number of floral buds and fruits retained, and final yield (McMichael and Hesketh 1982; Turner et al. 1986). Physiological differences between cotton genotypes which have been related to drought tolerance are reviewed below.

3.1 Traits Related to Drought Tolerance

3.1.1 Roots

Root growth has long been considered an important trait for drought tolerance. Water uptake capacity provided by soil exploration by roots and its adaptative responses in drying soils is a main conditioning factor for water-stress tolerance. At present, the recognition of the root as a sensor of water-deficit stress, and the first step in the chain reaction of the adaptative response, underlines its importance in drought-tolerance mechanisms. Research groups working in drought tolerance have been studying this trait in cotton germplasm (Quisenberry et al. 1981; de Souza et al. 1983; Leidi and Gutierrez 1993; Quisenberry and McMichael 1996). There is significant variability in cotton germplasm for partitioning of total biomass into shoot and root (McMichael and Quisenberry 1991). Deep rooting has been associated with higher yield under dryland conditions (Lopez et al. 1995). Genetic studies have shown heritability of 37% for root growth rate (de Souza et al. 1983) and additive, dominance, and additive by additive epistatic effects were present in characters like root length and relative root weight (Eissa et al. 1983).

Cotton has deep roots (>1.5 m), but there appears to have been little systematic investigation of variation for this character. Meyer and Ritchie (1980) reported on the effect of taproot length on plant water status in cotton. Rooting depth is affected by soil conditions and cultivation (Vepraskas and Guthrie 1992; Bland 1993; Aydin 1994; Mullins et al. 1994). Work in Israel suggests that the early establishment of a deep root system should be encouraged, and that regular drip irrigation left cotton plants susceptible to short-term water stresses (Carmi et al. 1992, 1993). Soil water extraction profiles in cotton have been described by Jacobson et al. (1994). McMichael et al. (1985, 1987) suggested that the number of vascular bundles in roots might be related to hydraulic conductivity – and hence water supply to the shoots. One mechanism for this could be the production of more lateral roots (McMichael et al. 1987), since other work suggested that water flux in the primary root was not related to the number of vascular bundles, at least in young seedlings (Oosterhuis and Wullschleger 1987b; Oosterhuis et al. 1988).

3.1.2 Photosynthesis and Related Traits

In cotton genotypes with different leaf morphology, differences in photosynthesis and water relations were observed in drought-stressed treatments (Karami et al. 1980). Differences in photosynthetic behavior under water stress may simply reflect the plants' capacity for water uptake (Lopez et al. 1993a). Under dryland conditions, a genotype with the highest photosynthesis under severe water stress was the one that had deeper roots (Lopez et al. 1995).

The inhibition of photosynthesis by stomatal closure shows genotypic variation (Cornish et al. 1991; Leidi et al. 1993; Lopez et al. 1993a; Pettigrew and Meredith 1994). This variability raises the possibility of improving water use efficiency (WUE) (Cornish et al. 1991) and of increasing productivity under conditions of limited water availability. On the other hand, work on Pima cotton lines (*G. barbadense*) indicates a relationship between high stomatal conductance and yield (Lu and Zeiger 1994). Maintenance of leaf turgidity under water stress, which is obviously related to maintenance of photosynthesis, has been suggested as a potential trait to select for drought tolerance (Quisenberry et al. 1985). Maintenance of turgidity was related to reduced leaf conductances and higher biomass accumulation (Quisenberry et al. 1985). However, in a previous report (Quisenberry et al. 1982), a positive correlation between growth and maintenance of transpiration at low relative water content was indicated. Radin (1992) reviewed the differences in stomatal response to water stress in cotton grown in the field and in controlled environments. Stomatal sensitivity to abscisic acid may change following water stress (Peng and Weyers 1994), again indicating the need to consider stress acclimation effects.

While individual leaf water loss can be estimated quickly and reasonably reliably, it is more difficult to integrate these measurements to quantify whole-plant water loss. Effects of leaf age, variable leaf damage, self-shading, plant architecture, etc. need to be considered. Whole-plant water loss may be measured using heat-pulse or heat-balance sap-flow measurements in the stems. Sap-flow measurements have been used to examine genetic variation in sugarcane (Saliendra and Meinzer 1992) and apple rootstocks (Hussein and McFarland 1994), and the technique has been used successfully with cotton (Dugas 1990; Petersen et al. 1992; Dugas et al. 1994). However, the expense and low number of observations which can be made limit the application of this technique in screening programs.

Assessments of water-saving and water-spending strategies need to be made in relation to the development of water deficit in different environments. The picture is further complicated by the relationship between water deficit and high-temperature stresses. At high temperatures (>45 °C) transpirational cooling of the leaves is an important factor, reducing leaf temperatures by 5–10 °C below ambient (see below). Stomatal closure induced by water deficit would therefore render cotton more susceptible to heat damage.

Genetic variation for water-use efficiency, measured as total biomass (shoot and root) produced per unit of water transpired, has been found in modern cultivars, primitive races, and different species (Quisenberry and McMichael 1991). At present, analysis of $^{13}C/^{12}C$ discrimination seems to be a suitable system for searching for higher WUE in cotton genotypes (Yakir et al. 1990; McDaniel and Dobrenz 1994; Gerik et al. 1996), although it still requires more evaluation.

Chlorophyll fluorescence measurements have been used to monitor water deficits in a variety of crops. In cotton, low day and night temperatures also affect chlorophyll fluorescence (Bilger and Bjorkman 1991; Winter and Koniger 1991; Warner and Burke 1993). Data from experiments in greenhouse and field indicated that Fv/Fm ratios were not affected until severe water deficit was experienced (Gorham et al. 1995). Genty et al. (1987) also reported that variable chlorophyll fluorescence, at least under some conditions, was relatively insensitive to mild drought stress. Atmospheric CO_2 concentrations also affect chlorophyll fluorescence in cotton (Betsche 1994).

3.1.3 Other Components of Drought Tolerance

Adaptation to drought by reducing leaf area and increasing specific leaf weight (SLW) might bring some advantages to the plant water economy (Fischer and Turner 1978). Thicker lamina (mainly by a taller palisade cells) and higher values of SLW (Fig. 3) were associated with higher yield under drought (Lopez et al. 1995). However, studying the variability for this trait in a wider range of genotypes, no correlation with yield was observed (E.O. Leidi, in prep.).

Growth habit also has an effect on cotton drought tolerance: indeterminate cultivars of cotton were the most productive under stress (Quisenberry and Roark 1976). This is related to the higher yielding potential of genotypes which produce a high number of reproductive nodes (Lopez et al. 1995). Another important effect of drought in cotton is the increase in earliness (Orgaz et al. 1991; Cook and El-Zik 1993). Under certain conditions, an increase in earliness could be useful as a drought-escape mechanism (Ludlow 1989). However, this trait has not been associated with higher yield under drought (Cook and El-Zik 1993; Lopez et al. 1995). The final yield depends on the number of fruits that are set and reach maturity. The relative sensitivity of yield to water stress changes during the reproductive phase (de Kock et al. 1990). Differences in boll abscission rates at different nodes, as well as in boll number and boll weight, have been observed among cotton genotypes under drought (Lopez et al. 1995).

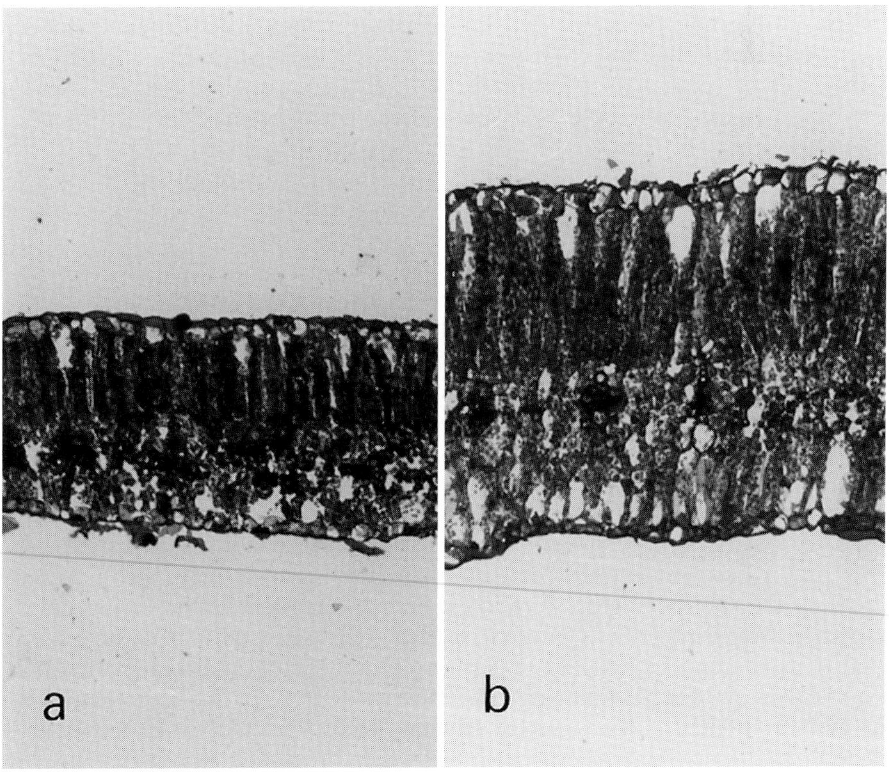

Fig. 3a,b. Transverse sections of leaves from **a** drought-sensitive and **b** drought-tolerant cotton cultivars grown under dryland conditions. Specific leaf weights ranged from 0.8 to 1.1 g/dm^2, respectively

Osmotic adjustment is an adaptative process which helps to maintain turgor at low water potential (Radin 1983). It has been suggested as a potential trait for selecting drought-tolerant crops (Ludlow and Muchow 1990), although in some studies negative correlations were found with growth in cotton genotypes (Quisenberry et al. 1984; E.O. Leidi, in prep.). Osmotic adjustment in the strictest sense involves the active accumulation or synthesis of solutes in response to water deficit; in the case of cotton these are mainly potassium and glycinebetaine. The resulting decrease in cellular solute potential helps to maintain turgor at low water potentials. The same result can also be achieved to some extent by lowering the water content of the plant. Osmotic adjustment of cotton leaves and roots has been reported (Cutler and Rains 1979, Oosterhuis 1985; Oosterhuis and Wullschleger 1987a). In cases where osmotic adjustment persists for a considerable time after rewatering, the plants are considered to be stress-adapted, and theoretically more resistant to subsequent stresses. Stress adaptation has been observed in cotton by several groups (Ackerson 1981, 1985; Ackerson and Hebert 1981; Oosterhuis and Wullschleger 1988). Genetic variation for osmotic adjustment in wild and cultivated *Gossypium* species was described by Oosterhuis et al. (1987).

Several other processes related to plant reaction to drought stress, like proline accumulation (Ferreira et al. 1979; Kumar et al. 1987; Singh and Sahay 1991; Lopez et al. 1993b) or susceptibility to drought-induced damage (de Souza et al. 1984) have been studied as possible indicators for drought tolerance.

3.2 Other Processes

Drought induces stomatal closure, which results in reduction in transpiration and an increase in leaf temperature. Leaf canopy temperature, and, more particularly, the difference between leaf canopy and air temperatures caused by transpirational cooling (and related to vapor pressure deficit in a crop water stress index (CWSI), have been used as indicators of crop water stress. Leaf canopy temperature has been used as an indicator of the need for irrigation in cotton (Stockle and Dugas 1992; Wanjura et al. 1992a,b; Wanjura and Mahan 1994). This increase in leaf temperature in water-stressed plants, which can be used to identify genotypes with differential water economy (Hatfield et al. 1987), may have profound effects on plant performance under stress by increasing the rate of photorespiration (Perry et al. 1983) and affecting other metabolic steps and growth patterns (Reddy et al. 1991). It has been suggested that the maintenance of leaf temperature within an optimal range (thermal kinetic window) for physiological functioning is important in cotton (Burke et al. 1988; Hatfield and Burke 1991). In many cotton-producing areas, drought is accompanied by high air temperatures which, in conjunction with reduced evaporative cooling by stomatal closure and high irradiance, can induce high-temperature stress in leaves. Cornish et al. (1991), Lu and Zeiger (1994), Lu et al. (1994), and Radin et al. (1994) studied the genetic variability in

G. barbadense for stomatal conductance in relation to heat adaptation, and developed higher-yielding varieties with greater heat tolerance. Inheritance of stomatal conductance responses was studied in *G. barbadense* (Percy et al. 1996). Selection for higher carbon isotope discrimination could increase yield in cotton grown in hot, high-irradiance, and well-irrigated environments (Lu et al. 1996).

At the molecular level, much attention has been paid to novel proteins which appear after a variety of stresses (heat shock proteins, HSP). HSP accumulation has been reported in field-grown cotton (Burke et al. 1985), and the HSP profiles are cultivar-dependent (de Ronde et al. 1993). Fender and O'Connell (1989) concluded that heritable thermotolerance in genetically characterized heat-tolerant and heat-sensitive cotton lines was not correlated with changes in HSP expression. The role of these HSPs in field resistance to high temperatures is still controversial, and work in Bangor and New Mexico failed to find a correlation between HSP accumulation and heat tolerance in cotton, at least in the short term. However, the HSP response is transient and affected by the thermal history of the plant, and it may be that the ability to continue producing HSPs after a series of heat shocks is more important than the initial response of plants not previously subjected to high temperatures. Heat shock appears to confer increased resistance to a subsequent salt shock (Kuznetsov et al. 1993). Cotton resistant to heat stress has been reported after in vitro selection and regeneration (Trolinder and Shang 1991). Pollen surviving a high temperature treatment has also been used to produce lines of *G. barbadense* with increased heat tolerance (Rodriguez-Garay and Barrow 1988). In cases where high temperatures coincide with early flowering, as in parts of India and Pakistan, setting and retention of early bolls has been used to screen for heat tolerance (Taha et al. 1981).

4 Summary and Future Outlook

Progress in obtaining stress-tolerant cotton can still be made by exploiting existing genetic variability. This progress may be speeded up by using techniques of molecular genetics such as DNA markers (Paterson et al. 1991). The location of genes associated with drought-related traits in rice (Champoux et al. 1995) and maize (Lebreton et al. 1995) by using molecular tools provides an example of an effective way for searching and selecting for such traits in breeding programs.

Somaclonal variation and selection of resistant cell lines is a valid approach for rapid selection for stress tolerance (Nabors 1990). Limitations to in vitro selection for salt tolerance can be overcome providing a useful technique for the development of tolerant plants if found. There are a number of examples of improvement of salt stress resistance by in vitro selection in various species (Bajaj and Gupta 1987; Nabors 1990; Watad et al. 1991; Ibrahim et al. 1992; Tal 1994).

Meanwhile, cloning of genes involved in transport processes in other organisms and their expression in transgenic cotton can simultaneously provide basic knowledge of the different mechanisms involved in tolerance. A gene encoding a transport protein from tonoplasts cloned in cotton (Wilkins 1993) might be related to cell adaptation to stress (Niu et al. 1995). Transgenic cotton cells overexpressing Mn-superoxide dismutase grew better than control cells when cultured in PEG-containing medium (Trolinder and Allen 1994). Gossett et al. (1994, 1996) also linked antioxidant enzyme activity to salt tolerance in cotton tissue culture.

Genetic variability exists in cotton for several drought tolerance-related traits. Interspecific hybridization can create additional genetic variability (Bajaj and Gill 1985; Stewart 1991; Dani 1992; Mirza and Sheikh 1994) for traits with narrow variability in *G. hirsutum*. Production of transgenic plants over-expressing genes induced by drought might be easily tested if more efficient protocols for regeneration were available.

Other approaches which could be tested (or are being tested at present) include the overexpression of late embryogenesis abundant protein (LEA) genes, the overproduction of proline, or synthesis of other organic osmolytes (like mannitol). However, the complex nature of the integrated response to stress, in which many genes are involved, requires pyramiding of individual traits before real progress in tolerance is obtained. Molecular approaches to improving stress tolerance need to be validated at the whole-plant level.

References

Abul-Naas AA, Omran MS (1974) Salt tolerance of seventeen cotton cultivars during germination and early seedling development. Z Acker Pflanzenb 140:229–236

Ackerson RC (1981) Osmoregulation in cotton in response to water stress. II. Leaf carbohydrate status in relation to osmotic adjustment. Plant Physiol 67:489–493

Ackerson RC (1985) Osmoregulation in cotton in response to water stress. III. Effects of phosphorus fertility. Plant Physiol 77:309–312

Ackerson RC, Hebert RR (1981) Osmoregulation in cotton in response to water stress. I. Alterations in photosynthesis, leaf conductance, translocation and ultrastructure. Plant Physiol 67:484–488

Aydin M (1994) Hydraulic properties and water-balance of a clay soil cropped with cotton. Irrig Sci 15:17–23

Bajaj YPS, Gill MS (1985) In vitro induction of genetic variability in cotton (*Gossypium* spp.). Theor Appl Genet 70:363–368

Bajaj YPS, Gupta RK (1987) Plants from salt-tolerant cell lines of napier grass *Pennisetum purpureum* Schum. Indian J Exp Biol 25:58–60

Ball RA, Oosterhuis DM, Mauromoustakos A (1994) Growth dynamics of the cotton plant during water-deficit stress. Agron J 86:788–795

Berlin J, Quisenberry JE, Bailey F, Woodworth M, McMichael BL (1982) Effect of water stress on cotton leaves. I. An electron microscopic stereological study of the palisade cells. Plant Physiol 70:238–243

Betsche T (1994) Atmospheric CO_2 enrichment – kinetics of chlorophyll a fluorescence and photosynthetic CO_2 uptake in individual, attached cotton leaves. Environ Exp Bot 34:75–86

Bilger W, Bjorkman O (1991) Temperature dependence of violaxanthin deepoxidation and nonphotochemical fluorescence quenching in intact leaves of *Gossypium hirsutum* L. and *Malva parviflora* L. Planta 184:226–234

Bland WL (1993) Cotton and soybean root system growth in three soil temperature regimes. Agron J 85:906–911

Bland WL, Dugas WA (1989) Cotton root growth and soil water extraction. Soil Sci Soc Am J 53:1850–1855

Burke JJ, Hatfield JL, Klein RR, Mullet JE (1985) Accumulation of heat shock proteins in field-grown cotton. Plant Physiol 78:394–398

Burke JJ, Mahan JR, Hatfield JL (1988) Crop-specific thermal kinetic windows in relation to wheat and cotton biomass production. Agron J 80:553–556

Carmi A, Plaut Z, Heuer B, Grava A (1992) Establishment of shallow and restricted root systems in cotton and its impact on plant response to irrigation. Irrig Sci 13:87–91

Carmi A, Plaut Z, Sinai M (1993) Cotton root growth as affected by changes in soil water distribution and their impact on plant tolerance to drought. Irrig Sci 13:177–182

Champoux MC, Wang G, Sarkarung S, Mackill DJ, O'Toole JC, Huang N, McCouch SR (1995) Locating genes associated with root morphology and drought avoidance in rice via linkage to molecular markers. Theor Appl Genet 90:969–981

Cook CG, El-Zik KM (1993) Fruiting and lint yield of cotton cultivars under irrigated and nonirrigated conditions. Field Crops Res 33:411–421

Cornish K, Radin JW, Turcotte EL, Lu ZM, Zeiger E (1991) Enhanced photosynthesis and stomatal conductance of Pima cotton (*Gossypium barbadense* L.) bred for increased yield. Plant Physiol 97:484–489

Cutler JM, Rains DW (1979) Effects of water stress and hardening on the internal water relations and osmotic constituents of cotton leaves. Physiol Plant 42:261–268

Dani RG (1992) Biotechnological research of cotton: two decades in Soviet retrospection. Adv Plant Sci 5:433–447

Davies WJ, Tardieu F, Trejo CL (1994) How do chemical signals work in plants that grow in drying soil? Plant Physiol 104:309–314

DeKock J, De Bruyn LP, Humann JJ (1990) The relative sensitivity to plant water stress during the reproductive phase of upland cotton (*Gossypium hirsutum* L.). Irrig Sci 11:239–244

de Ronde JA, van der Mescht A, Cress WA (1993) Heat-shock protein synthesis in cotton is cultivar-dependent. S Afr J Plant Soil 10:95–97

de Souza JG, Da Silva JBV, Neto MB, Giles JA (1983) Velocidade de crescimento da raiz como parámetro de resistencia à seca no algodoeiro. Pesqui Agropecu Bras 18:169–172

de Souza JG, Giles JA, Neto MB, Da Silva JV (1984) Selection for water stress tolerance in upland cotton in the northeast of Brazil. Trop Agric 61:2–4

Dugas WA (1990) Comparative measurement of stem flow and transpiration in cotton. J Appl Climatol 42:215–221

Dugas WA, Heuer ML, Hunsaker D, Kimball BA, Lewin KF, Nagy J, Johnson M (1994) Sap flow measurements of transpiration from cotton grown under ambient and enriched CO_2 concentrations. Agric For Meteorol 70:231–245

Eissa AM, Jenkins JN, Vaughan CE (1983) Inheritance of seedling root length and relative root weight in cotton. Crop Sci 23:1107–1111

Ephrath JE, Marani A, Bradvo BA (1990) Effects of moisture stress on stomatal resistance and photosynthetic rate in cotton (*Gossypium hirsutum*) 1. Controlled levels of stress. Field Crops Res 23:117–131

Ephrath JE, Marani A, Bradvo BA (1993) Photosynthetic rate, stomatal resistance and leaf water potential in cotton (*Gossypium hirsutum* L.) as affected by soil moisture and irradiance. Photosynthetica 29:63–71

Fender SE, O'Connell MA (1989) Heat-shock protein expression in thermotolerant and thermosensitive lines of cotton. Plant Cell Rep 8:37–40

Ferreira LGR, De Souza JG, Prisco JT (1979) Effects of water deficit on proline accumulation and growth of two cotton genotypes of different drought resistance. Z Pflanzenphysiol 93:189–199

Fischer RA, Turner NC (1978) Plant productivity in the arid and semiarid zones. Annu Rev Plant Physiol 29:277–317

Genty B, Briantais JM, Da Silva JBV (1987) Effects of drought on primary photosynthetic processes of cotton leaves. Plant Physiol 83:360–364

Gerik TJ, Gannaway JR, El-Zik KM (1996) Identifying high yield cotton varieties with carbon isotope analysis. In: Proc Beltwide Cotton Prod Res Conf, 9–12 Jan, Nashville, Tennessee, pp 595–596

Gorham J (1995) Betaines in higher plants – biosynthesis and role in stress tolerance. In: Wallsgrove RM (ed) Amino acids and their derivatives in higher plants – biosynthesis and metabolism. Soc Exper Biol Seminar Series, no 56. Cambridge University Press, Cambridge, pp 173–203

Gorham J (1996) Glycinebetaine is a major nitrogenous solute in the Malvaceae. Phytochemistry 43:367–369

Gorham J, Bridges J (1995) Effects of calcium on growth and leaf ion concentrations of *Gossypium hirsutum* grown in saline hydroponic culture. Plant Soil 176:219–227

Gorham J, Young EM (1996) Wild relatives of cotton and rice as sources of stress resistance traits. Proc Eucarpia Meeting on Tropical Plants. CIRAD, Montpellier pp 39–52

Gorham J, Bridges J, Mirza A (1995) Responses of wild cottons and hybrid lines to drought. Proc Interdrought 95, Montpellier, INRA IX 5

Gossett DR, Millhollon EP, Lucas MC, Banks SW, Marney MM (1994) The effects of NaCl on antioxidant enzyme activities in callus tissue of salt-tolerant and salt-sensitive cotton cultivars (*Gossypium hirsutum* L.). Plant Cell Rep 13:498–503

Gossett DR, Banks SW, Millhollon EP, Lucas MC (1996) Antioxidant response to NaCl stress in a control and an NaCl-tolerant cotton cell line grown in the presence of paraquat, buthionine sulfoximine, and exogenous glutathione. Plant Physiol 112:803–809

Hatfield JL, Burke JJ (1991) Energy exchange and leaf temperature behavior of three plant species. Environ Exp Bot 31:295–302

Hatfield JL, Quisenberry JE, Dilbeck RE (1987) Use of canopy temperature to identify water conservation in cotton germplasm. Crop Sci 27:269–273

Hussein IA, McFarland MJ (1994) Rootstock-induced differences in sap flow of Granny Smith apple. Hortscience 29:1120–1123

Hutmacher RB, Krieg DR (1983) Photosynthetic rate control in cotton. Stomatal and non-stomatal factors. Plant Physiol 73:658–661

Ibrahim KM, Collins JC, Collin HA (1992) Characterization of progeny of *Coleus blumei* following an in vitro selection for salt tolerance. Plant Cell Tissue Organ Cult 28:139–145

Jacobson BD, Murray DS, Stone JF (1994) Soil water extraction profiles of cotton (*Gossypium hirsutum*) and weed species. Weed Technol 8:190–198

Joham HE (1955) The calcium and potassium nutrition of cotton as influenced by sodium. Plant Physiol 30:4–10

Karami E, Krieg DR, Quisenberry JE (1980) Water relations and carbon-14 assimilation of cotton with different leaf morphology. Crop Sci 20:421–426

Kent LM, Läuchli A (1985) Germination and seedling growth of cotton: salinity-calcium interactions. Plant Cell Environ 8:155–159

Kerimov F, Kuznetsov VV, Shamina ZB (1993) Organism and cell levels of salt tolerance in two cotton cultivars (133 and Inebr-85). Russian Plant Physiol 40:111–114

Kumar V, Mehta NP, Gohil MD (1987) Investigation into some physiological parameters of cotton under water stress. Ann Arid Zone 26:267–271

Kuznetsov VV, Rakitin VY, Borisova NN, Rotschupkin BV (1993) Why does heat shock increase salt resistance in cotton plants? Plant Physiol Biochem 31:181–188

Läuchli A, Stelter W (1982) Salt tolerance of cotton genotypes in relation to K/Na-selectivity. In: San Pietro A (ed) Biosaline research. A look to the future. Plenum Press, New York, pp 511–514

Läuchli A, Kent LM, Turner JC (1981) Physiological responses of cotton genotypes to salinity. In: Proc Beltwide Cotton Prod Res Conf, 4–8 Jan, New Orleans, p 40

Lebreton C, Lazic-Jancic V, Steed A, Pekic S, Quarrie SA (1995) Identification of QTL for drought responses in maize and their use in testing causal relationships between traits. J Exp Bot 46:853–865

Leidi EO (1994) Genotypic variation of cotton in response to stress by NaCl or PEG. In: Peeters MC (ed) Cotton biotechnology. REUR Technical Series 32, FAO, Rome, pp 67–73

Leidi EO, de Castro A (1997) Inclusión de sodio: principal adaptación relacionada con tolerancia a salinidad en el algodonero. In: Sarmiento-Solís R, Leidi-Montes EO, Troncoso de Arce A (eds) Nutrición mineral de las plantas en la agricultura sostenible. Junta de Andalucía, Sevilla, pp 204–210

Leidi EO, Gutierrez JC (1993) Nota sobre la variabilidad en el crecimiento de raíces de algodón en estado de plántula: Efecto de genotipo y estrés osmótico. Invest Agric Prod Prot Veg 8:165–169

Leidi EO, Saiz JF (1997) Is salinity tolerance related to Na accumulation in Upland cotton (*Gossypium hirsutum*) seedlings? Plant Soil 190:67–75

Leidi EO, Nogales R, Lips SH (1991) Effect of salinity on cotton plants grown under nitrate or ammonium nutrition at different calcium levels. Field Crops Res 26:35–44

Leidi EO, Silberbush M, Soares MIM, Lips SH (1992) Salinity and nitrogen nutrition studies on peanut and cotton plants. J Plant Nutr 15:591–604

Leidi EO, Lopez JM, Lopez M, Gutierrez JC (1993) Searching for tolerance to water stress in cotton genotypes: photosynthesis, stomatal conductance, and transpiration. Photosynthetica 28:383–390

Lin H, Salus SS, Schumaker KS (1997) Salt sensitivity and the activities of the H^+-ATPases in cotton seedlings. Crop Sci 37:190–197

Lopez JM, Leidi EO, Lopez M, Gutierrez JC (1993a) Fotosíntesis, conductancia estomática, eficiencia en el uso del agua y temperatura foliar de cultivares de algodón en respuesta al estrés hídrico. Invest Agric Prod Prot Veg 8:17–27

Lopez M, Gutierrez JC, Leidi EO (1995) Selection and characterization of cotton cultivars for dryland production in the south-west of Spain. Eur J Agron 4:119–126

Lopez M, Leidi EO, Gutierrez JC (1993b) Respuesta de genotipos selecionados de algodón a condiciones de sequía terminal. In: Proc 1er Symp Hispano-Portugués Relaciones Hídricas en las Plantas, Palma de Mallorca, Spain, CSIC-UIB, pp 189–192

Lu ZM, Zeiger E (1994) Selection for higher yields and heat resistance in Pima cotton has caused genetically determined changes in stomatal conductances. Physiol Plant 92:273–278

Lu ZM, Radin JW, Turcotte EL, Percy RG, Zeiger E (1994) High yields in advanced lines of Pima cotton are associated with higher stomatal conductance, reduced leaf area and lower leaf temperature. Physiol Plant 92:266–272

Lu ZM, Chen J, Percy RG, Sharifi MR, Rundel PW, Zeiger E (1996) Genetic variation in carbon isotope discrimination and its relation to stomatal conductance in Pima cotton (*Gossypium barbadense*). Aust J Plant Physiol 23:127–132

Ludlow MM (1989) Strategies of response to water stress. In: Kreeb KH, Richter H, Hinckley TM (eds) Structural and functional responses to environmental stresses. SPB Academic Publ, The Hague, pp 269–281

Ludlow MM, Muchow RC (1990) A critical evaluation of traits for improving crop yields in water-limited environments. Adv Agron 43:107–153

Maas EV (1986) Salt tolerance of plants. Appl Agric Res 1:12–26

Malik RS, Dhankar JS, Turner NC (1979) Influence of soil water deficits on root growth of cotton seedlings. Plant Soil 53:109–115

Martinez V, Läuchli A (1994) Salt-induced inhibition of phosphate uptake in plants of cotton (*Gossypium hirsutum* L.). New Phytol 126:609–614

McDaniel RG, Dobrenz AK (1994) Water stress and carbon isotope discrimination in cotton. In: Proc Beltwide Cotton Prod Res Conf, 5–8 Jan, San Diego, pp 1297–1300

McMichael BL, Hesketh JD (1982) Field investigations of the response of cotton to water deficits. Field Crops Res 5:319–333

McMichael BL, Quisenberry JE (1991) Genetic variation for root-shoot relationships among cotton germplasm. Environ Exp Bot 31:461–470

McMichael BL, Burke JJ, Berlin JD, Hatfield JL, Quisenberry JE (1985) Root vascular bundle arrangement among cotton strains and cultivars. Environ Exp Bot 25:23–30

McMichael BL, Quisenberry JE, Upchurch DR (1987) Lateral root development in exotic cotton. Environ Exp Bot 27:499–502

Meyer WS, Ritchie JT (1980) Water status of cotton as related to taproot length. Agron J 72:577–580

Mirza MA, Sheikh AL (1994) Prospects of developing novel cottons through biotechniques. Pak J Agric Res 15:195–201

Mullins GL, Reeves DW, Burmester CH, Bryant HH (1994) In-row subsoiling and potassium placement effects on root growth and potassium content of cotton. Agron J 86:136–139

Nabors MW (1990) Environmental stress resistance. In: Dix PJ (ed) Plant cell line selection. Procedures and applications. VCH, Weinheim, Germany, pp 167–186

Nawar MT, Zaher AM, El-Sahhar K, Abdel-Rahim SA (1994) Influence of salinity on reproductive growth of three Egyptian cotton cultivars. In: Constable GA, Forrester NW (eds) Challenging the future. Proceedings of the World Cotton Research Conference-1, Brisbane, 14–17 Feb, CSIRO, Melbourne, pp 109–114

Neimann RH, Poulsen LL (1967) Interactive effects of salinity and atmospheric humidity on the growth of bean and cotton plants. Bot Gaz 128:69–73

Niu X, Bressan RA, Hasegawa PM, Pardo JM (1995) Ion homeostasis in NaCl stress environments. Plant Physiol 109:735–742

Oosterhuis DM (1985) Osmotic adjustment in water-stressed cotton leaves and roots. Arkansas Farm Res 34:5

Oosterhuis DM, Wullschleger SD (1987a) Osmotic adjustment in cotton (*Gossypium hirsutum* L.) leaves and roots in response to water stress. Plant Physiol 84:1154–1157

Oosterhuis DM, Wullschleger SD (1987b) Water flow through cotton roots in relation to xylem anatomy. J Exp Bot 38:1866–1874

Oosterhuis DM, Wullschleger SD (1988) Drought tolerance and osmotic adjustment. Arkansas Agric Exp Stn Sepc Rep 132:33–36

Oosterhuis DM, Wullschleger SD, Stewart JM (1987) Osmotic adjustment in commercial cultivars and wild types of cotton. Agron Abstr, p 97

Oosterhuis DM, Wullschleger SD, Stewart JM (1988) Diversity in cotton root xylem anatomy. Arkansas Farm Res 1988:19

Orgaz F, Bonachela S, Mateos L, Fereres E (1991) Desarrollo fenológico y producción de cultivares de algodón en el valle del Guadalquivir bajo diversos regímenes de riego. Invest Agr Prod Prot Veg 6:53–67

Paterson AH, Tanksley SD, Sorrells ME (1991) DNA markers in plant improvement. Adv Agron 46:39–90

Peng ZY, Weyers JDB (1994) Stomatal sensitivity to abscisic acid following water deficit stress. J Exp Bot 45:835–845

Percy RG, Lu Z, Radin JW, Turcotte EL, Zeiger E (1996) Inheritance of stomatal conductance in cotton (*Gossypium barbadense*). Physiol Plant 96:389–394

Perry SW, Krieg DR, Hutmacher RB (1983) Photosynthetic rate control in cotton. Photorespiration. Plant Physiol 73:662–665

Petersen KL, Moreshet S, Fuchs M (1991) Stomatal responses of field-grown cotton to radiation and soil moisture. Agron J 83:1059–1065

Petersen KL, Fuchs M, Moreshet S, Cohen Y, Sinoquet H (1992) Computing transpiration of sunlit and shaded cotton foliage under variable water stress. Agron J 84:91–97

Pettigrew WT, Meredith WR Jr (1994) Leaf gas exchange parameters vary among cotton genotypes. Crop Sci 34:700–705

Quisenberry JE, McMichael BL (1991) Genetic variation among cotton germplasm for water-use efficiency. Environ Exp Bot 31:453–460

Quisenberry JE, McMichael BL (1996) Screening cotton germplasm for root growth potential. Environ Exp Bot 36:333–337

Quisenberry JE, Cartwright GB, McMichael BL (1984) Genetic relationship between turgor maintenance and growth in cotton germplasm. Crop Sci 24:479–482

Quisenberry JE, Jordan WR, Roark BA, Fryrear DW (1981) Exotic cotton as genetic sources for drought resistance. Crop Sci 21:889–895

Quisenberry JE, Roark B (1976) Influence of indeterminate growth habit on yield and irrigation water-use efficiency in upland cotton. Crop Sci 16:762–765

Quisenberry JE, Roark B, McMichael BL (1982) Use of transpiration decline curves to identify drought-tolerant cotton germplasm. Crop Sci 22:918–922

Quisenberry JE, Wendt CW, Berlin JD, McMichael BL (1985) Potential for using leaf turgidity to select drought tolerance in cotton. Crop Sci 25:294–299

Radin JW (1983) Physiological consequences of cellular water deficits: Osmotic adjustment. In: Taylor HM, Jordan WR, Sinclair TR (eds) Limitations to efficient water use in crop production. ASA-CSSA-SSSA, Madison, USA, pp 267–276

Radin JW (1992) Reconciling water-use efficiencies of cotton in field and laboratory. Crop Sci 32:13–18

Radin JW, Hartung W, Kimball BA, Mauney JR (1988) Correlation of stomatal conductance with photosynthetic capacity of cotton only in a CO_2-enriched atmosphere: Mediation by abscisic acid? Plant Physiol 88:1058–1062

Radin JW, Lu ZM, Percy RG, Zeigler E (1994) Genetic variability for stomatal conductance in Pima cotton and its relation to improvements of heat adaptation. Proc Natl Acad Sci USA 91:7217–7221

Rathert G (1982) Influence of extreme K:Na ratios and high substrate salinity on plant metabolism of crops differing in salt tolerance. VI. Mineral distribution variability among different salt-tolerant cotton varieties. J Plant Nutr 5:183–193

Razzouk S, Whittington WJ (1991) Effects of salinity on cotton yield and quality. Field Crops Res 26:305–314

Reddy VR, Baker DN, Hodges HF (1991) Temperature effects on cotton canopy growth, photosynthesis, and respiration. Agron J 83:699–704

Rodriguez-Garay B, Barrow JR (1988) Pollen selection for heat tolerance in cotton. Crop Sci 28:857–859

Saiz JF, Leidi EO (1994) Caracteres fisiológicos relacionados con tolerancia a salinidad en algodón. In: Bases Fisiológicas, Bioquímicas y Moleculares de la Nutrición Mineral de las Plantas. Consej Agric Comercio J Extremadura, Badajoz, Spain, pp 38–45

Saliendra NZ, Meinzer C (1992) Genotypic, developmental and drought-induced differences in root hydraulic conductance of contrasting sugarcane cultivars. J Exp Bot 43:1209–1217

Singh D, Sahay RK (1991) Proline accumulation in relation to yield reduction and biomass recovery under stress conditions in cotton (*Gossypium* species). Indian J Agric Sci 60:739–741

Stewart J McD (1991) Biotechnology of cotton. CAB International, Wallingford, pp 1–50

Stockle CO, Dugas WA (1992) Evaluating canopy temperature-based indexes for irrigation scheduling. Irrig Sci 13:31–37

Taha MA, Malik MNA, Chaudhry FI, Makhdum MI (1981) Heat-induced sterility in cotton sown during early April in West Punjab. Exp Agric 17:189–194

Tal M (1994) In vitro selection for salt tolerance in crop plants. Theoretical and practical considerations. In Vitro Cell Dev Biol Plant 30P:175–180

Thomas JR (1980) Osmotic and specific salt effects on growth of cotton. Agron J 72:407–410

Trolinder NL, Allen RD (1994) Molecular approaches to stress tolerance in cotton. Int Cotton Res Conf, Brisbane, Australia, p 57

Trolinder NL, Shang X (1991) In vitro selection and regeneration of cotton resistant to high temperature stress. Plant Cell Rep 10:448–452

Turner NC, Hearn AB, Begg JE, Constable GA (1986) Cotton (*Gossypium hirsutum* L.): Physiological and morphological responses to water deficits and their relationship to yield. Field Crops Res 14:153–170

Vepraskas MJ, Guthrie DS (1992) Identifying soils where subsoiling can increase yields of tobacco, corn and cotton. J Prod Agric 5:482–488

Wanjura DF, Mahan JR (1994) Thermal environment of cotton irrigated using canopy temperature. Irrig Sci 14:199–205

Wanjura DF, Upchurch DR, Mahan JR (1992a) Automated irrigation based on threshold canopy temperature. Trans Am Soc Agric Eng 35:1411–1417

Wanjura DF, Upchurch DR, Mahan JR (1992b) Automated irrigation based on threshold canopy temperature. Trans Am Soc Agric Eng 35:153–159

Warner DA, Burke JJ (1993) Cool night temperatures alter leaf starch and photosytem II chlorophyll fluorescence in cotton. Agron J 85:836–840

Watad AA, Swartzberg D, Bressan RA, Izhar S, Hasegawa PM (1991) Stability of salt tolerance at the cell level after regeneration of plants from a salt-tolerant tobacco cell line. Physiol Plant 83:307–313

Wilkins TA (1993) Vacuolar H^+-ATPase 69-kilodalton catalytic subunit cDNA from developing cotton (*Gossypium hirsutum*) ovules. Plant Physiol 102:679–680

Winter K, Koniger M (1991) Dry matter production and photosynthetic capacity in *Gossypium hirsutum* L. under conditions of slightly suboptimum leaf temperatures and high levels of irradiance. Oecologia 87:190–197

Wyn Jones RG, Gorham J (1983) Osmoregulation. In: Lange OL, Nobel PS, Osmond CB, Ziegler H (eds) Encyclopedia of plant physiology (New Series) vol 12C, Physiological plant ecology III. Springer, Berlin Heidelberg New York, pp 35–58

Yakir D, DeNiro MJ, Ephrath JE (1990) Effects of water stress on oxygen, hydrogen and carbon isotope ratios in two species of cotton plants. Plant Cell Environ 13:949–955

IV.3 In Vitro Regeneration of High Temperature-Resistant Cotton Plants

N.L. TROLINDER[1], X. SHANG[2], and Y.P.S. BAJAJ[3]

1 Introduction

Cotton is grown in areas of the world that are subject to recurring periods of high temperature and drought stress. Drought is often associated with high air temperature, and high tissue temperature may result from drought due to loss of transpirational cooling. Pollen fertility, which is often reduced by excessive temperature and drought stress, affects both quantity and quality of the cotton fiber produced. Thus, in order to generate high temperature-resistant cotton plants in vitro, a culture approach was used.

The ability to recover whole plants from single cells has made possible the selection of variant cell lines and eventually whole plants resistant to the selection agent. *Gossypium hirsutum* can be readily regenerated from cell cultures (Trolinder and Goodin 1987). Earlier cadmium-resistant cell lines of tobacco were observed to show increased tolerance to heat shock (Huang and Goldsbrough 1988). This observation has been applied to cotton, and cell lines showing tolerance to high temperature were regenerated into complete plants (Trolinder and Shang 1991).

2 Methodology (for details see Trolinder and Shang 1991)

Plant Material. Callus cultures and cell suspensions of *Gossypium hirsutum* L. cv. Coker 312 were established as previously described (Trolinder and Goodin 1987). The suspension cultures were filtered and grown under control conditions, then exposed to the temperature treatments at the midlog stage of growth (Trolinder and Shang 1991).

[1] USDA-ARS, Cropping Systems Research Laboratory. *Present address*: BIOTEX, 599 E 50th, Lubbock, Texas 79404, USA
[2] Department of Agronomy, Horticulture, and Entomology, Texas Tech University, Lubbock, Texas 79401, USA
[3] Biotechnology in Agriculture and Forestry, A-137 New Friends Colony, New Delhi 110065, India

Biotechnology in Agriculture and Forestry, Vol. 42
Cotton (ed. by Y.P.S. Bajaj)

Effect of High Temperature on Cell Viability. Cell suspensions were exposed to 50 °C in a temperature-controlled incubator for 10h. Cultures were then returned to normal culture room conditions. The 50 °C exposure experiment was repeated as described except that samples were taken at 15, 30, and 45 min, in addition to 60 min.

Selection of Cells for High Temperature Resistance. At the midlog stage of growth, flasks were exposed to 45 °C for the following time periods: (1) 3 h/day to a total of 75 h exposure, (2) 3 h/day to a total of 81 h, then exposed continuously for 24 h, (3) 3 h/day to a total of 100 h, (4) 10 h continuous exposure.

To test for sustained resistance of surviving cells, the stress-selected cell lines were maintained at 30 °C for 6 months by monthly subculture to fresh medium. At the end of 6 months, the cultures were reexposed to the previously determined lethal time:temperature treatments (24 h at 45 °C or 1 h at 50 °C). Six replications of each selected cell line and control cell line were exposed to each lethal temperature treatment.

Cross-Resistance of Selected Cell Lines to PEG. Selected cell suspensions were transferred at the midlog stage of growth to liquid medium containing 150 g/l polyethylene glycol (PEG) 800 at a density of 2 ml cells/50 ml medium and incubated as before for one growth cycle (30 days). When the control cell line entered the stationary stage, cells were harvested by centrifugation and washed with MS medium to remove PEG. The cells were then dried at 80 °C for 5 days, and dry weight determined.

Regeneration of Plants. Selected cell lines were plated on solid medium for the induction of somatic embryos (Trolinder and Goodin 1987). Embryos that developed were selected for size more than 3 mm, and transferred to Petri dishes containing Stewart and Hsu (1977) basal salts (S) supplemented with 30 g/l glucose + 20 g/l agar at pH 6.8. The unsealed Petri dishes were incubated in the dark. Maturing embryos were removed and placed for germination in Petri dishes containing S medium supplemented with 5 g/l sucrose, 1.5 g/l Gelrite, 5 g/l agar at pH 6.8. The unsealed Petri dishes were placed in a controlled environment growth room at 30 °C. Germinated embryos with one true leaf were placed in pint Mason jars (clear plastic caps) containing the same medium as for germination for further growth.

As plants attained a height of 7–10 cm and a stem diameter ≥2 mm, the shoot tips (top three nodes) were cut and rooted in the same medium. All tips rooted within 7–9 days. Rooted shoots were transferred to 1-gal plastic pots containing 1:1 peat:vermiculite and watered with half-strength Rapid Grow. Sealed plastic bags were placed over the pots and the pots placed in the greenhouse under a shade cloth for several days. The plants were hardened by clipping the corners of the bags on day 2, cutting the top open on day 4, and removing the bag on day 6. On day 8 the plants were moved to full greenhouse conditions.

Testing for Increased Resistance to High Temperature. Two leaves were harvested from each regenerated plant. Each leaf blade and petiole was sectioned and placed on callus initiation medium (Trolinder and Goodin 1987) at 38 °C in an incubator under continuous light (20 μE/m^2/s). Six replicate plates (per cell line per tissue) were cultured, each containing ten sections from plants regenerated from the selected and control cell lines. Plates were evaluated visually for callus formation and sustained growth after 4 weeks.

3 Results and Discussion

3.1 Effect of Temperature: Time on Cell Viability

Cell viability immediately after exposure in liquid medium to 50 °C ranged from 75% after 1 h to 30% after 10 h (Fig. 1). When cell viability was determined 24 h later, viability of the 1-h treatments had decreased to 25% and after 2 weeks viability was 0. Delayed death was also observed when cells were exposed to 45 °C over an extended time or when exposed to 50 °C for a shorter period (less than 1 h). There was not always a linear relationship between the death of cultured cells and the time : temperature exposure periods (Figs. 2, 3). For example, exposure to 45 °C resulted in 20% cell death during the first 3 h but only an additional 10% died in the following 7 h. Between 10 and 16 h, cell death proceeded linearly with time and all cells were dead by 24 h. These data suggest that there are different populations of cells with variable ability to withstand prolonged high-temperature stress.

The criteria for selection of cotton cells resistant to high-temperature stress was based upon the temperature : time data and aspects of both the general selection scheme proposed by Handro (1981) and the successful

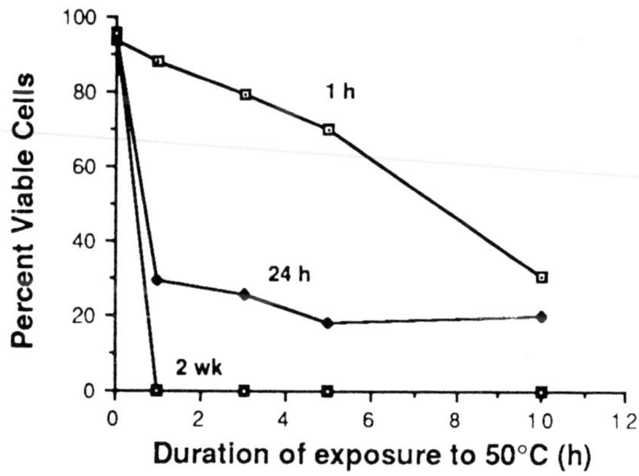

Fig. 1. Effect of length of exposure (*h*) to 50 °C on the viability of cotton cells grown in suspension culture – 1 h, 24 h, and 2 weeks after exposure

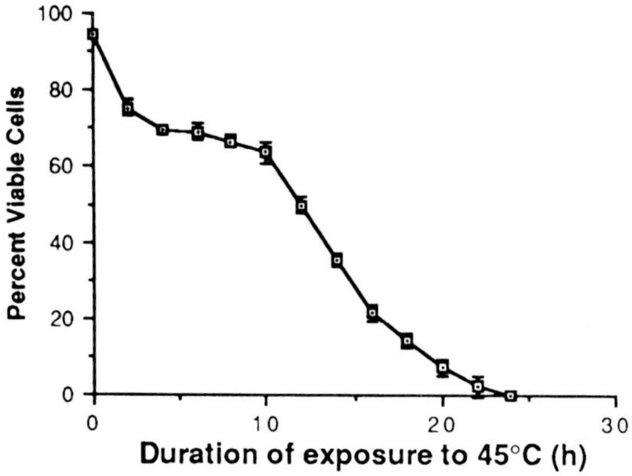

Fig. 2. Effects of length of exposure (h) to 45 °C on the viability of cotton cells grown in suspension culture – 2 weeks after exposure

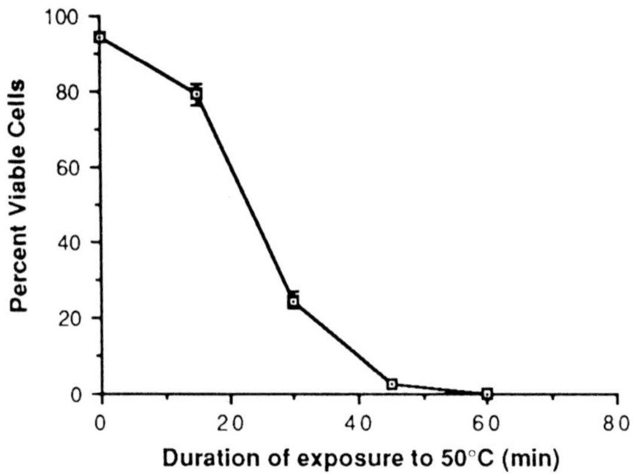

Fig. 3. Effects of length of exposure (min) to 50 °C on the viability of cotton cells grown in suspension culture – 2 weeks after exposure

stepwise selection used in animal cell cultures (Fougere-Deschatrette et al. 1984). Cotton leaf temperatures can daily reach close to 45 °C for short periods in some growing areas. The exposure of cells to a sublethal but highly damaging temperature treatment of 45 °C on a daily basis might allow the selection of cells better able to withstand repeated exposure to this temperature than wild-type cells. Three h was chosen as the stepwise selection treatment to correspond with the plateau area of cell viability between 3 and 10 h that suggests severe but not lethal stress. Ten h continuous exposure was chosen based upon the precipitous drop in cell viability beyond 10 h, another region of severe but not lethal stress. Two levels of pulsed stress accumulation were used in the selection treatments in addition to a third treatment in which the accumulated stress treatment was followed by a 24-h lethal temperature: time treatment. A large population of viable cells was obtained over the total accumulated temperature: time period; therefore, no attempt was made to ascertain frequency of mutation.

Table 1. Frequency of viable cells in selected and the control cell lines after reexposure to previously lethal temperature (FDA viability 2 weeks after exposure)

Cell line	Temperature treatment	
	45 °C: 24 h	50 °C: 1 h
Control	17.2	14.2
45 °C: 105 h	29.3[a]	46.7[a]
45 °C: 100 h	26.4[a]	43.2[a]
45 °C: 75 h	2.9[a]	42.9[a]
45 °C: 10 h	18.3	36.4[a]
Control[b]	0	0

[a] The selected cell lines were significantly different from the control cell line according to Fisher PLSD at $p < 0.0001$.
[b] Previous data for cell line prior to selection and extended culture period.

3.2 Resistance of Selected Cell Lines to High-Temperature Stress

After six passages of growth under normal temperature conditions, all selected and control cell lines were reexposed to the temperatures determined to be lethal to these lines prior to selection (Table 1). All cell lines, including the nonselected control, were more resistant than the initial nonselected control to the previously lethal 50 °C treatment. However, only two cell lines were more resistant than the control to the 45 °C treatment. The increased resistance of the control cell line might be attributed to somaclonal variation during the lengthy culture period.

3.3 Cross-Resistance of the Selected Cell Lines to PEG

Plants that are resistant to one type of environmental stress are often resistant to other types of stress as well (Huang and Goldsbrough 1988). In an agronomic situation, drought stress is often accompanied by high-temperature stress and vice versa. Therefore, it was logical to determine whether cells that have increased resistance to high temperature also show increased resistance to a component of drought stress. PEG has frequently been used to mimic drought stress in whole plants and cell cultures. The cell lines were grown with or without the addition of two levels of PEG in the culture medium. All selected cell lines were more resistant to the presence of 150 g/l PEG in the culture medium as compared to the control line (Fig. 4). Only the 105-h line showed a slight, but nonsignificant, increase over the control at the more severe stress induced by 250 g/l PEG.

3.4 Resistance of Leaf Explants from Ro Plants to 38 °C

At least ten plants per cell line, including the nonselected control line, were regenerated and tested for resistance of leaf and petiole tissue to high-

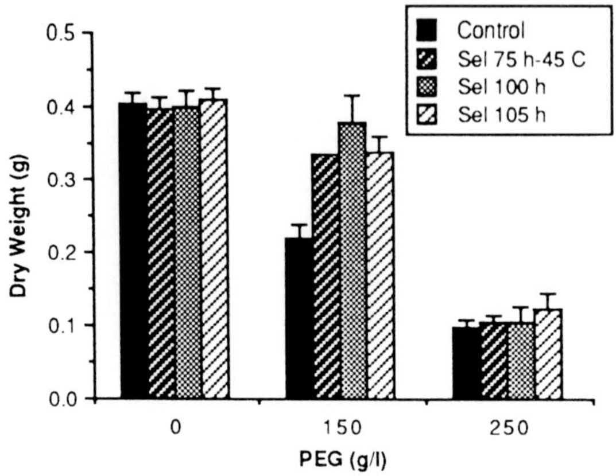

Fig. 4. Cross-resistance to polyethylene glycol of cotton cell lines selected for resistance to high temperature

Table 2. Percentage of cotton (Ro) leaf and petiole explants that formed calli and sustained growth under 38 °C

Cell line of Ro	% Explants forming callus	% calli that maintained growth
Control	8.7	0
45 °C: 105 h	30.3	25.7
45 °C: 100 h	28.7	25.3
45 °C: 75 h	14.5	11.0
45 °C: 10 h	12.7	5.3

temperature stress. Leaf and petiole explants were placed on callus induction medium and grown at 38 °C (Table 2). A high percentage of Ro leaf explants from the selected lines formed callus and maintained vigorous growth of that callus. Only a small percentage (8.7%) of explants from control Ro plants could form callus at the stress temperature, and the callus that formed became necrotic.

To conclude, it has been possible to select cotton cells in vitro that are resistant to both high temperature and 150 g/l of PEG. Sustained inheritance of a desirable trait by the progeny of the primary regenerants is essential for final proof that a selection strategy has been successful. However, most plants regenerated were infertile (male sterile). Cytological abnormalities and reduced fertility have also been noted in cotton regenerated from long-term cultures (Trolinder and Goodin 1987; Li et al. 1989). It may be possible to obtain progeny by cross-pollination with pollen from normal plants.

4 Summary

Cell suspension cultures of *Gossypium hirsutum* L. cv. Coker 312 were exposed to various temperature : time treatments in order to select cell lines resistant to high temperature stress. Cells were exposed to 45 °C for 3 h each day until the total accumulated hours of stress were 0, 10, 75, 100, or 105 h. After the stress treatments, the cells were plated onto embryo development medium and plants were recovered. The recovered embryogenic calli were subcultured monthly for 6 months and tested for increased resistance to the temperature : time treatments previously determined to be lethal and to water stress as imposed by PEG. All the selected cell lines were more resistant to both types of stress than the control cell lines. Leaf tissue of stress-selected plants (Ro) formed and maintained callus growth when incubated at 38 °C, whereas tissue excised from nonselected controls rarely formed callus, and calli which did form quickly became necrotic.

Acknowledgments. This chapter is a modified version of a paper published in Plant Cell Reports 1991. The original research was funded by the Texas Advanced Technology Program and USDA-ARS. Appreciation is expressed to Springer-Verlag for granting permission to reproduce data/figures.

References

Fougere-Deschatrette C, Schimke RT, Weil D, Weiss MC (1984) J Cell Biol 99:497

Handro W (1981) In: Thorpe TA (ed) Plant tissue culture. Academic Orlando pp 171–172

Huang B, Goldsbrough PB (1988) Plant Cell Rep 7:119–122

Li R, Stelly DM, Trolinder NL (1989) Cytogenetic abnormalities in cotton (*Gossypium hirsutum* L.) cell cultures. Genome 32:1128–1134

Stewart J McD, Hsu CL (1977) In ovulo embryo culture and seedling development of cotton (*Gossypium hirsutum* L.). Planta 137:113–117

Trolinder NL, Goodin JR (1987) Somatic embryogenesis and plant regeneration in cotton (*Gossypium hirsutum* L.). Plant Cell Rep 6:231–234

Trolinder NL, Shang Xaiomin (1991) In vitro selection and regeneration of cotton resistant to high temperature stress. Plant Cell Rep 10:448–452

IV.4 Nematode Resistance in Cotton

Roy G. Creech[1], J.N. Jenkins[2], G.W. Lawrence[3], and J.C. McCarty, Jr[2]

1 General Account

Nematode parasitism of cotton, has become increasingly important due to yield losses and increased distribution across cotton production regions. Yield losses attributed to nematodes in 1991 in the 14 cotton-producing states in the USA were estimated to be approximately 2.4%, or in excess of 476151 bales (Smith 1992). In 1992, plant-parasitic nematodes accounted for an estimated yield loss of 528000 bales valued at US$184951200 (Goodell 1993). The Cotton Disease Council reported yield losses in 1995 in Mississippi at about 4% or 82929 bales, and beltwide in the USA about 3.5%, or about 778703 bales. Decreases in fiber length and other adverse affects on fiber quality due to nematode infection were reported by Smith et al. (1991).

The most important widespread nematode species affecting cotton in the USA is root-knot nematode (RKN) race 3, *Meloidogyne incognita* (Kofoid & White) Chitwood (Smith 1992). RKN also may predispose plants to attack by other soil-borne organisms (Shepherd 1982a). Severe infestations are usually limited to well-drained soils. Other serious nematodes on cotton are reniform (*Rotylenchulus reniformis*, Linford and Oliveira 1940), lance (*Hoplolaimus columbus* Sher, *H. galeatus* (Thorne) Sher, and *H. magnistylus* Robbins (Robbins 1982). There are no commercial cultivars of *G. hirsutum* with resistance to RKN (Shepherd et al. 1988b; Jenkins et al. 1993).

The reniform nematode was first identified as a parasite of cotton in 1940 (Birchfield 1962). In many locations, the damage incurred by reniform nematode has surpassed the RKN in both the reduction of plant development and subsequent effect on suppression of cotton yields. The reniform nematode has a wide distribution and has been reported in over 38 countries. It occurs in many tropical and subtropical areas, but can also be found in the warmer

[1] Department of Plant and Soil Sciences, Mississippi State University, Mississippi State, Mississippi 39762, USA

[2] USDA/ARS Crop Science Research Laboratory, Mississippi State, Mississippi 39762, USA

[3] Department of Entomology and Plant Pathology, Mississippi State University, Mississippi State, Mississippi 39762, USA

Biotechnology in Agriculture and Forestry, Vol. 42
Cotton (ed. by Y.P.S. Bajaj)

regions of the temperate zones. Outside the tropics, the greatest concentration of the reniform nematode occurs in the southeastern part of the United States. The distribution of the reniform nematode is not limited by soil types, soil texture, soil pH, or the amount of rainfall. In general, the finer-textured silt soils seem to support larger nematode populations than the coarser-textured sandy soils (Heald and Robinson 1990; Lawrence et al. 1990, 1994). Reniform resistance has been identified in other *Gossypium* species including *G. stocksii*, *G. somalense*, *G. barbadense*, *G. longicalax*, and various plant introductions, but not *G. hirsutum* (Yik and Birchfield 1984).

Three species of lance nematodes are known to produce economic yield losses to cotton. *Hoplolaimus columbus* and *H. galeatus* were first identified in South Carolina in the 1950s (Heald et al. 1981; Watkins 1981). *H. magnistylus* was identified on cotton in Mississippi and Arkansas in 1982 (Robbins 1982). *H. columbus* occurs extensively in the Coastal Plains soils of North Carolina, South Carolina, and Georgia (Mueller and Sanders 1987). It has also been identified in one cotton-producing county in Alabama and in sugarcane in Louisiana. *H. galeatus* has a more extensive distribution and is found across the entire southeastern United States. The distribution of *H. magnistylus* is limited primarily to Arkansas, Louisiana, and Mississippi (Robbins 1982).

Efforts to breed cotton, *G. hirsutum*, for resistance to RKN, *M. incognita*, can be traced to the early 1900s (Ware 1936). No commercial cultivars that are resistant to this pest have been developed (Shepherd et al. 1988b; Jenkins et al. 1993). A series of highly resistant germplasm lines have been developed and released to the public (Shepherd 1974b, 1982b, 1983, 1987; Shepherd et al. 1988a,b, 1996).

Since no known sources of resistance to reniform and lance nematodes are known in *G. hirsutum*, even though searches are underway, this chapter will address efforts to develop RKN-resistant cotton cultivars and germplasm lines.

Conventional breeding for RKN resistance is time-consuming and labor-intensive, requiring both greenhouse and field evaluations. Techniques developed via molecular biology should become useful in transferring the resistance genes into cultivars available to the cotton growers once the genes are identified, isolated, and cloned. A good understanding of the biology of *M. incognita* in resistant and susceptible cotton lines is required to fully utilize these techniques. Recent biological and developmental studies of *M. incognita* on cotton partially provide the information needed to understand the mechanisms of resistance in cotton (Tang et al. 1994; Creech et al. 1995; Jenkins et al. 1995; McPherson et al. 1995).

Management tactics to maximize cotton yields and fiber quality on nematode-infested fields presently include use of less susceptible cultivars, crop rotation, and nematicides. The costs of effective nematicides, potential hazards of pesticides in the environment, and large-scale monoculture of cotton provide economic and environmental incentives for the development of genetically resistant cultivars for use as a strategic management tactic.

2 Origins of Genetic Resistance to Root-Knot Nematodes in Cotton

The history and problems associated with breeding cotton for resistance to RKN were reviewed by Fassuliotis (1982) and Sasser (1986). Most studies on the genetics of host-nematode interactions in cotton were primarily concerned with inheritance of plant resistance, and very few considered effects of resistant genotypes on nematode biology. A few studies have been conducted on the nature of resistance of cotton to *M. incognita*. Minton (1962) and McClure et al. (1974a,b), studying the Clevewilt-6-3-5 source of moderate resistance, indicated that resistance to *M. incognita* was due to physiological processes within the roots that prevented or delayed juvenile development, rather than to morphological factors that prevent nematodes from entering cotton roots. This was further supported by Tang et al. (1994) and Creech et al. (1995), working with the highly resistant Auburn 623 source of resistance. Auburn 623 was developed from a transgressive segregate in a cross of Clevewilt-6 with Mexican Wild. This germplasm was considerably more resistant than any other known sources at that time (Shepherd 1974a,b). McPherson (1993) and McPherson et al. (1995) provided evidence that highly resistant germplasm lines like Auburn 623, and its derivative germplasm lines, M-315 and M-240 being two examples, each possess at least two major genes for resistance, one dominant and the other quantitative. Recent investigations in this laboratory by J.C. McCarty and others have discovered resistance in several of the wild races of *G. hirsutum*, and research is in progress to determine if the gene(s) are allelic to those genes in Auburn 623 (J.C. McCarty et al., unpubl.).

3 Progress Toward Elucidating the Nature and Action of Cotton Root-Knot Nematode Resistance

Veech and McClure (1977) reported that methoxy-substituted terpenoid aldehydes are involved in resistance. Veech (1978, 1979) also showed that infection by *M. incognita* induced terpenoid aldehyde formation in the endodermis stele of cotton roots, with aldehyde formation more rapid and extensive in resistant germplasm than in susceptible germplasm. He further observed with in vitro bioassays that terpenoid aldehydes are toxic to this nematode. The natural mixture of terpenoids was more inhibiting to nematodes than gossypol.

Shepherd and Huck (1989) characterized the effects or RKN on root development and root damage in susceptible M-8 cotton roots and resistant Auburn 623 cotton roots. They did not, however, evaluate the effects of resistance on developmental stages of the nematode. Earlier research on postdevelopment of *M. incognita* on cotton had only the Clevewilt source of resistance available (McClure et al. 1974a,b).

Bird (1962) indicated that, with *M. javanica*, the lack of giant cell development in the host and failure of juveniles to develop were interrelated. We have confirmed this in our laboratory for *M. incognita* in highly resistant cotton (unpubl. data); however, some small giant cells developed in resistant cotton roots with a thick layer inside the cell wall and most died before the nematode developed to maturity.

Recent research efforts have focused on understanding the relationships between *M. incognita* and the cotton plant in germplasm lines with variable levels of plant resistance expressed as varying number of eggs and egg masses from inoculation to 40 days after inoculation (DAI). These investigations have provided basic and detailed information on the interaction between the cotton host plant and *M. incognita*, and have aided understanding of time of action of resistance at the physiological and developmental levels.

Penetration and reproduction or fecundity of the RKN, race 3, was compared on three cotton genotypes (*Gossypium hirsutum* L.) differing in level of resistance: M-8, highly susceptible to root-knot; M-78 (M-78RNR), moderately resistant; and M-315 (M-315RNR), highly resistant with the genes for resistance from Auburn 623 (Creech et al. 1995). These three lines were grown in a greenhouse in a Wickham sandy soil (a fine, loamy, mixed thermic, Typic Hapludult) and inoculated with second-stage juveniles. Seven developmental stages of RKN were counted every 2 days for 44 days. At 18 DAI, 70, 45, and 6% of RKN present were adults in M-8, M-78, and M-315, respectively. Responses were characterized by penetration of plant roots by root-knot nematode juveniles (Table 1) and by reproduction, measured as number of eggs (Table 2), egg masses per plant (Table 3) and eggs per egg mass (Table 4). Susceptible, moderately resistant, and highly resistant cotton genotypes were penetrated about equally by juveniles. Production of egg masses, eggs per egg mass, and eggs per plant by nematodes were different among the three cotton genotypes. The development of egg masses was delayed in M-78 and M-315 as compared to M-8, with the greatest delay and prevention of the formation of significant numbers of egg masses in the highly resistant M-315. Production of eggs and second-stage infective juveniles in the moderately resistant M-78 were sufficient to contribute significantly to the buildup of the next generation

Table 1. Mean numbers[a] of *Meloidogyne incognita* juveniles (Stage A J2, Tang et al. 1994) in the roots of three cotton genotypes from 2 to 8 days after inoculation (DAI). (Creech et al. 1995)

Days after inoculation (DAI)	Genotype			
	M-8	M-78	M-315	LSD
	— — — — No. — — — —			
2	88	100	94	83 n.s.
4	226	195	248	134 n.s.
6	148	162	188	121 n.s.
8	9	10	23*	10*

* Significantly different at $P = 0.05$.
[a] Mean of four replications of each genotype.

Table 2. Mean numbers of *Meloidogyne incognita* eggs per plant on cotton genotypes with different levels of resistance. (Creech et al. 1995)

Days after inoculation (DAI)	Genotype					
	M-8		M-78		M-315	
	Eggs per plant[a]	SE[b]	Eggs per plant	SE	Eggs per plant	SE
	No.					
18	47	35	24	24	0	–
20	1212	708	192	118	0	–
22	3522	1696	2162	809	4	4
24	3568	1889	5072	2412	86	36
26	30391	11708	7783	2628	250	124
28	40610	7308	14130	8211	542	124
30	57025	9247	18360	7214	822	262
32	83201	15593	26385	10582	1024	499
34	155638	24954	28782	10582	715	197
36	169775	26173	29465	7961	851	360
38	175200	11594	34245	12258	999	458
40	236850	52204	57450	17877	2025	1038
42	225600	20478	75900	38486	1256	377
44	230700	16377	77300	30843	1399	700

[a] Eggs were counted after removal from galls; therefore, the exact time of egg production would have been slightly earlier than 18 days after inoculation (DAI).
[b] Standard error of means of four replications of each genotype.

Table 3. Mean numbers of *Meloidogyne incognita* egg masses per plant on cotton genotypes with different levels of resistance. (Creech et al. 1995)

Days after inoculation (DAI)	Genotype					
	M-8		M-78		M-315	
	Egg masses per plant	SE[a]	Egg masses per plant	SE	Egg masses per plant	SE
	No.					
18	8.5	5.0	5.0	2.8	0	–
20	70.3	39.8	15.0	7.7	0	–
22	79.1	27.6	66.5	21.8	2.0	0.4
24	43.5	18.8	112.3	32.4	4.5	1.7
26	162.5	57.5	93.0	32.4	6.0	3.4
28	170.3	38.5	89.8	43.1	9.5	3.2
30	180.8	38.3	86.5	29.6	13.5	5.9
32	168.3	32.2	77.8	26.4	12.8	5.2
34	291.5	63.0	90.3	45.4	3.5	0.6
36	253.0	13.2	103.3	27.1	8.5	3.1
38	258.8	19.9	118.5	27.2	5.0	3.4
40	306.0	29.4	125.5	22.2	6.5	1.5
42	276.0	14.5	122.5	34.8	3.5	0.6
44	292.0	20.1	139.8	58.9	3.5	0.9

[a] Standard error of means of four replications of each genotype.

Table 4. Mean numbers of *Meloidogyne incognita* eggs per egg mass per plant on cotton genotypes with different levels of resistance. (Creech et al. 1995)

Days after inoculation (DAI)	Genotype					
	M-8		M-78		M-315	
	Eggs per egg mass	SE[a]	Eggs per egg mass	SE	Eggs per egg mass	SE
	No.					
18	5.0	1.4	9.0	–	0	–
20	14.7	3.4	14.5	3.9	0	–
22	38.5	16.4	30.5	7.0	3.0	–
24	74.3	19.6	36.5	11.6	22.0	6.6
26	151.3	46.9	72.5	18.5	53.5	18.6
28	249.5	23.6	119.0	37.9	78.3	33.2
30	343.8	48.2	186.8	31.4	102.3	55.9
32	499.0	54.7	314.3	77.5	79.5	22.8
34	565.0	68.1	312.5	35.5	192.0	23.0
36	666.3	77.4	290.5	36.6	114.8	34.6
38	693.7	89.5	280.0	83.1	221.0	93.8
40	755.5	108.0	486.0	136.9	261.8	81.4
42	818.2	70.1	574.5	187.9	330.9	87.1
44	800.5	75.9	579.5	27.9	317.3	88.8

[a] Standard error of the means of four replications of each genotype.

of root-knot populations in cotton production fields. Reproduction on M-315 was so low that nematode populations would be expected to decrease significantly under continuous culture.

Postpenetration development was also compared on plant roots of the three genotypes (Jenkins et al. 1995). In addition to slower development of RKN in M-315, resistance is expressed by significantly fewer nematodes developing to third and fourth stage juveniles at 6–8 DAI and fewer developing to mature females about 24 DAI (Fig. 1). Galling indices are presented in Table 5. The resistance gene in M-78 operates about 24 DAI. Most RKN that penetrated M-315 failed to establish and/or maintain giant cells. At 44 DAI, the numbers of egg-laying females on M-8, M-78, and M-315 were 299, 144, and 5, respectively. There were significantly fewer root galls on M-315 than on M-78 or M-8 beginning 10 DAI, and significantly fewer on M-78 than on M-8 at 40 DAI. Galls were significantly smaller on M-315 than on M-78 or M-8 beginning 8 DAI, and significantly smaller on M-78 than on M-8 beginning 18 DAI (Fig. 2). Thus, the post-penetration development of RKN was slower, fewer developed to adult females, and root galls were fewer and smaller on the resistant genotypes than on the susceptible M-8.

The above findings indicate that a critical stage for RKN development occurred in M-315 at 6 DAI and perhaps again at 24 DAI (Fig. 1). At each of these times, RKN numbers decreased markedly in M-315. Perhaps these are the stages when the two major resistance genes proposed by McPherson (1993) and McPherson et al. (1995) for M-315 are independently and actively expressed. These two stages are prime candidates for molecular genetic investigations in the resistant and susceptible genotypes.

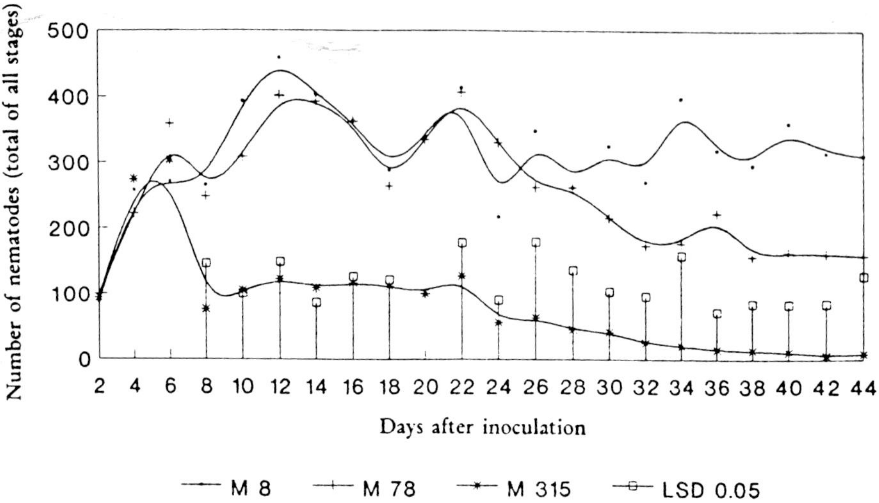

Fig. 1. Number of *Meloidogyne incognita* on three cotton genotypes differing in resistance at specified days after inoculation (Jenkins et al. 1995)

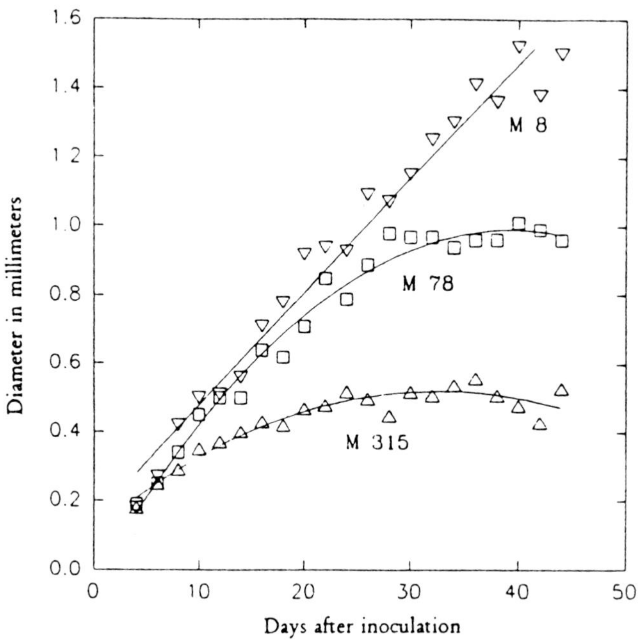

Fig. 2. Size of *Meloidogyne incognita* galls in relation to days after inoculation of three cotton genotypes with different levels of resistance (Jenkins et al. 1995)

Table 5. Root-knot nematode galling indices in relation to days after inoculation on three cotton genotypes with different levels of resistance. (Jenkins et al. 1995)

Germplasm	Days after inoculation				
	6	10	20	30	40
M-8	3.8a[a]	4.8a	4.8a	5.0a	5.0a
M-78	4.0a	3.8a	3.8a	4.0a	3.5b
M-315	3.5a	2.5b	2.3b	2.0b	1.0c
LSD 0.05	1.5	1.6	1.3	1.1	0.5

[a] Means in a column followed by the same letter are not significantly different at the 0.05 level with LSD.
1 = no galling; 2 = light galling; 3 = moderate galling; 4 = heavy galling; 5 = very heavy galling.

4 Summary

Research with all commercial cotton (*G. hirsutum*) cultivars offered for sale in Mississippi in 1993 found that all are susceptible to root-knot nematode race 3, reniform, and lance nematodes. Resistance is available in *G. hirsutum* only for RKN race 3 at this time. Incorporation of RKN resistance, which is available (joint release by USDA-ARS and MAFES 1989), into advanced breeding lines and cultivars, is of prime importance.

RKN race 3 is presently one of the most important and widespread pathogenic nematodes across the Cotton Belt in the USA, effecting significant yield losses. Our investigations elucidate ways of identifying genetic markers and other procedures for transfer of the root-knot-resistant genes to elite breeding lines and cultivars in an efficient manner. Our recent investigations have elucidated specific host-nematode interactions in giant cell formation and development in susceptible and resistant cotton (Tang et al. 1994, and unpubl. data in this laboratory).

Investigations have elucidated the biology of nematode development through various stages in susceptible, moderately resistant, and resistant cotton lines. These data also suggest the mode and time of action of the genes effecting nematode resistance in cotton plants subsequent to infection. Resistance management is also under study.

These investigations have contributed and are contributing to the use of RKN resistance in the development of new cultivars. Cotton-yield losses due to nematodes in 1992 across the Cotton Belt have been estimated and reported by the Cotton Disease Council to be about 2.4%, or in excess of 528000 bales, and 3.54%, or 82929 bales, in 1995. These represented increases over 1991 and previous years of the study.

The genes for resistance are being transferred to elite breeding lines, germplasm lines, and cultivars by our research team and others, including a few commercial breeders. Isogenic lines for molecular genetic and biochemical/physiological investigations are also under development.

References

Birchfield W (1962) Host-parasitic relations of *Rotylenchulus reniformis* or *Gossypium hirsutum.* Phytopathology 52:862–865

Bird AF (1962) The inducement of giant cells by *Meloidogyne javanica.* Nematologica 8:1–10

Creech RG, Jenkins JN, Tang B, Lawrence GW, McCarty JC (1995) Cotton resistance to root-knot nematode. I. Penetration and reproduction. Crop Sci 35:365–368

Fassuliotis G (1982) Plant resistance to root-knot nematodes. In: Riggs RD (ed) Nematology in the southern region of the United States. South Coop Ser Bull 267:33–49

Goodell PB (1993) Cotton nematodes. In: Cotton nematodes – our hidden enemies. Beltwide Cotton Nematode Survey and Education Committee, The Cotton Foundation, Memphis, TN, pp 1–3

Heald CM, Robinson AF (1990) Survey of current distribution of *Rotylenchulus reniformis* in the United States. J Nematol 22:695–699

Heald CM, Birchfield CW, Blackmon CW, Hussey RS, Orr CC, Shepherd RL, Veech J, Smith FH (1981) Nematodes. In: Watkins GM (ed) Compendium of cotton diseases. APS, St Paul, MN, pp 50–56

Jenkins JN, Creech RG, McCarty JC, McPherson R, Lawrence GW, Tang B (1993) Resistance of cotton cultivars to root-knot nematode. Mo Agric For Exp Stn Tech Bull 994:4

Jenkins JN, Creech RG, Tang B, Lawrence GW, McCarty JC (1995) Plant resistance in cotton to *Meloidogyne incognita.* II. Post-penetration development. Crop Sci 35:369–373

Lawrence GW, McLean KS, Batson WE, Miller D, Barbon JC (1990) Response of *Rotylenchulus reniformis* to nematicide applications on cotton. Ann Appl Nematol 22:707–711

Lawrence GW, McLean KS, Cornelius JJ (1994) Distribution of *Rotylenchulus reniformis* in a Mississippi cotton field. J Nematol 26:557

Linford MB, Oliveira JM (1940) *Rotylenchulus reniformis* Nov. gen. N. sp., a nematode parasite of roots. Proc Helminthol Soc Wash 7:35–42

McClure MA, Ellis KC, Nigh EL (1974a) Resistance of cotton to the root-knot nematode, *Meloidogyne incognita.* J Nematol 6:17–20

McClure MA, Ellis KC, Nigh EL (1974b) Post-infection development and histopathology of *Meloidogyne incognita* in resistant cotton. J Nematol 1:21–26

McPherson GR (1993) Inheritance of root-knot nematode resistance in cotton as determined by combining ability, generation means, and Mendelian analysis. PhD Diss, Mississippi State Univ Miss State, 135 pp

McPherson GR, Jenkins JN, McCarty JC Jr, Watson CE (1995) Combining ability analysis of root-knot nematode resistance in cotton. Crop Sci 35:373–375

Minton NA (1962) Factors influencing resistance of cotton to root-knot nematodes (*Meloidogyne* spp.). Phytopathology 52:272–279

Mueller JD, Sanders GB (1987) Control of *Hoplolaimus columbus* on late-planted soybean with aldicarb. Ann Appl Nematol 1:123–126

Robbins RT (1982) Description of *Hoplolaimus magnistylus* n. sp. (Nematoda: Hoplolaimidae). J Nematol 14:500–506

Sasser JN (1986) Half a century dynamics and control of cotton diseases: nematodes. In: Nelson CT (ed) Proc Beltwide Cotton Prod Res Conf, Las Vegas, NV, 4–9 Jan 1986, Natl Cotton Council, Memphis, TN, pp 48–51

Shepherd RL (1974a) Transgressive segregation for root-knot nematode resistance in cotton. Crop Sci 14:872–875

Shepherd RL (1974b) Registration of Auburn 623 RNR cotton germplasm. Crop Sci 14:911

Shepherd RL (1982a) Genetic resistance and its residual effects for control of the root-knot nematode-*Fusarium* wilt complex in cotton. Crop Sci 22:115–155

Shepherd RL (1982b) Registration of three germplasm lines of cotton. Crop Sci 22:692

Shepherd RL (1983) New source of resistance to root-knot nematode among primitive cottons. Crop Sci 23:999–1002

Shepherd RL (1987) Registration of three germplasm lines of cotton. Crop Sci 27:153

Shepherd RL, Huck MG (1989) Progression of root-knot nematode symptoms and infection on resistant and susceptible cotton. J Nematol 21:235–241

Shepherd RL, McCarty JC, Jenkins JN, Parrott WL (1988a) Registration of twelve non-photoperiodic lines with root-knot nematode resistant primitive cotton germplasm. Crop Sci 28:868–869

Shepherd RL, McCarty JC, Parrott WL, Jenkins JN (1988b) Resistance of cotton cultivars and elite breeding lines to root-knot nematodes. Mo Agric For Exp Stn Tech Bull 158:5

Shepherd RL, Parrott WL, McCarty JC, Jenkins JN (1996) Registration of nine cotton germplasm lines resistant to root-knot nematode. Crop Sci 36:820

Smith GR (1992) Report of the nematode management committee, 1991. In: Herber DJ, Richter DA (eds) Proc Beltwide Cotton Prod Res Conf, Nashville, Tennessee, 6–10 Jan 1992, Natl Cotton Council, Memphis, TN, 1:165

Smith GR, Veech JA, Gannaway JR (1991) The effect of *Meloidogyne incognita* on cotton development and fiber quality on the Texas high plains. In: Herber DJ, Richter DA (eds) Proc Beltwide Cotton Prod Res Conf, 8–12 Jan 1991, San Antonio, TX, Natl Cotton Council, Memphis, TN, 1:177–179

Tang B, Lawrence GW, Creech RG, Jenkins JN, McCarty JC Jr (1994) Post-infection development of *Meloidogyne incognita* in cotton roots. USDA and MO Agric For Exp Stn Tech Bull No 195:13

Veech JA (1978) An apparent relationship between methoxy-substituted terpenoid aldehydes and the resistance of cotton to *Meloidogyne incognita*. Nematologica 24:1, 81–87

Veech JA (1979) Histochemical localization and nematoxicity of terpene aldehydes in cotton. J Nematol 11:3, 240–246

Veech JA, McClure MA (1977) Terpenoid aldehydes in cotton roots susceptible and resistant to the root-knot nematode, *Meloidogyne incognita*. J Nematol 9:225–229

Ware JO (1936) Plant breeding and the cotton industry. In: US Dept Agric Yearbook of Agriculture. US Govt Printing Office, Washington, DC, 737 pp

Watkins GM (1981) Compendium of cotton diseases. APS Press, St Paul, Minnesota, 87pp

Yik Choi-Pheng, Birchfield W (1984) Resistant germplasm in *Gossypium* species and related plants to *Rotylenchulus reniformis*. J Nematol 16:146–153

Section V
Genetic Transformation; Transgenic Cotton Resistant to Insects and Herbicides; Field Trials; Risk Assessment Studies; Genetic Engineering of Fiber

V.1 Genetic Transformation of Cotton Through Particle Bombardment

D.E. McCabe, B.J. Martinell, and M.E. John[1]

1 Introduction

Plant transformation is a process by which an identifiable DNA segment is inserted into a plant genome. The introduced DNA is integrated into the chromosome, which may or may not be transcriptionally active, and is transmitted to the sexual offspring. Thus, transformation enables us to modify the genetic makeup of a plant, alter gene expression, and enhance agronomical or product traits. Although plant protoplasts allow free access to the DNA, the cells that have taken up the DNA have to be grown into a whole plant. However, since not all cells are totipotent, transformation methodologies that depend on regeneration of protoplasts through tissue culture are limited to certain cultivars. To overcome this shortcoming, direct DNA introduction into growing meristems has been developed.

Transformation requires a reporter (marker gene) or a selection process that distinguishes the transformed cells from background cells. Molecular biologists have identified several marker genes and developed corresponding assays for plant transformation work. A bacterial hydrolase gene, β-glucuronidase (*GUS; uidA*), that catalyzes the cleavage of many β-glucuronides, is an excellent reporter for plant transformation (Jefferson et al. 1987). A general description of transformations suitable for cotton follows.

One well-established method for DNA transfer into cotton is by the soil-borne bacterium *Agrobacterium tumefaciens*-mediated transformation. This bacterium infects a wound site in a plant, and transfers a segment of a plasmid (Ti) into a cell. The Ti plasmid contains several genes, including those encoding hormones, auxins, and cytokinins. The hormone genes are located in a portion of the Ti plasmid called T-DNA. The integration of T-DNA into the plant chromosome and subsequent expression of the hormone genes results in cell proliferation. Thus, each individual cell in the cancerous growth contains bacterial genes. When plants are regenerated from those individual cells, they carry *A. tumefaciens* DNA. The ability of *A. tumefaciens* to transfer DNA into plant cells can be exploited to transform plants with useful genes (Barton et al. 1983; Fraley et al. 1986). However, one of the major drawbacks of the *Agrobacterium*-mediated transformation of cotton is that only a limited number of cultivars can be regenerated into fertile plants through tissue cul-

[1] Agracetus, 8520 University Green, Middleton, Wisconsin 53562, USA

Biotechnology in Agriculture and Forestry, Vol. 42
Cotton (ed. by Y.P.S. Bajaj)

ture. Although *Agrobacterium* infects most varieties of cotton, the next step of regenerating cells from the callus has proven to be recalcitrant in the majority of them. At the present time, only Coker varieties have been consistently regenerated (Firoozabady et al. 1987; Trolinder and Goodin 1987; Umbeck et al. 1987; Bayley et al. 1992).

Agrobacterium-mediated transformation is being used to commercialize transgenic cotton (Perlak et al. 1990; Wilson et al. 1994; Stalker et al. 1996). The gene of interest is first introduced into a Coker variety and fertile transformants are selected. The transgene in the Coker transformant is then transferred into other commercial varieties of cotton through plant breeding. Several years of breeding and selection are required to identify lines suitable for commercialization, and fiber quality often suffers in this process.

A second method for genetic transformation of cotton is based on particle bombardment. This chapter deals with gene transfer to cotton meristems through particle bombardment.

2 Particle Bombardment

Rapid commercialization of genetically engineered cotton requires a reliable method of introducing genes directly into elite varieties of germplasm, circumventing the need for tissue culture, regeneration, and time-consuming plant breeding. Particle bombardment (gene gun), one method for physically transferring DNA through cell walls, meets the above objective. Klein et al. (1987) demonstrated that tungsten particles coated with DNA can be accelerated through *Allium cepa* (onion) cell walls, using a gunpowder-driven device. The epidermal cells expressed the introduced genes. This method was refined by McCabe and colleagues (Agracetus Inc., Middleton, WI) to introduce genes into a number of recalcitrant crops (McCabe et al. 1988; Christou et al. 1992; McCabe and Martinell 1993). Reports of cotton transformation through particle bombardment are summarized in Table 1. Gold particles (1 to 3 μm in diameter) were coated with DNA and accelerated toward plant tissues by an electric discharge gun. The extent of particle penetration to specific cell layers (L1, L2, and L3) and damage to the tissue could be controlled, since the electric discharge gun allowed fine tuning. The principle of the apparatus is shown in Fig. 1A. The microparticle carrying DNA is delivered to the cytoplasm or the nucleus when they come to rest in the tissue. The genes adhering on the microparticles are then taken up by the nucleus and transcribed within hours. This gene expression, termed transient expression, can be detected by means of a reporter gene (Fig. 1B). Several thousand copies of each plasmid (gene) can be resident on each microparticle. It appears that many of these copies contribute to transient expression. However, only one to several copies are eventually integrated into the plant genome (Cooley et al. 1995).

Particle bombardment transformation of elite cotton cultivars has been very successful (see Table 1). In short, the meristems of embryonic axes were bombarded with microparticles, containing plasmid DNA including a marker

Table 1. Particle bombardment-mediated cotton transformation

Reference	Observation/remarks
Finer and McMullen (1990)	Coker 310 embryogenic suspension culture was bombarded with hygromycin gene and placed under selection pressure Hygromycin-resistant plants were recovered and presence of input gene demonstrated by Southern blot analysis. Protocol limited to regenerable cultivars
John and Crow (1992)	Delta Pine 50 cultivar transformed with plasmid carrying carrot extensin gene and screenable marker gene GUS by meristem bombardment. Transformants identified through GUS screen and shown to express extensin gene by Southern and Northern blots analyses. The plants were generated by D.E. McCabe and B.J. Martinell
McCabe and Martinell (1993)	Described protocol for variety-independent transformation of cotton by meristem bombardment. Delta Pine 50 and 90, Sea Island and Pima S-6 were transformed with GUS gene with an electric discharge gun (Accell). Southern blot analysis demonstrated presence of GUS gene in R_0 and R_1 transformants. Progeny analysis established transmission of transgene in a Mendelian fashion. This is the first description of cultivar–independent transformation of cotton
Chlan et al. (1995)	Used biolistic method to introduce kanamycin gene into meristem of *G. hirsutum*. Plants were recovered under selection pressure. No gene integration data presented

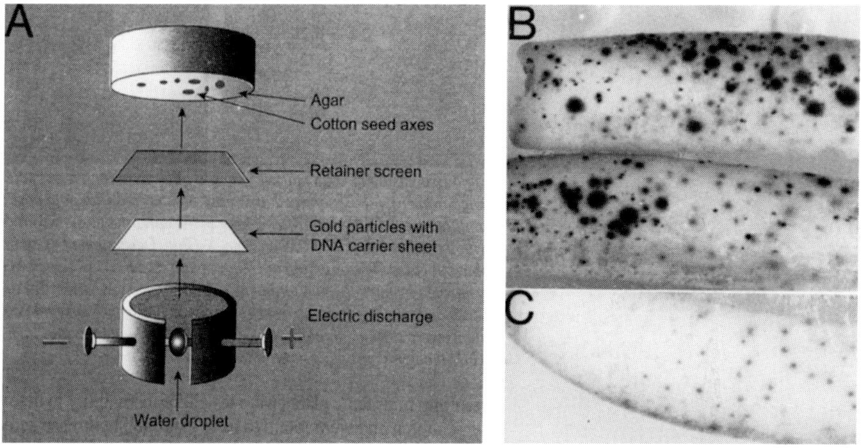

Fig. 1. A Schematic representation of particle bombardment principle. An electric current of 15 kV is applied between the electrodes to explode the water droplet. The resulting shockwave propels the carrier sheet towards the retainer screen, allowing the gold particles to move forward, but blocking the carrier sheet. The gold particles strike the plant tissue. DNA from gold particles are taken up by the cells. (After John 1994). **B** Cottonseed axes bombarded with a plasmid containing a *GUS* gene linked to cauliflower mosaic virus 35S promoter. The seed axes were assayed for GUS activity 24 h after bombardment by histochemical staining. The *blue staining* (dark here) indicates GUS activity. **C** Seed axis bombarded with a control plasmid without a *GUS* gene and subjected to histochemical staining for GUS

gene, *GUS*. After bombardment, the seed axes were allowed to develop for 2 to 3 weeks, and each leaf was tested for GUS activity by histochemical staining. Plants expressing *GUS* were mapped to identify nodes or axillary buds subtending the transformed leaves and pruned to induce the growth of the appropriate buds. The selective pruning resulted in plants that expressed *GUS* in each leaf (McCabe and Martinell 1993).

The mode of gene transfer in cotton results in two types of transformants, epidermal and germline. Various plant tissues arise from three physiological cell layers (L1, L2, and L3) of the meristem (Sussex 1989). Selective pruning and forcing of the axillary buds cause a few transformed cells from L1, L2, or L3 to gain ascendancy and populate an entire tissue layer of its origin (McCabe and Martinell 1993). If the transformed tissue layer is responsible for germline, the transgenes will be passed on to its progeny; these plants are referred to as germline transformants. Germline transformants are detected by histochemical staining for *GUS* in the vascular tissues and in pollen. In epidermal transformants, only the epidermal cell layers (L1) contain the transgene and therefore they do not pass on the transgene into their progeny. The epidermal transformants are useful for gene screens. For example, a herbicide- or insect-resistance gene can be inserted into cotton, and epidermal transformants can be tested for corresponding resistance in leaf biomass. Similarly, the effects of fiber modification genes can be tested in the fibers of epidermal transformants, as fibers are of epidermal origin (John 1996). Table 2A,B summarizes the timelines, the characteristics of germline and epidermal transformants, and their utilities. The following cotton cultivars have been transformed by particle bombardment: DP 50, DP 90, C312, Pima S6, Pima S7, Sea Island, Acala, and El Dorado (McCabe and Martinell 1993; unpubl.).

The heritability of the input genes was confirmed by Southern analysis of several generations. Each generation showed that the marker gene migrating as high molecular weight DNA when undigested genomic DNA was analyzed by gel electrophoresis suggesting integration of genes into the cotton genome. The analysis of progeny from various generations indicated normal Mendelian inheritance. A single cell, transformed in the original meristematic tissue of the bombarded shoot tip, ultimately gives rise to all of the germinal tissue for the whole shoot. The development of the clonal sectors in bombarded meristems is not induced by external hormonal treatment. The bombardment process itself must play a key role, as the meristem suffers some level of damage during bombardment. The random loss of cells from the three physiological layers responsible for the various tissue types of the plant allows individual cells to repopulate their original layer or even multiple layers. Each shoot is originally chimeric to a greater or lesser extent, depending upon the number of surviving meristematic cells within each physiological layer. As each chimera develops, the selective pruning and forcing of axillary buds causes subsets of the original population in the meristem to gain ascendancy and eventually populate the entire layer of its origin. Shoots from such meristems will then possess an entire tissue layer cloned from a single cell in the bombarded meristem. If that tissue layer is at least partially responsible for the germline, the new gene will be passed on to the progeny. The observation of chimeras, both germline and nongermline, further supports this interpretation of the data. Such chimeras would be expected from plants in which a transformed sector failed to dominate completely the meristematic layer of its origin.

The transformation frequency for cotton by particle bombardment is low. Only 0.1 to 0.2% of bombarded meristems are recovered as transgenic plants

Table 2. **A** Characteristics of particle bombardment mediated cotton transformation

Germline	Epidermal
Only L_2/L_3 layer transformed	Only L_1 (epidermis) layer transformed
Pollen is GUS-positive	Pollen is GUS-negative
Transgene transmitted through seeds (R_1)	Seeds are nontransgenic
Propagation through seeds	Vegetative propagation (R_0)
Transformation frequency is 0.08% (4/5000)	Transformation frequency is 0.19% (9/5000)
Homozygotes [R_1] in 11–12 months	Fiber tests in 6 months
Commercial development	Gene screen

Transformation frequency is based on the number of seed axes bombarded and total number of transformants recovered expressing GUS.

B Timeline for transgenic cotton development by particle bombardment

Stage	No. of plants	No. of months from start	Comments
1) Bombardment	10 000 explants	0	
2) 1st screen (GUS)	133 (1.33%)	1–1.5	GUS expression
3) 2nd screen (GUS)	27 (0.27%)	3	Insect/herbicide resistance test feasible on leaf biomass Germline identification (leaf)
4) Mature R_0 plant	25 (0.25%)	6	Insect/herbicide test Fiber quality test Germline confirmation (pollen)
5) Mature R_1 plant	8 (0.08%)	11	Fiber quality test Insect/herbicide test Field test

Assume bombardment of 10 000 seed axes (explants) with a *GUS* plasmid. The first screen in 1 to 1.5 months results in the identification of 133 putative transformants. The second screen reduces this number to 27 plants. In 6 months a total of 25 transformants (including 8 germlines) are identified.

(Table 3). However, the many positive attributes of this technology make it the premier method of obtaining transformed cotton. Further refinements, which would increase transformation frequency, would broaden its applications to routine cotton transformation in academic laboratories. Development of an antibiotic or herbicide selection regimen after particle bombardment may be useful in this regard (Chlan et al. 1995). A comparison of current cotton transformation methods and their characteristics is shown in Table 4.

3 Progress in Genetically Transformed Cotton

A number of genes with potential to confer agronomical advantages in cotton have been introduced into cotton through *Agrobacterium* or particle bombardment. Most progress has occurred in conferring insect resistance to cotton by insertion of the *Bacillus thuringiensis* (*B.t*) toxin gene (Perlak et al. 1990, 1991). Since 1990, extensive field tests have been conducted on transgenic *B.t* cotton and have shown that the *B.t* gene is effective against many late-season

Table 3. Particle bombardment-mediated cotton transformation frequencies

Experiment no.	Explants bombarded	Survived (%)	Total transgenics (%)	Epidermal frequencies	Germline frequencies
1	13292	6475 (49%)	33 (0.25%)	24 (0.18%)	9 (0.07%)
2	10913	6057 (56%)	45 (0.41%)	36 (0.33%)	9 (0.08%)
3	12278	5368 (44%)	17 (0.14%)	10 (0.08%)	7 (0.06%)
4	11378	4892 (43%)	27 (0.24%)	19 (0.17%)	8 (0.07%)
5	12930	6888 (53%)	43 (0.33%)	25 (0.19%)	18 (0.14%)
6	8799	3812 (43%)	19 (0.22%)	14 (0.16%)	5 (0.06%)
7	13159	7105 (54%)	41 (0.31%)	31 (0.24%)	10 (0.08%)
8	12587	6420 (51%)	35 (0.28%)	27 (0.21%)	8 (0.06%)
9	5013	2305 (46%)	12 (0.24%)	5 (0.1%)	7 (0.14%)
10	4993	2447 (49%)	10 (0.2%)	6 (0.12%)	4 (0.08%)
Total	105342	51769 (49%)	282 (0.27%)	197 (0.19%)	85 (0.08%)

Ten independent experiments are shown. Percent survival and transformation frequencies are based on the number of seed axes bombarded. In each experiment one or several plasmids were bombarded into seed axes. At least one *GUS* gene was included in all experiments.

Table 4. Comparison of cotton transformation methods

Characteristics	*Agrobacterium*	Protoplast gene transfter	Particle bombardment
Cycle time[a]			
Agronomical traits	2–6 months	2–6 months	3–4 months
Product traits	10–14 months	10–14 months	6 months
Cultivar	Dependent (Coker/Acala)	Dependent (Coker)	Independent (upland/*barbadense*)
Tissue culture/ regeneration/ Somoclonal Variations	Yes	Yes	No
Antibiotic selection	Yes	Yes	No
R_1 homozygote	14–18 months	14–18 months	11–12 months
Transgenic plant quality	Variable	Variable	Consistent (90 to 95% excellent)
Multigene transfer	Limited	Not limited	Not limited
Efficiency	High	(Limited data)	Low
Commercialization	7–10 years	(No data)	5–6 years[b]

[a] Cycle time refers to the earliest time or time period, after initiation of the experiment, when results can be assessed.

[b] Based on estimation (not actual data).

pests, including the tobacco budworm (Jenkins and McCarty 1995). Pink bollworm infestations of transgenic varieties were reduced by as much as 93 to 100%, and damage to crops by leaf-feeding Lepidoptera was also reduced without damage to beneficial insect species (Flint et al. 1995). *B.t* cotton is also not considered to have detrimental effects on soil ecosystems (Donegan et al. 1995). Monsanto Company (St. Louis, MO) received approval from federal agencies for commercial planting, and trangenic cotton seeds were marketed for the 1996 season. Transgenic cotton containing antielastase, antitrypsin, and antichymotrypsin protease inhibitor genes from *Manduca sexta* L. was shown to reduce the fecundity of the sweet potato whitefly (*Bemisia tabaci*) Genn.; Thomas et al. 1995).

New strategies based on genetically engineered cotton to confer resistance to herbicides that are effective against broadleaf weeds are being developed. Examples are 2,4-D (Bayley et al. 1992); bromoxynil (Stalker et al. 1996), and glyphosate (Shah et al. 1986). In addition, herbicides that inhibit acetolactate synthase (ALS or AHAS) are being investigated. A mutant tobacco, ALS, has been introduced to generate sulfonylurea-resistant cotton (Sarri and Mauvais 1996). Genes that confer bialaphos resistance (*bar* and *pat* from *Streptomyces hygroscopicus* and *S. viridochromogenes*, respectively) were also introduced into cotton (Keller et al. 1997). Herbicide-tolerant cotton varieties are expected in the market by 1998.

4 Summary

A particle bombardment protocol was developed as a variety-independent method to allow timely development of transgenic cotton. High-velocity gold beads coated with DNA were used to deliver genes directly into plants which were subsequently screened for marker gene expression. Buds in the axils of the transformed leaves were forced to develop into plants by pruning away nontransformed primary shoot tips. Plants derived from this process carried the transgene in one or more of their tissue layers. Transformation frequencies varied with genotype used, but all cultivars attempted to date have yielded transgenic progeny. The transgenes in transformants were stably integrated and transmitted in a Mendelian fashion.

5 Experimental Protocol

5.1 Preparation of DNA and Particle Delivery

Five micrograms of DNA were mixed with 100 μl of 0.1 M spermidine (free base) and used to suspend 10 mg of 1.5–3.0-μm gold beads (Alfa Chemical Co.). One hundred microliters of 2.5 M $CaCl_2$ were then added with agitation to precipitate the DNA onto the gold. The gold DNA complex was allowed to settle and the pellet[1] resuspended in 100% ethanol at a rate of 0.5 mg gold per ml. The ethanolic suspension was then pipetted onto 18 × 18-mm squares of aluminized Mylar (DuPont 50 MMC) at a rate of 320 μl per sheet. The gold was allowed to settle and the alcohol removed. The resulting loading rate was 0.05 mg gold + DNA complex per cm^2. The device used to accelerate the carrier sheets and gold was the electric discharge gun (Accell) described by Christou et al. (1990). The carrier sheets were accelerated toward the target tissues by the discharge of an 18–20-kV arc.

5.2 Explant Preparation and Plant Regeneration

Cottonseed from the commercial varieties Pima S-6, Sea Island, Delta Pine 90, and Delta Pine 50 were surface sterilized by soaking for 3 min in 2.5% sodium hypochlorite. Seeds were rinsed in sterile distilled water, then soaked for an additional 24 h at 28 °C in fungicide suspension containing 30 ppm each of Bravo, Captan, and Benomyl, plus 125 mg/l cefotaxime and 200 mg/l carbenicillin. Following surface sterilization, the seed was drained and either used immediately or incubated at 15 °C in the dark for up to 48 h. Embryonic axes were removed from the germinated seed and dissected to expose the meristem. The axes were then laid on modified MS medium containing 3 mg/l benzylaminopurine (BAP) and incubated overnight at 15 °C in the dark. These explants were then oriented so that their meristems would be accessible to bombardment (meristem in the line of flight of the beads). Following bombardment, the axes were replated on the modified MS medium plus BAP for an additional 24 h at 28 °C in darkness. After this healing period, the explants were transferred to Plantcons containing McCown's woody plant medium with 20 g/l glucose instead of sucrose. The axes were allowed to develop at 28 °C under a 16-h photoperiod.

5.3 Screening

Screening for transformants made use of a previously described GUS staining protocol. When explants had developed two to three leaves (2–3 weeks postbombardment), a piece of each leaf was sampled for GUS activity. Plants showing activity were then mapped to identify nodes or axillary buds subtending the transformed leaves. These plants were transferred to the greenhouse and pruned to induce growth of the appropriate bud. Once selective pruning produced a plant in which every leaf had GUS activity, the potential for germline transformation could be evaluated. Characterization of the transformants for this purpose relied on the observations of D.E. McCabe, P. Christou, and B.J. Martinell. They observed that whenever any part of the

[1] Footnote: Accell® is the trademark of the electric discharge gun of Agracetus, Inc.

vascular system of a soybean leaf produced from meristem bombardment expressed the reporter gene, it was able to pass that gene to its progeny at a very high frequency. Cross-sections of either a leaf or petiole from the forced cotton shoots were therefore stained to identify the specific tissue types expressing the *GUS* gene. Hand-cut sections of a fresh leaf or petiole were incubated under appropriate conditions for visualization of GUS activity. If one or more vascular bundles expressed GUS, the shoot was assumed to be a germline transformant. Plants showing only epidermal or cortical GUS activity were classified as nongermline events. Both germline and nongermline transformants were evaluated further at anthesis by staining pollen for GUS. Free pollen was shaken into a Petri dish containing the GUS buffer and substrate. Great care was taken to prevent any extraneous maternal tissue, i.e., anther, petal, etc., from contaminating the pollen assay. Any such contamination could yield false positive results. Pollen was incubated at 37 °C for up to 4 h, then scored for the presence of GUS activity.

5.4 Progeny Analysis

R_1 seeds recovered from pollen-positive R_0 plants were germinated and grown in the greenhouse. Leaf punches were taken from each plant and analyzed for GUS activity to determine segregation patterns. Up to ten positive plants from each family were then allowed to develop to anthesis. At this point, pollen staining for GUS permitted selection of homozygous individuals for the production of R_2 plants for further analysis.

Acknowledgment. The authors thank Cheryl Scadlock and Peggy Wagner for editorial assistance. This study was partially funded by the Department of Commerce, National Institute of Technology, Advanced Technology Program Grant JONANB5H1061.

References

Barton KA, Binns AN, Matzke AJM, Chilton MD (1983) Regeneration of intact tobacco plants containing full-length copies of genetically engineered T-DNA and transmission of T-DNA to R1 progeny. Cell 32:1033–1043

Bayley C, Trolinder N, Ray C, Morgan M, Quisomberry JE, Ow DW (1992) Engineering 2,4-D resistance into cotton. Theor Appl Genet 83:645–662

Chlan CA, Lin J, Cary JW, Cleveland TE (1995) A procedure for biolistic transformation and regeneration of transgenic cotton from meristematic tissue. Plant Mol Biol Rep 13:31–37

Christou P, McCabe DE, Martinell BJ, Swain WF (1990) Soybean genetic engineering – commercial production of transgenic plants. TIBTECH 8:145–151

Christou P, Ford TL, Kofron JM (1992) The development of a variety-independent gene-transfer method for rice. TIBTECH 10:239–246

Cooley J, Ford T, Christou P (1995) Molecular and genetic characterization of elite transgenic rice plants produced by electric-discharge particle acceleration. Theor Appl Genet 90:97–104

Donegan KK, Palm CJ, Fieland VJ, Porteous LA, Ganio LM, Shaller DL, Bucao LQ, Seidler RJ (1995) Changes in levels, species, and DNA fingerprints of soil microorganisms associated with cotton expressing the *Bacillus thuringiensis* var. Kurstaki endotoxin. Appl Soil Ecol 2:111–124

Finer JJ, McMullen MD (1990) Transformation of cotton (*Gossypium hirsutum* L.) via particle bombardment. Plant Cell Rep 8:586–589

Firoozabady E, DeBoer DL, Merlo DJ, Halls E, Halk EL, Amerson LN, Rashka KE, Murray EE (1987) Transformation of cotton (*Gossypium hirsutum* L.) by *Agrobacterium tumefaciens* and regeneration of transgenic plants. Plant Mol Biol 10:105–116

Flint HM, Henneberry TJ, Wilson FD, Holquin E, Parks N, Buchler RE (1995) The effect of transgenic cotton. (*Gossypium hirsutum* L.) containing *Bacillus thuringiensis* toxin genes for the control of the pink bollworm *Pectinophora gossypiella* and other arthropods. Southwest Entomol 20:281–292

Fraley RT, Rogers SG, Horsch RB (1986) Genetic transformation in higher plants. Crit Rev Plant Sci 4:1–46

Jefferson RA, Kavanagh TA, Bevan MW (1987) GUS fusions: β-glucuronidase as a sensitive and versatile gene fusion marker in higher plants. EMBO J 6:3901–3907

Jenkins JN, McCarty JC (1995) B.t. cotton: a new era in cotton production. In: Proc Beltwide Cotton Conf, vol 1. National Cotton Council, Memphis, TN, pp 171–173

John ME (1994) Re-engineering cotton fibre. Chem Ind N17, pp 676–679

John ME (1996) Structural characterization of genes corresponding to cotton fiber mRNA, E6: reduced E6 protein in transgenic plants by antisense gene. Plant Mol Biol 30:297–306

John ME, Crow LJ (1992) Gene expression in cotton (*Gossypium hirsutum* L.) fiber: Cloning of the mRNAs. Proc Natl Acad Sci USA 89:5769–5773

Keller G, Spatola L, McCabe D, Martinell B, Swain W, John ME (1997) Transgenic cotton resistant to herbicide bialaphos. Transgenic Res (in press)

Klein TM, Wolf ED, Wu R, Sanford JC (1987) High-velocity microprojectiles for delivering nucleic acids into living cells. Nature 327:70–73

McCabe DE, Martinell BJ (1993) Transformation of elite cotton cultivars via particle bombardment of meristems. Bio/Technology 11:596–598

McCabe DE, Swain WF, Martinell BJ, Christou P (1988) Stable transformation of soybean (*Glycine max*) by particle acceleration. Bio/Technology 6:923–926

Perlak FJ, Deaton RW, Armstrong TA, Fuchs RL, Sims SR, Geenplate JT, Fischhoff DA (1990) Insect-resistant cotton plants. Bio/Technology 8:939–943

Perlak FJ, Fuchs RL, Dean DA, McPherson SL, Fischhoff DA (1991) Modification of the coding sequence enhances plant expression of insect control protein genes. Proc Natl Acad Sci USA 88:3324–3328

Saari LL, Mauvais CJ (1996) Sulfonylurea herbicide resistant crops. In: Duke SO (ed) Herbicide-resistant crops: agricultural, environmental, enconomic, regulatory, and techchical aspects. Lewis Publishers, New York, pp 127–142

Shah DM, Horsch RB, Klee HJ, Kishore GM, Winter JA, Tumer NE, Hironaka CM, Sanders PR, Gasser CS, Aykent S, Siegel NR, Rogers SG, Fraley RT (1986) Enginering herbicide tolerance in transgenic plants. Science 233:478–481

Stalker DM, Kiser JA, Baldwin G, Coulombe B, Houck CM (1996) Cotton weed control using the BXN system. In: Duke SO (ed) Herbicide-resistant crops: agricultural, environmental economic, regulatory, and technical aspects. Lewis Publishers, New York, pp 93–105

Sussex IM (1989) Developmental programming of the shoot meristem. Cell 56:225–230

Thomas JC, Adams DG, Keppenne VD, Wasmann CC, Brown JK, Kanost MR, Bohnert HJ (1995) Protease inhibitors of *Manduca sexta* expressed in transgenic cotton. Plant Cell Rep 14:758–762

Trolinder NL, Goodin JR (1987) Somatic embryogenesis and plant regeneration in cotton (*Gossypium hirsutum* L.). Plant Cell Rep 6:231–234

Umbeck P, Johnson G, Barton K, Swain W (1987) Genetically transformed cotton (*Gossypium hirsutum* L.) plants. Bio/Technology 5:263–266

Wilson FD, Flint HM, Deaton WR, Buehler RE (1994) Yield, yield components, and fiber properties of insect-resistant cotton lines containing a *Bacillus thuringiensis* toxin gene. Crop Sci 34:38–41

V.2 Insect-Resistant Transgenic Cotton

J.C. THOMAS[1]

1 Introduction

Insect-ravaged cotton crops are a dominant concern for the global cotton agricultural community. Pivotal factors leading to extensive cotton damage include discontinued use of many effective pesticides (due to environmental concerns) and rapidly arising insect resistance to the remaining pesticides. To meet the reccurring insect challenge, advances in newer and safer chemical pesticides, pheromones, insect parasites, plant breeding, modified crop cultivation practices, and integrated pest management (IMP) have helped reduce crop destruction. Despite these protective strategies, cotton pests can still cause large crop losses, up to 50% (Schwartz 1983). Insect damage to cotton has been estimated to cost consumers over US$600 million per year (Perlak et al. 1990). Adding to the field of cotton predators, the appearance of the new and polyphagous superbug, or *Bemisia tabaci* (Genn.) (whitefly) has caused even more extensive damage (Henneberry and Toscano 1993). By direct feeding, honeydew contamination, and damage inflicted by viruses, *B. tabaci* alone caused over US$500 million in crop damage in 1991 (Brown and Bird 1992; Perring et al. 1993).

The need to enhance insect resistance in cotton has been initiated with transgenic *Bacillus thuringensis* (B.t.)-Bollgard cotton (Monsanto). Target insects for Bollgard cotton are bollworm complex plus pink bollworm and cabbage looper, because existing methods are costly and may not be completely effective. Bollworm, *Heliothis zea* (Boddie) (corn earworm), and tobacco budworm, *H. virescens* (Fabricius), are very destructive to cotton. Larvae destroy several fruiting forms, thus low insect populations can cause enormous economic damage. Early use of insecticides helps control these insects, especially small second-generation bollworms. Pink bollworm, *Pectinophora gossypiella* (Saunders) also devastates cotton. Control measures include releasing sterilized adult moths to lower arising insect populations. The cabbage looper, *Trichoplusia ni* (Hubner), is primarily a problem in the south, damaging cotton foliage usually after early insecticide applications. In cotton, cabbage looper populations usually remain low due to a viral disease.

[1] Department of Natural Sciences, University of Michigan-Dearborn, Dearborn, Michigan 48128-1491, USA

Biotechnology in Agriculture and Forestry, Vol. 42
Cotton (ed. by Y.P.S. Bajaj)

In contrast, many insect pests of cotton are unaffected by B.t. toxin. Whitefly *Bemisia tabaci* and *B. argentifolii* cause cotton losses from US$500 million to US$1 billion annually, damaging leaves, vectoring plant viruses, and producing honeydew that complicates fiber processing. To control whitefly, careful planning of planting sites, planting and harvesting dates, and use of smooth-leaf cotton varieties are suggested. Scouting, control with insect growth regulators, and postponed and/or limited pyrethroid insecticides have been advised. Thrips represent a diverse insect group which causes damage to seedling cotton. Severe thrips infestations may require insecticides, especially early in the season. Boll weevil (*Anthonomous grandis grandis* Boheman) damage is caused by oviposition on squares and bolls, the young feeding within and causing square shedding. Insect control depends upon early planting, adequate fertilization, weed control, use of early-maturing cotton varieties, and insecticides during fruit set and boll maturity. Tarnished plant bug, *Lygus lineolaris* (Palisot de Beauvois), or *Lygus hesperus* pierce terminal growth, squares, flowers and bolls, and extract plant juices. Punctured squares and terminals drop off (blast), causing plants to remain in a vegetative state, delaying fruiting and fiber maturity. Lygus bug is now considered the number one problem insect on cotton in Arizona in 1996–1997 (L. Antilla, pers. comm.). Insecticides are used to control *Lygus*; however, lowering *Lygus* populations may stimulate increases in pink bollworm and white fly. Because B.t. toxins currently available are ineffective against whitefly, thrips, boll weevil, and lygus bug, novel in planta insecticides are needed for the future protection of cotton. Tactics for insect pest management including the introduction and expression of novel antiinsect genes in transgenic plants are currently poised to enter the worldwide cotton market. While progress towards plant protection has been achieved with *Bacillus thuringiensis* (B.t.) toxin expression in cotton, questions regarding insect specificity and resistance persist. This chapter reviews recent advances in insect-tolerant cotton and outlines some viable, to date noncommercial genes and strategies to be considered in the future.

2 Genes for Antiinsect Activities

2.1 *Bacillus thuringiensis* B.t. Toxin

Toxins produced in the bacterium *Bacillus thuringiensis* (B.t.) have been the focus of many antiinsect strategies. Early studies focused on exogenous application of the bacterially produced toxin on plants to thwart insect induced damage (Ali and Watson 1982). While current transgenic technology strives to express B.t. toxins in cotton, the mechanism of B.t. action remains unclear.

The advantages of using B.t.s are that they break down quickly and safely in the environment and they lack toxicity in mammals, including man. B.t. δ endotoxins kill only at the larval stage and they are insect-specific. Interestingly, nearly 40% of 1000 known B.t. isolates have no measurable activity on insects (Martin and Travers 1989), suggesting rapid evolution of the toxins in the environment. Most B.t. Cry toxins contain three N-terminal domainess. Domaine I is characterized by six amphipathic α-helices surrounded by a central hydrophobic region, thought to participate in membrane insertion. Domaine II is composed of three β-sheets with terminal loops which, according to domane swapping and mutagenesis experiments, appear to be responsible for specificity of ligand-receptor binding. Initially, the prototoxin is ingested and the alkaline pH of the insect gut solubilizes the protein crystal. Next, endogenous insect proteases cleave the larger prototoxin into the smaller and biologically active (55–65 kDa) peptide. The toxin binds an insect-specific receptor and then inserts itself into the plasma membrane of the columnar cells, creating a pore of about 0.6 nm. Pore formation causes a breakdown in K^+ transport (Knowles and Dow 1993). Eventually, loss of the K^+ gradient results in osmotically induced cell lysis, septicemia, and starvation. Further discussion of the B.t. mechanism can be found in several reviews (Gill et al. 1992; Knowles and Dow 1993).

2.1.1 Mechanism of B.t. Action

Toxins of *B. thuringiensis* have been used commercially for many years (Aronson et al. 1986). Furthermore, introduction and transgenic plant expression of B.t. toxins have provided some degree of plant protection from insects (Barton et al. 1987; Fischhoff et al. 1987). In early studies, the level of B.t. that accumulated in these transgenic plants was relatively low, approximately 0.001% of total protein. Initial field tests indicated that native B.t. production and accumulation in transgenic cotton did not protect the plants from insect damage when compared to the nontransformed controls (Jenkins et al. 1990). Improved B.t. accumulation and efficacy in cotton tissues was needed. By removing all but the biologically active N-terminal encoding region of the B.t. gene, increased B.t. levels were observed (Fischoff et al. 1987). An in-frame fusion of the NPT gene and B.t. toxin also improved protein accumulation sufficiently to provide plants some protection from insects (Vaeck et al. 1987). Without changing the amino acid sequence, Monsanto workers modified codon usage of the CryIA (b) and CryIA (c) B.t. genes to utilize abundant tRNAs in plants. Additionally, the scientists substituted many high AT-rich regions within the native B.t. gene with GC-rich regions (Perlak et al. 1991). Using this gene redesign approach, up to 100-fold expression of B.t. toxins, compared to the native genes, was achieved in transgenic plants, and these modified B.t. toxins encoded by the CryIA (b) and CryIA (c) genes accumulated to 0.05 and 0.1% of the total protein, respectively, in transgenic cotton (Perlak et al. 1990).

2.1.2 Applications with B.t. Toxins in Cotton

1. *Insect/Field Tests.* Independent of transgenic protein expression and accumulation, actual field studies with native insect species were crucial to the commercialization process of B.t. cotton. Cotton plants that expressed the

genetically altered and "optimized" CryIA (b) gene were greatly protected against the cabbage looper (Perlak et al. 1990). In similar tests on identical B.t. cotton lines, beet army worm mortality was lower on nontransformed plants compared to cotton producing the "optimized" CryIA (b) protein. Cotton plants accumulating either the CryIA (b) and CryIA (c) B.t. toxins were either lethal to cotton bollworm or lengthened this insect's developmental rate when compared to controls (Perlak et al. 1990). Genetically engineered B.t. toxins have also been expressed in other species including rice (Fujimoto et al. 1993) and corn (Koziel et al. 1993).

Prior to release, the impact of residual transgenic cotton plant matter on the resident soil ecosystem was examined. Two of three transgenic (B.t.) plants stimulated microbial populations in the surrounding soil (Donegan et al. 1995). These results were not directly attributable to B.t. toxin itself, but were thought due to the host plant, itself a product of somaclonal variation from tissue culture-derived plants (Donegan et al. 1995). These important and significant results have helped establish the feasibility of insecticide-producing cotton, paving the way for the B.t. cotton program at Monsanto and other companies.

Commercialization of insect-protected transgenic crops producing B.t. have been undertaken by several companies, including Monsanto, Mycogen, Calgene, and Ciba-Geigy. Extensive field testing of cotton expressing B.t. proteins has indicated that some of the most damaging insect pests to cotton can be controlled using this approach (Jenkins et al. 1991a,b; Wilson et al. 1994). For example, insect damage to bolls of B.t. expressing cotton was between 2 and 12% of all bolls examined, when contrasted to the 29% boll damage recorded for control plants. In addition, boll size was smaller in the B.t. cotton varieties, and one line contained a higher percentage of lint than controls (Jenkins et al. 1991a), underlining the importance of a structured breeding program to introgress transgenic lines into cultivated varieties.

One direct result of extensive field studies was the rapid Environmental Protection Agency (EPA) approval of B.t. cotton registration for Monsanto, the first insect-protection gene to be approved for this species. This announcement followed the successful completion of regulatory reviews by the Food and Drug Administration (FDA) and the US Department of Agriculture (USDA). The B.t.-cotton product provides a built-in pesticide that protects against lepidopterans: cotton bollworm, tobacco budworm, and pink bollworm. The EPA also granted permission to Monsanto to make large-scale plantings of cotton varieties containing the B.t. gene to produce seed for 1996 sales to America's cotton farmers (Anonymous 1996). In March 1995, the US Patent and Trademark Office granted a composition patent (No. 5500365) for a class of genes encoding B.t. proteins from *Bacillus thuringiensis* to Monsanto (Anonymous 1996). Due to increased public concern and the development of pyrethroid resistance in insects, B.t. cotton may also be a viable alternative in Australian-grown cotton if costs are favorable for the growers (Forrester 1994).

Recently, Monsanto has agreed to purchase the plant biotechnology assets (genetic transformation methods, cotton fiber improvement, proteins in pharmaceutical and industrial applications) of Agracetus. Further partnerships and acquisitions including the Zeneca Group PLC (Biopolymers), the Delta and Pine Land Company (the leading US cotton seed company), its wholly owned Paymaster Technology Corporation, and the Jacob Hartz Seed Company are all indicative of Monsanto's intention to remain an industry leader. Future plans are in place to produce biodegradable polyhydroxyalkanoic acid (PHA) polymers in transgenic plants for use in consumer products and packaging (Anonymous 1996). PHA polymers are already known to be synthesized by some microorganisms. One advantage of PHA production in plants is that availability is not dependent on fossil fuels. Secondly, plants accumulating PHAs may be less nutritious to insects than non-PHA plants, lending a value added aspect to transgenic plants producing biopolymers.

2. *B.t. Resistance in Insects.* B.t. resistance has been suggested as possibly arising in some insect species (Gould et al. 1992), particularly the boll worm (Bradley 1996). In addition, high levels of B.t. expression in transgenic plants may actually promote the development of insect resistance (Harris 1991). Susceptible refuge plants within a B.t. field is one approach to solve the developing resistance threat (Deaton 1996). Another view would be to introduce or stagger different B.t. genes (encoding distinct toxins) together with other antiinsect genes, work that is ongoing to date (Lambert et al. 1996). Others have pointed out the importance of tissue-specific expression of Bt and other antiinsect compounds to better combat Bt resistance (Harris 1996). These and similar approaches are currently underway and being evaluated.

2.2 Alternative Antiinsect Strategies

1. *Protease Inhibitors.* To combat insect predation, natural defense mechanisms in many plants include endogenous inducible protease inhibitors (PIs). These PIs are thought to arrest digestive enzymes in the insect gut and limit essential amino acid availability to the developing insect (Ryan 1990). The net effect of increased PI accumulation is a decrease in value of the plant material when used as an insect food source (Broadway and Duffey 1986; Broadway et al. 1986). Addition of PIs to insect diets has been shown to limit insect development (Broadway and Duffey 1986). Overexpression of introduced PIs has protected plants against damage caused by some insects (Hilder et al. 1987; Johnson et al. 1989).

Toxicity of PIs may not follow a simple dose and response curve. Some PIs are more effective when presented to insects in a low protein diet, while other inhibitors act to retard insect growth in either high or low protein-containing artificial diets (Burgess et al. 1991). Trypsin inhibitors supplemented into the insect diet reduced growth rate of *Heliothis zea*, but measurable trypsin activity in the insect gut increased (Broadway and Duffey 1986; Burgess et al. 1991). When expressed under the control of the CaMV 35S promoter, the rice-encoded cystatin (a cysteine proteinase inhibitor) accumulated to 0.5% of total soluble protein in transgenic tobacco, with demonstrable activity against papain (Masoud et al. 1993). However, resistance of one line of cowpea seeds to bruchid beetles was not correlated necessarily to levels of cysteine proteinase inhibitors (Fernandes et al. 1993). Hyperproduction of digestive enzymes in response to a diet containing protease inhibitors could result in decreased concentrations of sulfur-containing amino acids, thus indirectly inhibiting insect growth (Ryan 1990). If this is the case, then lower toxicity of PI should be observed in more adult developmental stages, and suggesting other types of approaches should be used in concert with PIs.

Protease inhibitors from a variety of sources are thought to protect plants against insect feeding (Ryan 1990). Modifications of the alaserpin cDNA encoding an anti-elastase protease inhibitor were described previously (Kanost et al. 1989). The P1 alanine residue to alaserpins defines the primary selectivity of serpin-1B for elastase-like enzymes; however, the selectivity can be altered by mutation (Jiang et al. 1995). We have used PCR-based in vitro mutagenesis to alter the active site, creating chymotrypsin-specific or trypsin-specific inhibitors (Thomas et al. 1995a).

Antitrypsin, antichymotrypsin, and antielastase protease inhibitor (PI) genes from *Manduca sexta* were easily expressed in transgenic alfalfa and tobacco (Thomas et al. 1994, 1995b). When expressed in alfalfa plants, up to 0.125% of the total soluble protein was accounted for as the engineered *M. sexta* protease inhibitor (Thomas et al. 1994). Noticeable damage to transgenic PI-expressing alfalfa plants was delayed nearly three times as much as in controls (Thomas et al. 1994). Transgenic tobacco plants accumulated 0.1% of the total soluble protein as PI and the plants were poor hosts in supporting whitefly reproduction, compared to controls (Thomas et al. 1995b). In *G. hirsutum*, selection of 35 elite transgenic cotton lines from 198 independent kanamycin-resistant transformants found that the same *Manduca sexta* encoded PIs accumulated to the greatest levels in bracts and mature leaves when transcription was under the control of the 35S promoter of CaMV (Thomas et al. 1995a). Overall, PI accumulated to approximately 0.1% of total protein in cotton, depending on the tissue analyzed (boll, stem, leaf, etc.). Compared to controls, whitefly bioassays on four kanamycin-resistant progeny (T-1) plants from independent *G. hirsutum* var. Coker 312 PI-expressing transgenic lines demonstrated a lower degree of reproduction of *Bemisia tabaci* (Genn.) (Thomas et al. 1995a). In addition, one very active protease inhibitor known to have antiinsect activity from taro, *Alocasia macrorrhiza*, has been recommended as potentially useful against insects if expressed in transgenic cotton (Llewellyn et al. 1994).

2. *Lectins/α-Amylase.* Orally administered insecticides also include lectins and amylase inhibitors. An α-amylase inhibitor of the common bean (*Phaseolus vulgaris*) inhibits insect and mammalian α-amylases but not plant α-amylases (Moreno and Chrispeels 1989). When transgenic pea seeds were engineered to express α-amylase inhibitor of bean, the plants became more resistant to bruchid beetles than controls (Shade et al. 1994). Other plant-encoded lectins have also been shown to be effective in controlling insects (Boulter et al. 1990). In addition, N-acetylglucosamine-specific lectins from

Griffonia simplicifolia (Leguminosae) have recently been identified as having anti-insect activities (Zhu et al. 1996).

3. *Chitinase.* An extracellular amino sugar, chitin, is found in insect cuticle and egg shells and has long been considered a selective target for pesticide action, particularly by enzymes known as chitinases (Cohen 1993). While effective against fungi, chitinase expression in transgenic plants and subsequent protection against insects is still in development (Ding et al. 1995).

4. *Cholesterol Oxidase. Streptomyces*-produced cholesterol oxidase was shown to disrupt the midgut epithelial membrane and significantly inhibit larvael and ovary development in *Anthonomus grandis grandis* Boheman (Purcell et al. 1993; Greenplate et al. 1995; Cho et al. 1996). The mechanism of action of cholesterol oxidase has yet to be elucidated; however, lytic activity has been associated with excessive cholesterol or lipid in membranes (Lange et al. 1984; Berheimer et al. 1987). When the gene encoding the enzyme cholesterol oxidase is expressed, the transgenic plants are protected from weevils and worms (Cho et al. 1996). This *Streptomyces*-encoded enzyme may be quite effective in transgenic cotton and other crops.

5. *Tryptamine/Glucoscinalates.* Tryptophan decarboxylase (TdC) expression in transgenic tobacco plants impairs whitefly reproduction when compared to tobacco previously transformed with the *uidA* (GUS) gene (Jefferson et al. 1987; Thomas et al. 1995d). Accumulation of tryptamine in leaf and phloem was observed, perhaps due to the CaMV 35S promoter activity in phloem as well as in other tissues (Benfey et al. 1990). TDC expression was demonstrated immunologically and the presence of TDC and tryptamine in tobacco plants was coincident with a decrease in whitefly pupae emergence, as much as 97% compared to control plants (Thomas et al. 1995d). Our efforts are currently directed towards the introduction and expression of TdC in transgenic cotton.

Exactly how tryptamine may act against insects is unknown. One thought is that TdC may lead to decreased levels of tryptophan, an amino acid needed for insect oogenesis (Chapman 1969). Our results indicated that tryptophan levels were similar in control and TdC plants, suggesting that TdC may act via another antiinsect mechanism (Thomas et al. 1995d). Tryptamine may exert antiwhitefly effect(s) directly by slowing insect development or indirectly by decreasing aspects of insect reproduction, such as decreased oviposition rate (Thomas et al. 1996, 1997). Tryptamine also inhibits serotonin-mediated brain monoamine oxidase (Vehovszky and Walker 1991), and can be converted to other indole-alkaloid derivatives which may subsequently affect insect feeding (Niemeyer 1990; Aerts et al. 1991). Surface glucosinolates, which greatly influence insect host selection (Niemeyer 1990), decreased dramatically when TDC was expressed in transgenic plants (Chavadej et al. 1994). Clearly, more effort towards understanding the mechanism of tryptamine action is needed.

2.3 Mobilizing Endogenous Defenses Against Insects

2.3.1 Pigments and Other Alleochemicals of Cotton

Historically, combinations of toxic insecticidal, structural, and genetic components have been part of antiinsect strategies in cotton (Bailey et al. 1980). Many structural characteristics that are deleterious to insects already exist in some cotton varieties. The nectarless trait of okra-leaf cotton has demonstrable resistance to pink bollworm (Flint et al. 1992). The open-bract trait allows improved boll penetration of applied insecticides (Smith 1989). Biosynthesis and deposition of epicuticular waxes and *trans*-aconitic acid in plants are also known to discourage insect feeding (Eigenbrode and Espelie 1995). Glandular trichomes containing toxic compounds may also adversely affect feeding, growth, and survival of insects (Jackson et al. 1986).

Cotton is known to contain a variety of alleochemicals that inhibit insect growth, including gossypol. This sesquiterpene aldehyde has been shown to adversely affect growth of tobacco budworm and cotton leafworm (Lukefahr and Martin 1966; Meisner et al. 1977). Hemigossypols (or heliocides) have also been described (Stipanovic et al. 1977), as well as other terpenoids (Stipanovic et al. 1986). Tannins and gossypol, together with leaf hairiness and moisture content, can greatly affect insect oviposition rates (Sharma and Agarwal 1984). Isley (1928) found that the red leaf color of some cotton varieties had a negative effect on adult boll weevils (*Anthonomous grandis grandis* Boheman). Due to color or other associated alleochemical qualities, accessory plant pigments may play a significant role in insect damage to cotton (Lukefahr and Martin 1966). By combining feeding, visual, and structural cues, pigments and other alleochemicals already present in cotton germplasm may discourage insect feeding.

There are many known antiinsect compounds that could be applied and/or engineered for cotton expression in combinatorial insecticidal strategies (or gene stacking) in *Gossypium*. For example, chlorogenic acid and polyphenol oxidase have been found to enhance B.t. toxicity in *Heliothis* (Ludlum et al. 1991). L-canavine plus B.t. toxins were shown to be more effective against *Manduca sexta* than either applied alone (Felton and Dahlman 1984). Flavonoid content and composition may also play a role in conferring insect resistance (Hedin et al. 1992). On cotton, additive resistance to tobacco budworm was conferred by the introduction of large numbers of both gossypol glands and nectarless traits (Wilson and George 1983). Cotton-produced tannins are very effective, especially against early-staged larvae of *H. virescens* (Chan et al. 1978; Navon et al. 1993). However, caution must be exercised when combining tannins together with B.t., as B.t. effects are masked rendering this combined strategy ineffective (Navon et al. 1993). Somewhat surprisingly, gossypol itself did not affect larval growth in artificial diet experiments. Remarkably, when gossypol was tested in combination with B.t. toxin, larvae actually grew faster (Meade and Felton 1994). These results underline the somewhat unpredictable nature of antiinsect compounds and demonstrate the need for further study into plant-insect interactions.

2.3.2 Role of Plant Growth Regulators on Insect Resistance

In response to damage, wound responses are thought to contribute to a plant's natural defense strategy (McGurl et al. 1994). The "conditioning" of plants, or the induction of already existing genetic traits conferring insect resistance, has been demonstrated in cotton. Exposure to *Tetranychus turkestani* was shown to later confer resistance to *T. urticae* (Karaban and Carey 1984). Because *T. urticae* injects saliva into cotton during the course of feeding, it has been hypothesized that resistance may be acquired as a result of earlier exposure and a plant-based wound response (Stroms 1971). Mechanical abrasion has also been shown to induce resistance toward mites (Karaban 1985), lending support to the idea that induction of genes, or wound responses, are then able to combat insect attack.

Growth rates of plants are also known to affect the extent of damage caused by insects (Hinds 1906). Perhaps by affecting endogenous plant wound responses and/or growth rates, plant growth regulators can also have an influence on insect resistance in cotton. N-N-dimethylpiperidinium (PIX) applications increased production of gossypol, making plants less acceptable to *Heliothis virescens*, but had no effect on tobacco budworm (Mulrooney et al. 1985). Similarly, *Manduca sexta* larvae and *Myzus persicae* development were delayed on transgenic tobacco plants overproducing cytokinin (Smigocki et al. 1993). Because cytokinins are thought to induce many stress and wound responses (Thomas et al. 1995c), plant-encoded defenses may be mobilized upon increased cytokinin biosynthesis. In addition, gibberellin application also exerts a great influence on endogenous proteinase inhibitors accumulation (Jacobsen and Olszewski 1996).

Genetically controlled signal molecules such as systemin may stimulate widespread and intrusive genetic pathways in plants and may protect crops like cotton. Systemin is a peptide produced in response to wounding and thought to mediate transduction of wound and stress responses in plants. When systemin is overproduced in a constitutive transgenic background, proteinase inhibitor synthesis is also induced, up to 1 mg/g of dry weight (McGurl et al. 1994). In addition, systemin activates several secondary pathways, thought to contribute to alleochemical biosynthesis (Constabel et al. 1995). It has recently been suggested that specific systemin-DNA interactions in the promoter region of proteinase inhibitors and other wound-inducible genes could explain the biological activity of systemin (Slosarek et al. 1995). Some plant responses to environmental stress may be mediated by systemin, cytokinins, or other plant growth regulators.

3 Tissue-Specific Expression

Improved antiinsect activity may be achieved by tissue specific expression of a particularly toxic compound. Tissue-specific expression is governed, in part, by the type of promoter used to express the *trans*gene. A large variation in the expression of *trans*genes has been observed when expression was under the control of the strong and constitutive 35S promoter of CaMV (Benfey et al. 1990). In cotton, three-quarters of transgenic plants containing the 35S promoter with PI constructions were kanamycin-resistant and also accumulated PI, as detected by Western blot analysis (Thomas et al. 1995a). Both single and double 35S promoter types led to similar levels of PI accumulation in alfalfa, cotton, and tobacco (Thomas et al. 1995a,b). Leaf-specific promoters have been used to optimize B.t. expression in rice (Kyozuka et al. 1993). The stress-inducible and strong promoter of phosphenolpyuvate carboxylase (*Ppc-1*) from *M. crystallinum* (Schaffer et al. 1995) led to little PI protein accumulation in either stressed or unstressed plants (Thomas et al. 1994, 1995a,b). Thus, use of tissue-specific promoters originally isolated from cotton may help alleviate nonuniform expression from transgenic plant to transgenic plant, sometimes ascribed to "position effects" or differences as a result of species (Narvaéz-Vásquez et al. 1992). This approach would be particularly effective for boll-specific promoters, as low activity of the 35S promoter was observed in cotton bolls (Thomas et al. 1995a).

Some lepidopterans may be easily controlled by plant-encoded toxins expressed throughout the plant; however, other phloem-feeding pests such as

Fig. 1. Somatic embryos of Coker 312 previously transformed with the β glucuronidase *uidA* (GUS) gene under the control of the 35S promoter of CaMV. Cotyledon tissues were cocultivated with pBI121.1 in LBA4404 (see Jefferson et al. 1987) and transformed calli selected on 50 μg/ml of kanamycin (Thomas et al. 1995a). After several subcultures, embryos were recovered and stained with XGLUC, a colorless substrate that, when cleaved by *uidA* (GUS), results in a blue precipitate

aphids and the whitefly may be more efficiently controlled by targeting gene expression to the phloem and companion cells. The whitefly is thought to feed on phloem tissues, after inserting its proboscis through the epidermis and several layers of mesophyll (Rosell et al. 1995). The promoter often used to express insecticidal proteins in transgenic plants is the 35S promoter from CaMV, also known to allow transgene expression in the phloem, presumably via companion cell production and transport to the phloem (Benfey et al. 1990). Phloem exudate naturally contains several proteins including phloem-abundant proteins (Bostwick et al. 1992). Promoters of these phloem-localized

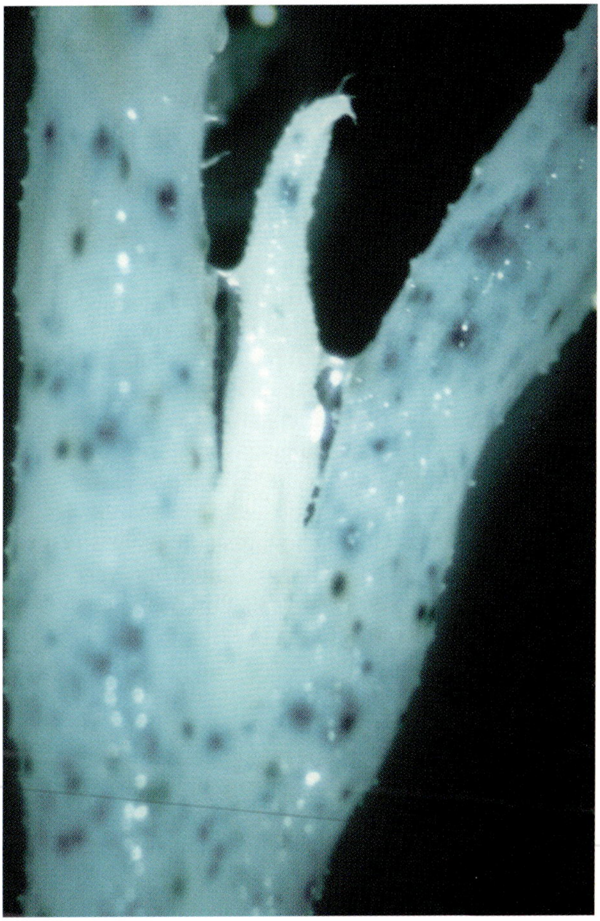

Fig. 2. DNA bombardment approach to transient expression in *Gossypium hirsutum* shoot tips. Coker 312 seeds were surface sterilized and germinated in standard MS medium (Thomas et al. 1995a). After 7 days, shoots were dissected and used in bombardment experiments. DNA plasmid pBI121.1 was precipitated on tungsten particles and bombarded onto shoot tips, as described previously (Klein et al. 1987). *UidA* (GUS) staining and destaining was according to Jefferson et al. (1987)

proteins may provide an excellent means of targeting insecticidal proteins to the phloem.

Epidermal tissue is also a likely target to present antiinsect products near the surface where much plant/insect interaction occurs (e.g., the plant recognition sites) (Clark et al. 1992). A differentiated cotton epidermal (fiber) promoter was shown to direct the expression of a transgene in a developmentally regulated and tissue-specific fashion in transgenic cotton plants (John and Crow 1992). Promoter fusions to the reporter gene β-glucuronidase *uidA* (GUS) and the subsequent GUS expression in transgenic cotton will adequately describe the tissue-specific expression of transcriptionally induced mRNA (Figs. 1 and 2). Caution must be applied, however, to toxin expression in the pollen, as this is an important site of bee-mediated cotton fertilization. Tissue-specific expression of antiinsect genes will be a future research goal, to increase effectiveness against insect feeding while decreasing the likeliness of arising insect resistance (Harris 1991).

4 Blending Biopesticides with Plant Breeding

When resistance to a normally toxic pesticide occurs, the passage of the resistance trait is implied to continue throughout the population. One potential problem with the singular antiinsect gene expression approach may be illustrated by the adaptive resistance to protease inhibitors by *Spodoptera* larvae, suggesting that other strategies may be necessary to insure antiinsect activity for several cropping seasons (Jongsma et al. 1995). Similar problems have been outlined above for B.t. toxin expression in transgenic plants. Expressing multiple insecticide genes in plants has been suggested to improve effectiveness and decreasing resistance in the target insect (Boulter et al. 1990). Several serine protease inhibitors were found to enhance insecticidal activity of *Bacillus thuringiensis* (B.t.) toxins, even without detectable PI toxicity alone (MacIntosh et al. 1990). Considering the arising resistance, similar combinatorial strategies have been recently outlined to combat arising B.t. resistance (McGaughey and Whalon 1992). Gene expression is a principle component in controlling arising resistance. Combinations of high, low, and moderate doses of the genes encoding antiinsect proteins in concert with targeted (tissue specific) gene expression, together with field management tactics such as gene rotation, mosaic planting, and providing refuges, will help minimize insect resistance (McGaughey and Whalon 1992).

5 Conclusion

Current agricultural methods of insect control can no longer exclusively rely on pesticides to protect the cotton fields. Today, a central goal of crop improvement is to develop novel strategies which utilize natural and environmentally acceptable methods to protect crops through a combination of chemical, biological, and behavioral strategies. Insects such as *Bemisia tabaci* (whitefly), lepidopterans (*Heliothis virescens*), pink bollworm (*Pectinophora gossypiella*), boll weevil (*Anthonomous grandis grandis* Boheman), and others continue to attack cotton crops. An average annual expenditure of over US$200 million is necessary to protect cotton from insects (Jenkins et al. 1991b). Nevertheless, insecticide resistance is a formidable opponent, arising after a short time (McGaughey and Whalon 1993). A coherent plan, employing traditional measures together with the creation of novel biopesticides which are expressed in a combined and targeted fashion directly in cotton crops, will inhibit the development of insects and insect resistance.

In 1996, Monsanto's Bollgard cotton (containing the B.t. gene) comprised 13% of the US cotton (over 1.8 million acres). About 60% of Bollgard cotton growers did not spray insecticides for tobacco budworms, bollworms, or pink bollworms. To control the additional insect pests, a single spray was usually sufficient. Bollgard cotton had an average yield improvement of 7% (56 pounds per acre yield advantage), which plus yield (fiber) increase, provided an economic advantage (about US$33 per acre) compared to non-Bollgard varieties (Anonymous 1997). Elimination of insecticide spraying for budworm/bollworm will allow beneficial insects to increase and help control aphids, whiteflies, mites, and armyworms. However, decreased spraying will permit more damage by B.t.-tolerant pests such as boll weevil, whitefly, tarnished plant bug, and fall armyworm. This fact presents a new challenge to the biotechnology industry.

Acknowledgments. Portions of this work were funded by the Center for Plant/Insect Interaction and the Michigan Bedding Plants Society (to J.C. Thomas), and grants by the Arizona Cotton Growers Association and the University of Arizona College of Agriculture (to J.C. Thomas and H.J. Bohnert). The author would also like to thank Drs. H.J. Bohnert, J. Endrizzi, F. Katterman, and D. Ray for their many years of support and encouragement in researching the cotton plant. Finally, Liz Davies is thanked for her help in writing and proofing this manuscript.

References

Aerts RJ, Snoeijer W, Aerts-Teerlink O, van der Meijden E, Verpoorte R (1991) Control and biological implications of alkaloid synthesis in *Cinchona* seedlings. Phytochemistry 30(11):3571–3577

Ali AA, Watson TF (1982) Effects of *Bacillus thuringiensis* var. *kurstaki* on tobacco budworm (Lepidoptera: Noctuidae) adult and egg stages. J Econ Entomo 75:596–598

Anonymous (1996) [Online] Available http://www.monsanto.com/crop_protection/html_files/news.html

Anonymous (1997) Monsanto in the news – additional reading. Bollgard Cotton Update, Jan 6, 1997 http://www.monsanto.com/MonPub/InTheNews/Articles/97-01-06BollgardUpdate.html

Aronson AI, Beckman W, Dunn P (1986) *Bacillus thuringiensis* and related insect pathogens. Microb Rev 50:1–24

Bailey JC, Hanny BW, Meredith WR Jr (1980) Combinations of resistant traits and insecticides: effect on cotton yield and insect populations. J Econ Entomol 73:58–60

Barton KA, Whitely HR, Yang KS (1987) *Bacillus thuringiensis* endotoxins expressed in transgenic *Nicotiana tabacum* provides resistance to lepidopteran insects. Plant Physiol 85:1103–1109

Benfey PN, Ren L, Chua N-H (1990) Tissue-specific expression from CaMV 35S enhancer subdomains in early stages of plant development. EMBO J 9:1677–1684

Berheimer AW, Robinson WG, Linder R, Mullins D, Yip YK, Cooper NS, Seidman I, Uwajima T (1987) Toxicity of enzymatically-oxidized low-density lipoprotein. Biochem Biophys Res Commun 148:260–266

Bostwick DE, Dannenhoffer JM, Skaggs MI, Lister RM, Larkins BA, Thompson GA (1992) Pumpkin phloem lectin genes are specifically expressed in companion cells. Plant Cell 4:1539–1548

Boulter D, Edwards G, Gatehouse A, Gatehouse J, Hilder V (1990) Additive protective effects of different plant derived insect resistance genes in transgenic tobacco plants. Crop Protect 9:351–354

Bradley JR (1996) Controlling cotton pests, Letter to the Editor. Science 273:1642

Broadway RM, Duffey SS (1986) Plant protease inhibitors: mechanism of action and effect on the growth and digestive physiology of larvae *Heliothis zea* and *Spodoptera exiqua*. J Insect Physiol 32:827–833

Broadway RM, Duffey SS, Pearce G, Ryan CA (1986) Plant protease inhibitors: a defense against herbivorous insects? Entomol Exp Appl 41:33–38

Brown JK, Bird J (1992) Whitefly transmitted geminiviruses in the Americas and the Caribbean Basin: past and present. Plant Dis 76:220–225

Burgess EPJ, Stevens PS, Keen GK, Laing WA, Christeller JT (1991) Effects of protease inhibitors and dietary protein level on the black field cricket *Teleogryllus commodus*. Entomol Exp Appl 621:123–130

Chan BG, Waiss AC Jr, Lukefahr M (1978) Condensed tannins: an antibiotic chemical from *Gossypium hirsutum*. J Insect Physiol 24:113–118

Chapman RF (1969) The insects: structure and function. The English Universities Press, London

Chavadej S, Brisson N, McNeil JN, DeLuca V (1994) Redirection of tryptophan leads to production of low indole glucosinolate canola. Proc Natl Acad Sci USA 91:2166–2170

Cho H-J, Choi KP, Yamashita M, Morikawa H, Murooka Y (1996) Introduction and expression of the *Steptomyces* cholesterol oxidase gene (ChoA), a potent insecticidal protein active against boll weevil larvae, into tobacco cells. Appl Microbiol Biotechnol 44:133–137

Clark AM, Verbeke JA, Bohnert HJ (1992) Epidermis-specific gene expression in *Pachyphytum*. Plant Cell 4:1189–1198

Cohen E (1993) Chitin synthesis and degradation as targets for pesticide action. Arch Insect Biochem Physiol 22:245–261

Constabel CP, Bergey DR, Ryan CA (1995) Systemin activates synthesis of wound-inducible tomato leaf polyphenol oxidase via the octadecanoid defense signaling pathway. Proc Natl Acad Sci USA 92:407–411

Deaton WR (1996) Controlling cotton pests, Letter to the Editor. Science 273:1641

Ding X, Johnson L, White F, Gopalakrishnan B, Kramer K, Muthukrishnan S (1995) Insect chitinase-mediated resistance to tobacco budworm (*Heliothis virescens*) in transgenic tobacco plants. In Vitro Cell Dev Biol 31:71A

Donegan KK, Palm CJ, Fieland VJ, Porteous LA, Ganion LM, Schaller DL, Bucao LQ, Seidler RJ (1995) Changes in levels, species and DNA fingerprints of soil microorganisms associated with cotton expressing the *Bacillus thuringiensis* var. *kurstaki* endotoxin. Appl Soil Ecol 2:111–124

Eigenbrode SD, Espelie KE (1995) Effects of plant epicuticular lipids on insect herbivores. Annu Rev Entomol 40:171–194

Felton GW, Dahlman DL (1984) Alleochemical-induced stress: Effects of L-canavanine on the pathogenicity of *Bacillus thuringiensis* in *Manduca sexta*. J Invert Pathol 44:187–191

Fernandes KV, Sabelli PA, Barratt DH, Richardson M, Xavier-Filho J, Shewry PR (1993) The resistance of cowpea seeds to bruchid beetles is not related to levels of cysteine proteinase inhibitors. Plant Mol Biol 23:215–219

Fischhoff DA, Bowdish HS, Perlak FJ, Marrone PG, McCormick SM, Niedermeyer JG, Dean DA, Kusano-Kretzmer K, Mayer EJ, Rochester DE, Royers SG, Fraley RT (1987) Insect tolerant transgenic tomato plants. Bio/Technology 5:807–813

Flint HM, Wilson FD, Parks NJ, Reynoso RY, Stapp BR, Szaro JL (1992) Suppression of pink bollworm and effect on beneficial insects of a nectariless okra-leaf cotton germplasm line. Bull Entomol Res 82:379–384

Forrester NW (1994) Use of *Bacillus thuringiensis* in integrated control, especially on cotton pests. Agric Ecosyst Environ 49:77–83

Fujimoto H, Itoh K, Yamamoto M, Kyozuka J, Shimamoto K (1993) Insect resistant rice generated by introduction of a modified δ-endotoxin gene of *Bacillus thuringiensis*. Bio/Technology 11:1151–1155

Gill SS, Cowles EA, Pietrantonio PV (1992) The mode of action of *Bacillus thuringiensis* endotoxins. Annu Rev Entomol 37:615–636

Gould F, Martinez-Ramirez A, Andersen A, Ferre J, Silva F, Moar WJ (1992) Broad-spectrum resistance to *Bacillus thuringiensis* toxins in *Heliothis virescens*. Proc Natl Acad Sci USA 89:7896–7900

Greenplate JT, Duck NB, Pershing JC, Purcell JP (1995) Cholesterol oxidase: an oöstatic and larvicidal agent active against the cotton boll worm, *Anthonomus grandis*. Entomol Exp Appl 74:253–258

Harris MK (1991) Bt and pest control. Science 253:1075

Harris MK (1996) Controlling cotton pests, Letter to the Editor. Science 273:1641–1642

Hedin PA, Jenkins JN, Parrott WL (1992) Evaluation of flavonoids in *Gossypium arboreum* (L.) cottons as potential source of resistance to tobacco budworm. J Chem Ecol 18(2):105–114

Henneberry TJ, Toscano NC (1993) First Annual Review of the 5-year National Research and Action Plan for Development of Management and Control of the Sweetpotato Whitefly. USDA-ARS Bulletin, Washington, DC

Hilder VA, Gatehouse AMR, Sheerman SE, Barker RF, Boulter D (1987) A novel mechanism of insect resistance engineered into tobacco. Nature 330:160–163

Hinds WE (1906) Proliferation as a factor in the natural control of the Mexican cotton boll weevil. US Dept Agric Bur Entomol Bull 59:45

Isley D (1928) The relation of leaf color and leaf size to boll weevil infestation. J Econ Entomol 21:553–559

Jackson DM, Severson RF, Johnson AW, Herzog GA (1986) Effects of cuticular duvane diterpenes from green tobacco leaves on tobacco budworm (Lepidopera: Noctuidae) oviposition. J Chem Ecol 12:1349

Jacobsen SE, Olszewski NE (1996) Gibberellins regulate the abundance of RNAs with sequence similarity to proteinase inhibitors, dioxygenases and dehydrogenases. Planta 198:78–86

Jefferson RA, Kavanaugh TA, Bevan MW (1987) *uidA* (GUS) fusions: β-glucuronidase as a sensitive and versatile gene-fusion marker in higher plants. EMBO J 6:3903–3907

Jenkins JN, Parrott WL, Umbeck PF, Barton KA (1990) Field and laboratory evaluations of transgenic cotton strains containing a gene from *Bacillus thuringiensis kurstaki* strain HD-1. In: Brown JM (ed) Proc Beltwide Cotton Prod Res Conf Las Vegas, NV 9–14 Jan 1990, Natl Cotton Council, Memphis, TN, 635 pp

Jenkins JN, Parrott WL, McCarty JC (1991a) Field performance of transgenic cotton containing the B.t. gene. In: Herber DJ, Richter DA (eds) Proc Beltwide Cotton Prod Res Conf San Antonio, TX 8–12 Jan 1991, Natl Cotton Council, Memphis, TN, 576 pp

Jenkins JN, Parrot WL, McCarty JC Jr, Barton KA, Umbeck PF (1991b) Field test of transgenic cottons containing a *Bacillus thuringiensis* gene. Mo Agric For Exp Stn Tech Bull 174

Jiang H, Mulnix AB, Kanost MR (1995) Expression and characterization of recombinant *Manduca sexta* serpin-1B and site-directed mutants that change its inhibitory selectivity. Insect Biochem Mol Biol 25:1093–1100

John ME, Crow LJ (1992) Gene expression in cotton (*Gossypium hirsutum* L.) fiber: cloning of the mRNAs. Proc Natl Acad Sci USA 89:5769–5773

Johnson R, Narvaéz J, An G, Ryan CA (1989) Expression of proteinase inhibitors I and II in transgenic tobacco plants: effects on natural defense against *Manduca sexta* larvae. Proc Natl Acad Sci USA 86:9871–9875

Jongsma MA, Bakker PL, Peters J, Bosch D, Stiekema WJ (1995) Adaptation of *Spodoptera exigua* larvae to plant proteinase inhibitors by induction of gut proteinase activity insensitive to inhibition. Proc Natl Acad Sci USA 92:8041–8045

Kanost MR, Sarvamangala VP, Wells MA (1989) Primary structure of a member of the serpin superfamily of proteinase inhibitors from an insect *Manduca sexta*. J Biol Chem 264:965–972

Karaban R (1985) Resistance against spider mites in cotton induced by mechanical abrasion. Entomol Exp Appl 37:137–141

Karaban R, Carey JR (1984) Induced resistance of cotton seedlings to mites. Science 225:53–54

Klein TM, Wolf ED, Wu R, Sanford JC (1987) High-velocity microprojectiles for delivering nucleic acids into living cells. Nature 327:70–73

Knowles BH, Dow JAT (1993) The crystal δ-endotoxins of *Bacillus thuringiensis*: models for their mechanism of action on the insect gut. BioEssays 15:469–476

Koziel MG, Beland GL, Bowman C, Carozzi NB, Crenshaw R, Crossland L, Dawson J, Desai N, Hill M, Kadwell S, Launis K, Lewis K, Maddox D, McPherson K, Meghji E, Merlin MR, Rhodes R, Warren GW, Wright M, Evola SV (1993) Field performance of elite transgenic maize plants expressing an insecticidal protein derived from *Bacillus thuringiensis*. Bio/Technology 11:194–200

Kyozuka J, McElroy D, Hayakawa T, Xie Y, Wu R, Shimamoto K (1993) Light-regulated and cell-specific expression of tomato *rbc-gusA* and rice *rbcS-gusA* fusion genes in transgenic rice. Plant Physiol 102:991–1000

Lambert B, Buysse L, Decock C, Jansens S, Piens C, Saey B, Seurinck J, Van Audenhove K, Van Rie J, Van Vliet A, Peferoen M (1996) A *Bacillus thuringiensis* insecticidal crystal protein with a high activity against members of the family Noctuidae. Appl Environ Microbiol 62:80–86

Lange Y, Matthies H, Steck TL (1984) Cholesterol oxidase susceptibility of the red cell membrane. Biochim Biophys Acta 769:551–562

Llewellyn D, Cousins Y, Mathews A, Hartweck L, Lyon B (1994) Expression of *Bacillus thuringiensis* insecticidal protein genes in transgenic crop plants. Agric Ecosyst Environ 49:85–93

Ludlum CT, Felton GW, Duffey SS (1991) Plant defenses: chlorogenic acid and polyphenol oxidase enhance toxicity of *Bacillus thuringiensis* subsp. *kurstaki* to *Heliothis zea*. J Chem Ecol 17:217–237

Lukefahr MJ, Martin DF (1966) Cotton-plant pigments as a source of resistance to the bollworm and tobacco budworm. J Econ Entomol 59:176–179

MacIntosh SC, Kishore GM, Perlak FJ, Marrone PG, Stone TB, Sims SR, Fuchs RL (1990) Potentiation of *Bacillus thuringiensis* insecticidal activity by serine protease inhibitors. J Agric Food Chem 38:1145–1152

Martin PAW, Travers RS (1989) Worldwide abundance and distribution of *Bacillus thuringiensis* isolates. Appl Environ Microbiol 55:2437–2442

Masoud SA, Johnson LB, White FF, Reeck GR (1993) Expression of a cysteine proteinase inhibitor (oryzacystatin-I) in transgenic tobacco plants. Plant Mol Biol 21:655–663

McGaughey WH, Whalon ME (1992) Managing insect resistance to *Bacillus thuringiensis* toxins. Science 258:1451–1455

McGurl B, Orozco-Cardenas M, Pearce G, Ryan CA (1994) Overexpression of the prosystemin gene in transgenic tomato plants generates a systemic signal that constitutively induces proteinase inhibitor synthesis. Proc Natl Acad Sci USA 91:9799–9802

Meade T, Felton GW (1994) Combined effects of gossypol and *Bacillus thuringiensis* on the bollworm and the tobacco budworm. P-42 11th Biennial Workshop, International Plant Resistance to Insects, Feb 27–March 3, 1994, Stillwater Oklahoma

Meisner J, Zur M, Kabonci E, Ascher KRS (1977) Influence of gossypol content of leaves of different cotton strains on the development of *Spodoptera littoralis* larvae. J Econ Entomol 70:714–716

Moreno J, Chrispeels MJ (1989) A lectin gene encodes the alpha-amylase inhibitor of the common bean. Proc Natl Acad Sci USA 86:7885–7889

Mulrooney JE, Hedin PA, Parrott WL, Jenkins JN (1985) Effects of PIX, a plant growth regulator, on allelochemical content of cotton and growth of tobacco budworm larvae (Lepidoptera: Noctuidae). J Econ Entomol 78:1100–1104

Narvaéz-Vásquez J, Orozco-Cárdenase MI, Ryan CA (1992) Differential expression of a chimeric CaMV-tomato proteinase inhibitor I gene in leaves of transformed nightshade, tobacco and alfalfa plants. Plant Mol Biol 20:1149–1157

Navon A, Hare JD, Federici BA (1993) Interactions among *Heliothis virescens* larvae, cotton condensed tannin and the CryIA(c) δ-endotoxin of *Bacillus thuringiensis*. J Chem Ecol 19:2485–2499

Niemeyer HH (1990) The role of secondary plant compounds in aphid-host interactions. In: Campbell RK, Eikenbary RD (eds) Aphid-Plant Genotype Interactions. Elsevier, Amsterdam, pp 187–205

Perlak FJ, Deaton RW, Armstrong TA, Fuchs RL, Sims SR, Greenplate JT, Fischoff DA (1990) Insect resistant cotton. Bio/Technology 8:939–943

Perlak FJ, Fuchs RL, Dean DA, McPherson SL, Fischoff DA (1991) Modifications of the coding sequence enhances plant expression of insect control proteins. Proc Natl Acad Sci USA 88:3324–3328

Perring TM, Cooper AD, Rodriguez RJ, Farrar CA, Bellows TS (1993) Identification of a whitefly species by genomic and behavioral studies. Science 259:74–77

Purcell JP, Greenplate JT, Jennings MG, Ryerse J, Pershing JC, Sims SR, Prinsen MJ, Corbin DR, Tran M, Sammons RD, Stonard RJ (1993) Cholesterol oxidase: a potent insecticidal protein active against boll weevil larvae. Biochem Biophys Res Commun 196:1406–1413

Rosell RC, Lichty JE, Brown JK (1995) Ultrastructure of the mouthparts of adult sweetpotato whitefly, *Bemisia tabaci* Gennadius (Homopteran: Aleyrodidae). Int J Insect Morphol Embryol 24:297–306

Ryan CA (1973) Proteolytic enzymes and their inhibitors in plants. Annu Rev Plant Physiol 24:173–196

Ryan CA (1990) Protease inhibitors in plants: genes for improving defenses against insects and pathogens. Annu Rev Phytopathol 28:425–449

Schaffer HJ, Forsthoefel NR, Cushman JC (1995) Identification of enhancer and silencer regions involved in salt-responsive expression of Crassulacean acid metabolism (CAM) genes in the facultative halophyte *Mesembryanthemum crystallinum*. Plant Mol Biol 28:205–218

Schwartz RH (1983) Cotton insect management with special reference to boll weevil. In: Ridgway RL, Loyd EP, Cross WH (eds) Agricultural Handbook 589. USDA, ARS, Washington, DC

Shade RE, Schroeder HE, Pueyo JJ, Tabe LM, Murdock LL, Higgins TJV, Chrispeels MJ (1994) Transgenic pea seeds expressing α amylase inhibitor of the common bean are resistant to bruchid beetles. Bio/Technology 12:793–796

Sharma HC, Agarwal RA (1984) Factors imparting resistance to stem damage by *Earias Vittella* F. (Lepidoptera: Noctuidae) in some cotton phenotypes. Protect Ecol 6:35–42

Slosarek G, Kalbitzer HR, Mucha P, Rekowski P, Kupryszewski G, Giel-Pietraszuk M, M. Szymanski M, Barciszewski J (1995) Mechanism of the activation of proteinase inhibitor synthesis by systemin involves beta-sheet structure, a specific DNA-binding protein domain. J Struct Biol 115:30–36

Smigocki A, Neal JW, McCanna I, Douglass L (1993) Cytokinin-mediated insect resistance in *Nicotiana* plants transformed with the ipt gene. Plant Mol Biol 23:325–335

Smith CM (1989) Use of plant resistance in insect pest management systems. In: Smith CM (ed) Plant resistance to insects: a fundamental approach. John Wiley, New York, pp 243–268

Stipanovic RD, Bell AA, O'Brien DH, Lukefahr MJ (1977) Heliocide H2: An insecticidal sesquiterpenoid from cotton (*Gossypium*). Tetrahedron Lett 6:567–570

Stipanovic RD, Williams HJ, Smith LA (1986) Cotton terpernoid inhibition of *Heliothis virescens* development. In: Green MB, Hedin PA (eds) Natural resistance of plants to pests. ACS Symposium Series 296. Am Chem Soc, Washington, DC, pp 79–94

Stroms JJH (1971) Some physiological effects of spider mite infestation of bean plants. Neth J Plant Pathol 77:154–167

Thomas JC, Wasmann CC, Echt C, Dunn R, Bohnert HJ, McCoy TC (1994) Introduction and expression of an insect proteinase inhibitor in alfalfa (*Medicago sativa* L.). Plant Cell Rep 14:31–36

Thomas JC, Adams DG, Keppenne VD, Wasmann CC, Brown JK, Kanost MR, Bohnert HJ (1995a) Protease inhibitors of *Manduca sexta* expressed in transgenic cotton. Plant Cell Rep 14:758–762

Thomas JC, Adams DG, D. Keppenne VD, Wasmann CC, Brown JK, Kanost MR, Bohnert HJ (1995b) *Manduca sexta* encoded protease inhibitors expressed in *Nicotiana tabacum* provide protection against insects. Plant Physiol Biochem 33(5):611–614

Thomas JC, Smigocki AC, Bohnert HJ (1995c) Light induced expression of *ipt* from *Agrobacterium tumefaciens* results in cytokinin accumulation and osmotic stress symptoms in transgenic tobacco. Plant Mol Biol 27:225–235

Thomas JC, Adams DG, Nessler CL, Brown JK, Bohnert HJ (1995d) Tryptophan decarboxylase, tryptamine and reproduction of the whitefly (*Bemisia tabaci*). Plant Physiol 109:717–720

Thomas JC, Saleh EF, Akroush AM (1996) Tryptamine: plant-synthesized compounds for self-defense against insects. Plant Physiol 111:87 (Abstr)

Thomas JC, Saleh EF, Alammar N, Akroush AM (1997) The indole alkaloid tryptamine impairs reproduction in *Drosophila melanogaster*. J Econ Entomol (in press)

Vaeck M, Reynaerts A, Höfte H, Jansens S, de Beukler M, Dean C, Zabeau M, Van Montagu M, Leemans J (1987) Transgenic plants protected from insect attack. Nature 328:33–37

Vehovszky A, Walker RJ (1991) An analysis of the 5-hydroxytryptamine (serotonin) receptor subtypes of central neurons of *Helix aspersa*. Comp Biochem Physiol C 100(3):463–476

Wilson FD, George BW (1983) A genetic and breeding study of pink bollworm resistance and agronomic properties in cotton. Crop Sci 23:1–4

Wilson FD, Flint HM, Deaton WR, Buehler RE (1994) Yield, yield components, and fiber properties of insect-resistant cotton lines containing a *Bacillus thuringiensis* toxin gene. Crop Sci 34:38–41

Zhu K, Huesing JE, Shade RE, Bressan RA, Hasegawa PM, Murdock LL (1996) An insecticidal N-acetylglucosamine-specific lectin gene from *Griffonia simplicifolia* (Leguminosae). Plant Physiol 110:195–202

V.3 Transgenic Cotton Resistant to the Herbicide 2,4-D

D.W. Ow[1] and J.E. Quisenberry[2]

1 Introduction

For close to half a century, 2,4-dichlorophenoxyacetic acid (2,4-D) has been a popular herbicide for controlling broadleaf weeds. The synthetic auxin has high selectivity on dicotyledonous plants, making it ideal for use among the monocotyledonous cereals, sugarcane, and turf lawn. The herbicide is also relatively inexpensive, and offers a cost-effective means for controlling weeds over vast land areas, such as forest undergrowth. The one drawback with the use of this herbicide is its volatility. Since it drifts with the prevailing winds, nearby dicotyledonous crops can be severely damaged. This is particularly true for cotton, where extremely low levels of 2,4-D drift can cause severe damage and yield reductions. In the US Southern High Plains, these annual events result in unnecessary lawsuits and increases in insurance premiums. Several years ago, cotton was engineered to tolerate 2,4-D drift (Bayley et al. 1992; Lyon et al. 1993). The approach taken was to transfer into cotton a gene that can degrade the herbicide. In this chapter, our work on engineering of 2,4-D resistance into cotton is summarized.

A number of soil bacterial strains have been described which contain plasmid-borne genes that confer the ability to catabolize 2,4-D (Don and Pemberton 1981; Amy et al. 1985). The gene for the first step of 2,4-D degradation was initially isolated from pJP4, a plasmid found in *Alcaligenes eutrophus* (Streber et al. 1987). This gene, *tfdA*, was shown to encode 2,4-D monooxygenase, an α-ketoglutarate-dependent dioxygenase which converts 2,4-D into 2,4-dichlorophenol and glyoxylate (Fukumori and Hausinger 1993). Lyon et al. (1989) and Streber and Willmitzer (1989) reported successful transfer of modified versions of this gene into tobacco, resulting in transgenic plants that tolerated the herbicide. To enable cotton to tolerate 2,4-D, this gene was expressed in cotton.

[1] Plant Gene Expression Center, USDA/ARS, 800 Buchanan St., Albany, California 94710, USA
[2] Cropping Systems Research Laboratory, USDA/ARS, Route 3, Box 213, Lubbock, Texas 79401, USA

Biotechnology in Agriculture and Forestry, Vol. 42
Cotton (ed. by Y.P.S. Bajaj)

2 2,4-D Monooxygenase (*tfd*A) Gene from *Alcaligenes eutrophus* Plasmid pJP5

Our source of the *tfdA* gene was pJP5, a variant of the pJP4 plasmid also found in *A. eutrophus*. Starting with synthetic oligonucleotide probes corresponding to the sequence of the pJP4-derived *tfd*A gene (Streber et al. 1987), a pJP5 fragment of ~2 kb that has *tfdA* homology was cloned (Fig. 1). Partial sequence of the 5′ and 3′ ends of this fragment revealed sequence similarity with the pJP4-derived *tfd*A coding region. As in the case of pJP4, GTG appeared to be the translation start codon of the *tfdA* gene from pJP5.

A modified *tfd*A gene was constructed for expression in plant cells. The putative GTG start codon was changed to ATG with the use of a synthetic oligonucleotide adapter (Fig. 1). This adapter contains the first 15 basepairs of the *tfd*A coding region including the substitution of ATG in place of GTG. Preceding this 15-bp coding sequence is a 32-bp sequence to serve as an untranslated mRNA leader. The 5′ and 3′ ends of this adapter have XbaI and FokI sticky ends, respectively. The FokI end of this adapter was ligated to the

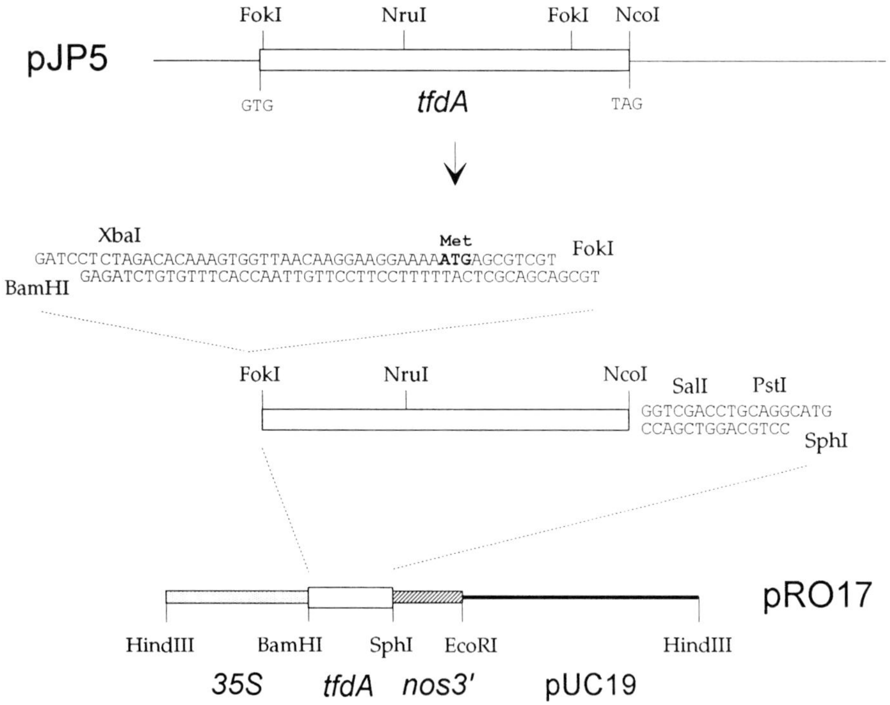

Fig. 1. Schematic representation of the *tfd*A gene from pJP5 leading to constructs pRO17 (shown) and pBIN19::pRO17 (not shown). Restriction sites relevant to the text are indicated. Abbreviations: *35S* cauliflower mosaic virus 35S RNA promoter; *nos3′* polyadenylation signal-containing fragment from nopaline synthase gene. (After Bayley et al. 1992)

```
           10         20         30         40         50         60
BamHI       |          |          |          |          |          |
001 GGATCCTCTA GACACAAAGT GGTTAACAAG GAAGGAAAAA TGAGCGTCGT CGCAAATCCC
                                               MET SerValVal AlaAsnPro

061 CTTCATCCTC TTTTCGCCGC AGGGGTCGAA GACATCGACC TTCGAGAGGC CTTGGGTTCG
    LeuHisProLeu PheAlaAla GlyValGlu AspIleAspLeu ArgGluAla LeuGlySer

121 ACCGAGGTCC GAGAGATCGA ACGGCTAATG GACGAGAAGT CGGTGCTGGT GTTCCGGGGG
    ThrGluValArg GluIleGlu ArgLeuMet AspGluLysSer ValLeuVal PheArgGly

181 CAGCCCCTGA GTCAGGATCA GCAGATCGCC TTCGCGCGCA ATTTCGGGCC ACTCGAAGGC
    GlnProLeuSer GlnAspGln GlnIleAla PheAlaArgAsn PheGlyPro LeuGluGly

241 GGTTTCATCA AGGTCAATCA AAGACCTTCG AGATTCAAGT ACGCGGAGTT GGCGGACATC
    GlyPheIleLys ValAsnGln ArgProSer ArgPheLysTyr AlaGluLeu AlaAspIle

301 TCGAACGTCA GTCTCGACGG CAAGGTCGCG CAACGCGATG CGCGCGAGGT GGTCGGGAAC
    SerAsnValSer LeuAspGly LysValAla GlnArgAspAla ArgGluVal ValGlyAsn

361 TTCGCGAACC AGCTCTGGCA CAGCGACAGC TCCTTTCAGC AACCTGCTGC CCGCTACTCG
    PheAlaAsnGln LeuTrpHis SerAspSer SerPheGlnGln ProAlaAla ArgTyrSer

421 ATGCTCTCCG CGGTGGTGGT TCCGCCGTCG GGCGGCGACA CCGAGTTCTG CGACATGCGT
    MetLeuSerAla ValValVal ProProSer GlyGlyAspThr GluPheCys AspMetArg

481 GCGGCATACG ACGCGCTGCC TCGGGACCTC CAATCCGAGT TGGAAGGGCT GCGTGCCGAG
    AlaAlaTyrAsp AlaLeuPro ArgAspLeu GlnSerGluLeu GluGlyLeu ArgAlaGlu

541 CACTACGCAC TGAACTCCCG CTTCCTGCTC GGCGACACCG ACTATTCGGA AGCGCAACGC
    HisTyrAlaLeu AsnSerArg PheLeuLeu GlyAspThrAsp TyrSerGlu AlaGlnArg

601 AATGCCATGC CGCCGGTCAA CTGGCCGCTG GTTCGAACCC ACGCCGGCTC CGGGCGCAAG
    AsnAlaMetPro ProValAsn TrpProLeu ValArgThrHis AlaGlySer GlyArgLys

661 TTTCTCTTCA TCGGCGCGCA CGCGAGCCAC GTCGAAGGCC TTCCGGTGGC CGAAGGCCGG
    PheLeuPheIle GlyAlaHis AlaSerHis ValGluGlyLeu ProValAla GluGlyArg

721 ATGCTGCTTG CGGAGCTTCT CGAGCACGCG ACACAGCGGG AATTCGTGTA CCGGCATCGC
    MetLeuLeuAla GluLeuLeu GluHisAla ThrGlnArgGlu PheValTyr ArgHisArg

781 TGGAACGTGG GAGATCTGGT GATGTGGGAC AACCGCTGCG TTCTTCACCG CGGACGCAGG
    TrpAsnValGly AspLeuVal MetTrpAsp AsnArgCysVal LeuHisArg GlyArgArg

841 TACGACATCT CGGCCAGGCG TGACGTGCGC CGGGCGACCA CCCTGGACGA TGCCGTCGTC
    TyrAspIleSer AlaArgArg AspValArg ArgAlaThrThr LeuAspAsp AlaValVal

901 TAGCGCACGC CATGGGTCGA CCTGCAGGCA TGC
    ---
                                 SphI
```

Fig. 2. DNA sequence of the BamHI to SphI fragment shown in Fig. 1 and which contains the modified *tfdA* coding region. Compared to the pJP4-derived *tfdA* sequence (Streber et al. 1987), there is a GC to CG change at positions 864 to 865 with corresponding amino acid substitutions of Glu-Leu to Asp-Val

corresponding end of an ~860 bp FokI to NcoI *tfdA* gene fragment obtained through cleavage with NcoI and partial cleavage with FokI. After adding SalI linkers to the ends of this fragment, and subsequent cleavage with XbaI and SalI, the resulting XbaI to SalI fragment was inserted into the corresponding sites of pUC19. This allowed retrieval of the *tfdA* coding region as a BamHI to SphI fragment. We have recently characterized this 933-bp fragment and the sequence from this pJP5-derived gene is essentially identical to the one isolated from pJP4 (Streber et al. 1987). Only two base changes were found, a CG in place of GC at positions 864 to 865, as shown in Fig. 2, which would translate to two amino acid substitutions, from Glu-Leu to Asp-Val. This BamHI to SphI framgnet was inserted into the corresponding sites in place of the luciferase gene of pDO432 (Ow et al. 1986). Figure 1 shows the resulting construct, pRO17, which contains the *tfdA* coding region between a 1585-bp fragment containing the cauliflower mosaic virus 35S RNA promoter (*35S*) and a 754-bp fragment of the nopaline synthase gene termination region (*nos3′*).

To transfer the *tfd*A gene into an *Agrobacterium* gene transfer vector, pRO17 was linearized by HindIII and ligated to HindIII-cleaved pBIN19 (Bevan 1984) to form a cointegrate plasmid, with transcription of the *tfd*A gene towards the left T-DNA border. The pBIN19::pRO17 cointegrate was transferred into the "disarmed" *A. tumefaciens* strain GV3111 harboring the helper Ti plasmid pTiB6S3S3. As the 35S RNA promoter has activity in bacteria (Assaad and Signer 1990), *tfdA* function in pRO17 was initially tested in *E. coli*. The *tfdA* gene product converts 2,4-D to 2,4 dichlorophenol and glyoxylate, the latter of which can be acid-hydrolyzed to CO_2. Activity of 2,4-D monooxygenase in *E. coli* was determined by measuring the conversion of ^{14}C-labeled 2,4-D into $^{14}CO_2$ (Amy et al. 1985). As with *A. eutrophus* harboring pJP5, *E. coli* carrying pRO17 was efficient in releasing $^{14}CO_2$, showing that the modified *tfd*A gene in pRO17 encodes functional 2,4-D monooxygenase (Fig. 3).

3 Transgenic Plants

Tobacco was initially used to evaluate the effectiveness of this chimeric construct. Kanamycin-resistant plants were obtained following *Agrobacterium* infection of tobacco leaves. Shoots were regenerated in the presence of 2 to 6 mg/l of 2,4-D from the leaves of kanamycin-resistant plants. Vegetatively derived clones of the transformants were transferred to the greenhouse and sprayed with varying doses of the herbicide. Whereas nontransformed plants were sensitive at 4 kg/ha, transformed lines were unaffected at levels up to 20 kg/ha. Seeds of representative transgenic lines were examined in a germination assay. Nontransformed seeds germinated only at 10^{-7} M 2,4-D, whereas seeds derived from two primary transformants (nos. 3 and 4) germinated in 10^{-3} M 2,4-D with frequencies of 77% (62/80) and 74% (73/99), respectively. The ratio of three resistant to one sensitive progeny suggested a single 2,4-D resistance locus in the parents.

The same pBIN19::pRO17 construct was introduced into cotton (*Gossypium, hirsutum* L., cv. Coker 312) by *Agrobacterium*-mediated transformation. Shoots of putative transformed plants were placed on rooting medium containing 0.05 mg/l 2,4-D and/or meristems were treated with

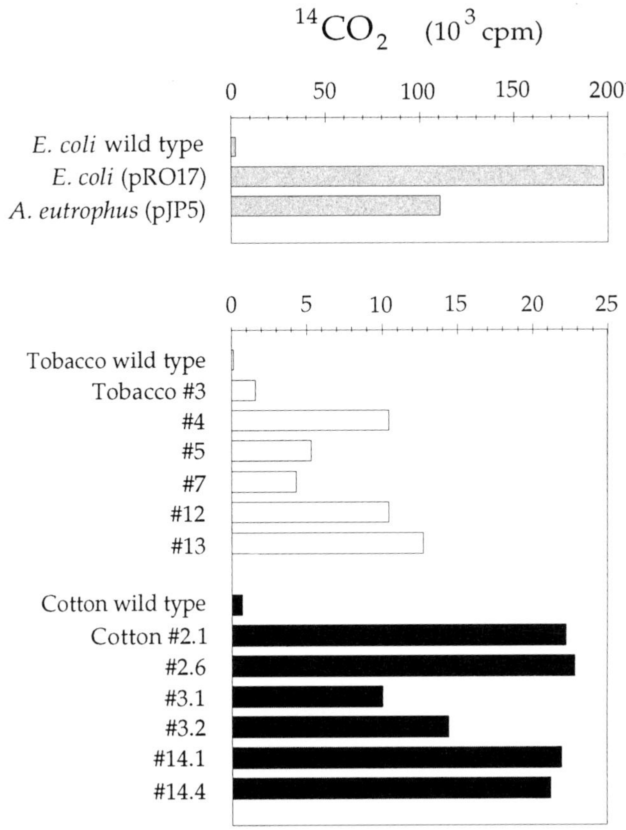

Fig. 3. 2,4-D monooxygenase activity in bacteria (*top panel*), tobacco (*open bars*) and cotton (*filled bars*). For tobacco, leaf mesophyll protoplasts from primary transformants were used. For cotton, protoplasts from cotyledons and true leaves were isolated from selfed seedlings derived from plant nos. *2*, *3*, and *14*, and previously determined by PCR analysis to harbor the *tfd*A gene. Activity is defined as counts per minute (cpm) of $^{14}CO_2$ released after incubation in medium containing ^{14}C-labeled 2,4-D. (After Bayley et al. 1992)

0.01 mg/l 2,4-D. Approximately 300 transformants from 30 cell lines rooted normally and/or were unaffected by the meristem treatment. Vegetatively derived clones of the primary transformants were transferred to the greenhouse and sprayed with various doses of herbicide. Transformants were unaffected at up to three times the recommended spray rate of 2,4-D, whereas nontransformed plants were severely affected at 300 times less of the herbicide. In a germination assay, progeny from wild-type plants were unable to form roots on medium containing 0.1 mg/l 2,4-D. In contrast, nearly three-quarters of the germinating progeny seedlings from each of three transformed lines (nos. 2, 3, 14, as in Fig. 3) were unaffected by the herbicide. The 3:1 segregation pattern is indicative of a single 2,4-D resistance locus in the parental lines. Direct spraying of the herbicide was conducted on the progeny of a representative primary transformant (no. 2). Germinated in the absence of

2,4-D, plants at the two-leaf stage were sprayed with a commercial preparation of the herbicide at a dosage equivalent of 1.5 kg/ha. This is approximately three times the manufacturer's highest recommended rate for wheat, corn, sorghum, and pastures (0.2, 0.2, 0.2, and 0.6 kg/ha, respectively). Whereas all 18 wild-type plants were killed within 5 days, 83 out of 107 progeny plants derived from the transformant were unaffected by the herbicide treatment.

Molecular analysis confirmed the presence of the *tfdA* transgene in representative cotton lines. DNA primers that anneal to the 35S RNA promoter and the carboxyl end of the *tfd*A gene were used in a polymerase chain reaction assay to amplify a fragment with an expected size of 940 bp. This 940-bp PCR fragment has the characteristic restriction pattern and could indeed be cleaved by endonuclease NruI to yield two fragments close to the predicted sizes of 513 and 427 bp. For biochemical confirmation that 2,4-D resistance in the transgenic plants was due to degradation of the herbicide, we measured the $^{14}CO_2$ released from acid hydrolysis of glyoxylate, an enzymatic product of 2,4-D monooxygenase action on ^{14}C-labeled 2,4-D. Since plant cells can metabolize glyoxylate, excess unlabeled glyoxylate was supplemented to the incubation solution to reduce metabolism of the ^{14}C-labeled product. In a sampling of six tobacco lines, $^{14}CO_2$ production was 11- to 85-fold higher than that from non-transformed material (Fig. 3), indicating that resistance to 2,4-D was due to herbicide degradation. Similarly, two seedlings from each of three different *tfd*A transgenic cotton lines showed the presence of 2,4-D monooxygenase with activity 17- to 38-fold above that of the nontransformed control.

4 Conclusion and Prospects

The desire to minimize annual losses of cotton caused by 2,4-D drift prompted us to isolate and engineer for expression in plants a bacterial gene encoding 2,4-D monooxygenase. Some of the transgenic cotton lines obtained by us can tolerate up to three times the recommended 2,4-D level for cereal crops. This degree of tolerance would be more than adequate to protect cotton from drift levels of the herbicide.

Since our initial report (Bayley et al. 1992), transgenic cotton lines harboring a single copy of the *tfdA* construct have been released to a number of investigators for field trials, and the results obtained substantiated our greenhouse-based conclusions. Figure 4 shows one transgenic line sprayed with the herbicide at a high rate of 2.2 kg/ha. A side-by-side comparison to a row of nontransgenic cotton clearly illustrates high level resistance to the herbicide. Cotton breeders who received the seeds are currently using a backcrossing method to move the transgene into varieties adapted to the different production climates of the US Cotton Belt. At present, most of these companies only consider the gene as an insurance against the accidental drift of 2,4-D onto cotton, as 2,4-D has not been cleared for over-the-top application on cotton.

Fig. 4. Field test of wild-type and *tfdA* transgenic cotton plants. Photo shows 10-week-old nontransgenic (*left row*) and transgenic (*right row*) plants sprayed with 2.2 kg/ha of 2,4-D at week 8

High-level tolerance conferred by pRO17 may allow use of the *tfdA* gene itself as a selectable marker for future cotton transformation. This would eliminate the need for a cotransformed kanamycin resistance gene. Reducing the number of antibiotic or herbicide resistance markers incorporated into crop plants would minimize concerns on the widespread dissemination of these potentially controversial genes (Dale and Ow 1991; Yoder and Goldsbrough 1994). Tolerance to field applications of 2,4-D also entertains its possible use as a herbicide for cotton. Whether this will prove to be the case in the years to come is now outside the realms of research and into the realm of business and government.

Acknowledgments. The authors are grateful to various colleagues involved in the various aspects of this work: C. Bayley, N. Trolinder, C. Ray, G. Cartwright, and M. Oliver, and to J. Becton (United Agricultural Products, Greeley, Colorado) for allowing us to photograph his test plot as shown in Fig. 4. This work was funded by the Agricultural Research Service, US Department of Agriculture.

References

Amy PS, Schulke JW, Frazier LM, Seidler RJ (1985) Characterization of aquatic bacteria and cloning of genes specifying partial degradation of 2,4-dichlorophenoxyacetic acid. Appl Environ Microbiol 49:1237–1245

Assaad FF, Signer ER (1990) Cauliflower mosaic virus P35S promoter activity in *Escherichia coli*. Mol Gen Genet 223:517–520

Bayley C, Trolinder N, Ray C, Morgan M, Quisenberry JE, Ow DW (1992) Engineering 2,4-D resistance into cotton. Theor Appl Genet 83:645–649

Bevan M (1984) Binary *Agrobacterium* vectors for plant transformation. Nucleic Acids Res 12:8711–8721

Dale E, Ow DW (1991) Gene transfer with subsequent removal of the selection gene from the host genome. Proc Natl Acad Sci USA 88:10558–10562

Don RH, Pemberton JM (1981) Properties of six pesticide degradation plasmids isolated from *Alcaligenes paradoxus* and *Alcaligenes eutrophus*. J Bacteriol 145:681–686

Fukumori F, Hausinger RP (1993) *Alcaligene eutrophus* JMP134 2,4-dichlorophenoxyacetate monooxygenase is an α-ketoglutarate-dependent dioxygenase. J Bacteriol 175:2083–2086

Lyon BR, Llewellyn DJ, Huppatz JL, Dennis ES, Peacock WJ (1989) Expression of a bacterial gene in transgenic tobacco plants confers resistance to the herbicide 2,4-dichlorophenoxyacetic acid. Plant Mol Biol 13:533–540

Lyon BR, Cousins Y, Llewellyn DJ, Dennis ES (1993) Cotton plants transformed with a bacterial degradation gene are protected from accidental spray drift damage by the herbicide 2,4-dichlorophenoxyacetic acid. Transgen Res 2:162–169

Ow DW, Wood KV, DeLuca M, deWet JR, Helinski DR, Howell SH (1986) Transient and stable expression of the firefly luciferase gene in plant cells and transgenic plants. Science 234:856–859

Streber WR, Willmitzer L (1989) Transgenic tobacco plants expressing a bacterial detoxifying enzyme are resistant to 2,4-D. Bio/Technology 7:811–816

Streber WR, Timmis KN, Zenk MH (1987) Analysis, cloning, and high-level expression of 2,4-dichlorophenoxyacetate monooxygenase gene *tfd*A of *Alcaligenes eutrophus* JMP134. J Bacteriol 169:2950–1955

Yoder JI, Goldsbrough AP (1994) Transformation systems for generating marker-free transgenic plants. Bio/Technology 12:263–267

V.4 Effect of Transgenic Cotton Expressing the *Bacillus thuringiensis* var. *kurstaki* Endotoxin on Soil Microorganisms – Risk Assessment Studies

K.K. Donegan[1] and R.J. Seidler[2]

1 Introduction

The genetic engineering of plants has facilitated the production of agronomically desirable crops that exhibit increased resistance to pests, herbicides, pathogens, and environmental stress, and enhancement of qualitative and quantitative crop traits (Gasser and Fraley 1992). Along with these many benefits, however, comes the potential for adverse ecological effects because of the often sustained expression in the genetically engineered (transgenic) plant of the engineered trait(s) and the persistence of the transgenic plant or plant residue in the environment. Consequently, we have undertaken research to evaluate the potential ecological effects of transgenic plants and their products.

Some of our research has included microcosm studies with cotton plants that are genetically engineered to produce the *Bacillus thuringiensis* var. *kurstaki* (*B.t.k.*) endotoxin (Donegan et al. 1995). Many agriculturally important plants have been engineered to produce endotoxins from different subspecies of the bacterium *Bacillus thuringiensis* (*B.t.*) (Vaeck et al. 1987; Delanney et al. 1989; Perlak et al. 1990; Lundstrum 1992; Koziel et al. 1993). The endotoxin of *B. thuringiensis* var. *kurstaki* (*B.t.k.*) has demonstrated insecticidal activity against lepidopterans (Hofte and Whiteley 1989). Although high specificity has been assumed for most *B.t.* endotoxins, their effects on nontarget organisms have not been fully evaluated. Studies have been performed exposing nontarget invertebrates to various *B.t.*-producing bacterial strains and have demonstrated such detrimental effects as mortality and reduced fecundity (Ali et al. 1973; Tolstova and Ionova 1976; Molloy and Jamnback 1981; Mulla et al. 1982; Flexner et al. 1986; Miller 1990; James et al. 1993). In preliminary experiments where transgenic *B.t.k.* cotton plants were placed in natural soils and decomposed (Pratt et al. 1993; Palm et al. 1994), we discovered that the *B.t.k.* endotoxin persisted and retained its immunological and biological activity at levels similar to those observed with microbially produced *B.t.* endotoxins. Therefore, we considered it important to determine the impact of the *B.t.k.* endotoxin in decomposing transgenic plants on soil

[1] Dynamac Corporation NHEERL-WED, 200 S.W. 35th Street, Corvallis, Oregon 97333, USA
[2] U.S. Environmental Protection Agency, NHEERL-WED, 200 S.W. 35th Street, Corvallis, Oregon 97333, USA

Biotechnology in Agriculture and Forestry, Vol. 42
Cotton (ed. by Y.P.S. Bajaj)

microorganisms because of the ubiquity of microorganisms in soil and the crucial role they play in soil processes.

Most concern about the environmental release of plants containing *B.t.* endotoxins has been for the development of resistance in the target pests (Fox 1991; Johnson and Gould 1992; USDA 1992) or for gene flow and plant invasiveness (Umbeck et al. 1991; Manasse 1992; Crawley et al. 1993; Kareiva et al. 1994; Klinger and Elstrand 1994). Some studies have considered nontarget effects of the *B.t.* endotoxin but have used microbial *B.t.* strains rather than plants that produce *B.t.* toxins (Molloy and Jamnback 1981; Flexner et al. 1986; Miller 1990). Only a few studies have used transgenic plants to assess the potential direct or indirect effects of *B.t.* endotoxins on plant and soil ecosystems (Donegan et al. 1995, 1996, 1997).

In addition to the potential effects of *B.t.* endotoxins produced by transgenic plants, there is the possibility that other plant characteristics may be unintentionally altered during the insertion of the transgene (Lange 1990; MacKenzie 1990; The Economist 1990; Jenkins et al. 1991; The Gene Exchange 1992; Yamada 1992). These types of alterations in plant characteristics caused by genetic manipulation (e.g., changes in plant enzyme production and biomass), that are independent of expression of the inserted gene, may also produce ecological effects.

In this chapter four experiments are described that investigated the biological and molecular changes in microbial populations following the incorporation of purified *B.t.k.* endotoxin or *B.t.k.*-producing cotton into natural soils. Microbial populations were monitored for changes in the total numbers and species composition of culturable bacteria and fungi, in the substrate utilization of the bacterial community and in the total DNA content and DNA fingerprints of the eubacteria.

2 Experimental Studies

2.1 Plant Propagation, Sample Preparation, and Experimental Design

Seeds of parental and transgenic cotton were provided by the Monsanto Company (St. Louis, MO). The parental cotton line was Coker 312. The transgenic cotton lines producing *B.t.k.* endotoxin were line 81, HD-1, CryIA(b); line 247, HD-73, CryIA(c); and line 249, HD-73, CryIA(c) (Perlak et al. 1990). Cottonseeds were propagated in a microcosm and harvested when flowers began forming. Only cotton leaves were used in these experiments because studies showed they have the highest expression level of the *B.t.k.* endotoxin (Ream et al. 1992). For treatments containing plant material, parental or transgenic cotton leaves were incorporated at 1:3 by weight (leaves:soil) into a fine sandy loam soil (Wasco) that was obtained from the USDA Cotton Research Station in Shafter, CA. In Experiment 4 only, an additional soil, a clay loam soil (Panoche) that was received from the USDA Agricultural

Research Station in Fresno, CA, was used. For the purified endotoxin treatments, the HD-1 or HD-73 endotoxin provided by the Monsanto Company was added at a concentration of 0.05 μg toxin/g soil; calculated as the equivalent of the endotoxin concentration in the transgenic plant treatments. Sterile water was added to all treatments to bring the soil to 45% water-holding capacity. Three replicates were prepared for each treatment and sampled on days 0, 7, 14, 21, and 28 in experiments 1, 2, and 3, and on days 0, 7, 14, 28, and 56 in experiment 4.

Four experiments with the following combinations of treatments and bioassays were conducted:

2.2 Determination of Endotoxin Concentration, Microbial Populations, Bacterial Species Composition, and DNA Fingerprints

Endotoxin concentrations in the samples were determined immunologically by enzyme-linked immunoabsorbent assay (ELISA) according to Palm et al. (1994), and by determining bioactivity using bioassays with *Heliothis virescens* according to Pratt et al. (1993).

Population levels of total culturable bacteria and fungi were determined by plating samples on selective media. Numbers of protozoa (amoebae, ciliates, and flagellates), performed only in experiment 4, were determined at the Microbial Biomass Service at Oregon State University, Corvallis, OR, USA, according to the method of Darbyshire et al. (1974).

Table 1. Experiments with combinations of treatments and bioassays

Experiment no.	Treatments	Assays
1 and 2	Soil only (Wasco sandy loam)	Bacterial populations
	+Purified HD-73 toxin	Fungal populations
	+Parental cotton	ELISA
	+Transgenic 249 cotton	
3	Soil only (Wasco sandy loam)	Bacterial populations
	+Purified HD-1 toxin	Fungal populations
	+Parental cotton	ELISA
	+Parental cotton + HD-1 toxin	
	+Transgenic 81 cotton	
4	Soil only (Wasco sandy loam)	Bacterial populations
	+Purified HD-73 toxin	Fungal populations
	+Parental cotton	Amoebae populations
	+Parental cotton + HD-73 toxin	Ciliate populations
	+Transgenic 247 cotton	Flagellate populations
	+Transgenic 249 cotton	Target insect bioassay
	Soil only (Panoche clay loam)	Biolog/Bacterial community
	+Purified HD-73 toxin	Biolog/Species ID
	+Parental cotton	ELISA
	+Parental cotton + HD-73 toxin	Total DNA content
	+Transgenic cotton 247	DNA fingerprints
	+Transgenic cotton 249	

In experiment 4, for the parental, transgenic 247, and transgenic 249 treatments in Wasco and Panoche soil on sample days 0, 14, and 56, bacterial species identification was performed. Bacterial colonies were subcultured, Gram-stained, and identified biochemically based on substrate utilization of 95 carbon sources in Biolog Gram-negative or Gram-positive microtiter plates (Biolog 1992). In all of the experiments, fungal colonies were examined visually for qualitative morphological differences.

In experiment 4, the parental, transgenic 247, and transgenic 249 cotton treatments in Wasco soil were evaluated for differences in bacterial community composition. Diluted soil samples that were used for the total bacteria and fungi plating were also placed in Biolog Gram-negative microtiter plates to determine bacterial community utilization of 95 carbon sources (Garland and Mills 1991). Three replicate samples of Wasco and Panoche soil treatments were analyzed for total DNA content. The method of Porteous et al. (1994) was used for DNA extraction, polymerase chain reaction (PCR), amplification, and DNA fingerprinting.

2.3 Statistical Analyses

Total bacterial, fungal, and protozoan population levels were analyzed in SAS with ANOVA and repeated measures analysis to determine significant ($p < 0.05$) differences among treatments. The log CFU/g means of the treatments were compared within each sample day with Tukey's Studentized Range Test (SAS Institute 1989).

Principal components analysis (PCA) was used to analyze the bacterial community substrate utilization assays for each sample day in order to distinguish differences in patterns of carbon source utilization among the parental cotton treatment and the transgenic 247 and 249 cotton treatments (Garland and Mills 1991).

3 Results

3.1 Enzyme-Linked, Insect, and Protozoan Assays

Both immunological and biological activity of the *B.t.k.* toxin persisted in the purified toxin treatments and in the transgenic plant treatments. Purified *B.t.k.* HD-1 and HD-73 toxins and *B.t.k.* toxin in the transgenic lines 81 and 279 cotton plants were determined by ELISA to persist at detectable levels (the ELISA detection limit is 0.5 ng toxin/g soil) for up to 28 days in experiments 1, 2, and 3. In experiment 4, the purified *B.t.k.* toxin was detectable by ELISA up to 28 days and the purified toxin + parental plant, and the *B.t.k.* toxin in the transgenic 247 and 249 cotton plants, were detectable by ELISA up to 56 days in the Wasco soil. In experiment 4, the purified *B.t.k.* toxin and the *B.t.k.* toxin in the transgenic 247 and 249 cotton plants were detectable by ELISA up to 28 days and the purified toxin + parental cotton plant was detectable by ELISA up to 56 days in the Panoche soil. The target insect bioassays (performed only in experiment 4) indicated biological activity of the *B.t.k.* toxin remained up to 28 days in the purified *B.t.k.* toxin treatment and in the transgenic 247 and 249 plant treatments (day 56 samples were not assayed) in the Wasco soil and in the Panoche soil. Further discussion of the persistence of the immunological and biological activity of the *B.t.k.* toxins is provided in publications on ELISA studies (Palm et al. 1994) and target insect bioassay studies (Pratt et al. 1993).

No significant differences occurred on any sample day in the number of amoebae, ciliates, or flagellates per g dry weight of soil among the transgenic 247, transgenic 249, and parental cotton treatments for the protozoan assays that were performed in experiment 4.

3.2 Culturable Bacteria and Fungi Counts

Results were compared among treatments without plant material (with the soil-only treatment as the control) and among treatments with plant material (with the soil + parental cotton treatment as the control). This was done because the addition of the cotton leaves, whether they were parental or transgenic, nearly always caused a significant increase in bacterial and fungal numbers. This effect was due to the nutrients from the added cotton leaves promoting microbial growth; plant material added to soil often results in high microbial counts (Parr and Papendick 1978; Broder and Wagner 1988).

The addition of purified *B.t.k.* toxin never caused a detectable effect on microbial populations. No significant differences in bacterial and fungal population levels were observed between the soil-only treatment and the purified HD-1 toxin treatment or purified HD-73 toxin treatment on any of the sample days in all four experiments.

In contrast, the addition of two of the three lines of transgenic cotton caused detectable, and often significant, changes in microbial population levels. In experiment 1, bacterial population levels on sample days 7, 14, and 21, and fungal population levels on sample days 7 and 14 were significantly higher in the transgenic 249 cotton treatment than in the parental cotton treatment (Fig. 1). In experiment 2, bacterial population levels on sample days 7 and 14 and fungal population levels on sample days 7, 14, 21, and 28 were significantly higher in the transgenic 249 cotton treatment than in the parental cotton treatment (Fig. 2). In experiment 3, there were no significant differences in bacterial or fungal population levels on any sample days between the transgenic 81 cotton treatment and the parental cotton treatment, or the parental cotton treatment + purified HD-1 toxin treatment. In the experiment 4 treatments with Wasco soil, bacterial populations in the transgenic 247 cotton treatment were significantly higher than in the parental cotton + purified HD-73 toxin treatment on sample days 7, 14, and 28. The transgenic 249 cotton treatment was significantly higher than the parental cotton treatment in

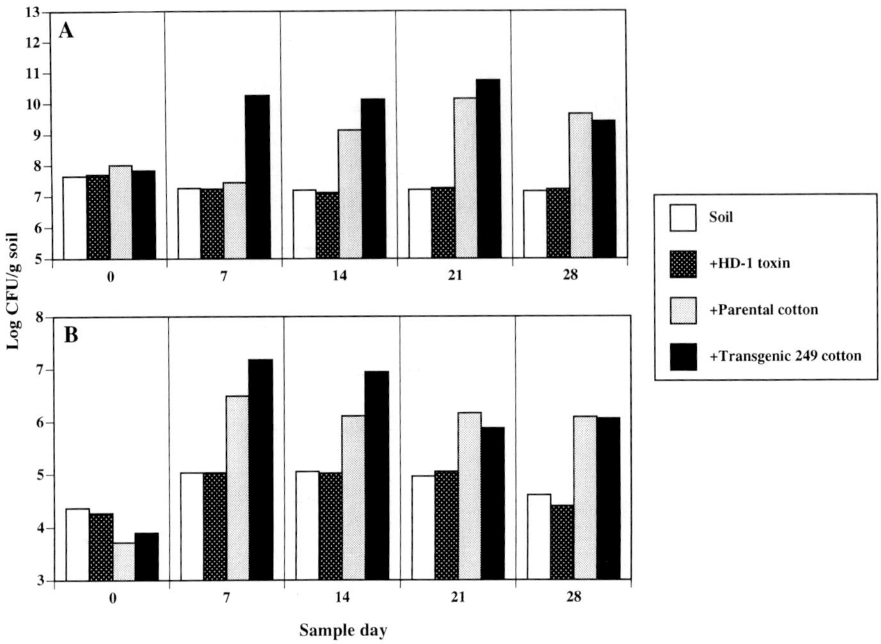

Fig. 1A,B. Numbers of culturable bacteria (**A**) and fungi (**B**) in experiment 1, averaged from duplicate plates of three treatment replicates per sample day, of soil alone, soil + purified HD-73 toxin, soil + parental cotton, and soil + transgenic 249 cotton. Significant ($P < 0.05$) differences among treatments were determined with Tukey's Studentized Range Test. The minimum significant differences in log values calculated for bacterial populations were 0.42 (day 0), 0.29 (day 7), 0.95 (day 14), 0.42 (day 21), 0.66 (day 28), and for fungal populations were 0.44 (day 0), 0.14 (day 7), 0.19 (day 14), 0.28 (day 21), 0.34 (day 28)

bacterial population levels on sample day 56. The transgenic 249 cotton treatment was also significantly higher than the parental cotton + purified HD-73 toxin treatment in bacterial population levels on sample days 7 and 56 and in fungal population levels on sample days 7 and 14. For the treatments in experiment 4 with Panoche soil, bacterial populations in the transgenic 247 cotton treatment were significantly higher than those in the parental cotton treatment on sample day 7.

3.3 Bacterial Species Identification

Identification of subcultures from the plates used for the bacterial counts in experiment 4 indicated that differences in bacterial species' composition developed among the transgenic 247 cotton treatment, transgenic 249 cotton treatment, and the parental cotton treatment over the course of the experiment in the Wasco soil and to a lesser extent in the Panoche soil. Differences among the treatments in bacterial species composition were not observed at the start of the experiment; on sample day 0, all colonies subcultured from the

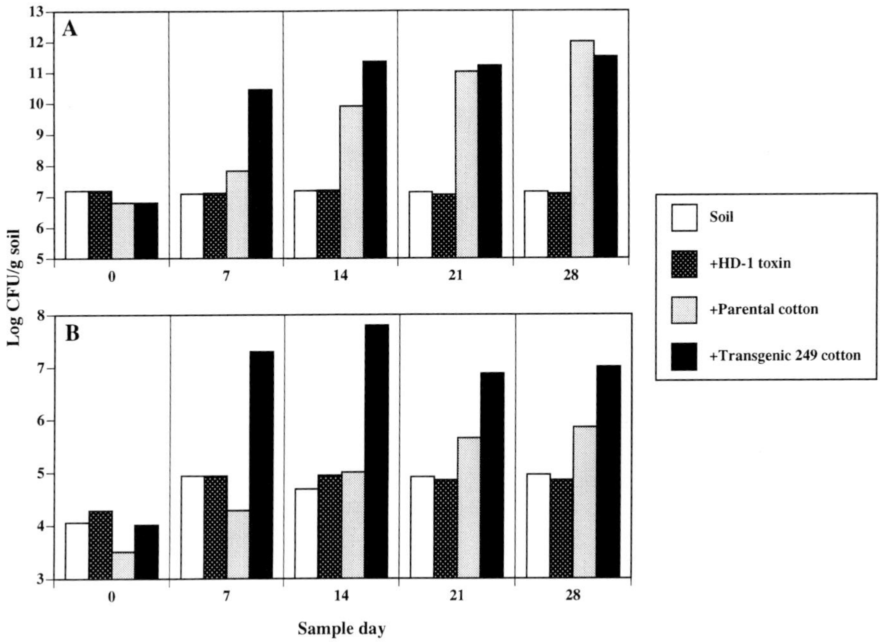

Fig. 2A,B. Numbers of culturable bacteria (**A**) and fungi (**B**) in experiment 2, averaged from duplicate plates of three treatment replicates per sample day, of soil alone, soil + purified HD-73 toxin, soil + parental cotton, and soil + transgenic 249 cotton. Significant ($P < 0.05$) differences among treatments were determined with Tukey's Studentized Range Test. The minimum significant differences in log values calculated for bacterial populations were 0.23 (day 0), 1.40 (day 7), 0.87 (day 14), 1.07 (day 21), 0.40 (day 28), and for fungal populations were 0.51 (day 0), 0.91 (day 7), 0.79 (day 14), 0.36 (day 21), 0.32 (day 28)

parental, transgenic 247 and transgenic 249 cotton treatments in Wasco or Panoche soil were Gram-positive and there were few differences in genus or species composition (colonies were mainly *Actinomycete* or *Bacillus* spp. in the Wasco soil and mainly *Streptococcus* or *Bacillus* spp. in the Panoche soil).

The greatest differences in Gram identification and species composition in the Wasco soil were observed on sample day 14. The parental cotton treatment had a larger number of Gram-positive bacteria (34%) than the 247 transgenic treatment (0%) and the 249 transgenic treatment (4%), and had mainly *Bacillus* spp. (25% of the identified Gram-positive colonies), *Enterobacter cloacae B* (25% of the identified Gram-negative colonies) and *Enterobacter* spp. (25% of the identified Gram-negative colonies). In the transgenic 247 cotton treatment, the predominant species were *E. cloacae B* (23% of the identified Gram-negative colonies) and *Pseudomonas corrugata* (35% of the identified Gram-negative colonies) and in the transgenic 249 cotton treatment *Enterobacter asburiae* (86% of the identified Gram-negative colonies) was the dominant species. On sample day 56 in the Wasco soil, bacterial composition was most similar between the parental and transgenic 249 cotton treatments; both had a predominance of Gram-positive bacteria (64% for the parental treatment and 75% for the transgenic 249 treatment) that were mainly *Bacillus* spp. (36% for the parental treatment and 39% for the transgenic 249 treatment of the identified Gram-positive colonies) and also relatively high numbers of *Pseudomonas diminuta* (23% for the parental treatment and 8% for the transgenic 249 treatment of the identified Gram-negative colonies). In contrast, in the transgenic 247 cotton treatment on day 56, the majority of colonies were Gram-negative and were identified as *Alcaligenes denitrificans* (20% of the identified Gram-negative colonies) and *P. corrugata* (25% of the identified Gram-negative colonies).

In the Panoche soil on sample day 14, the parental, transgenic 247, and transgenic 249 cotton treatments were similar and had a fairly equal distribution of Gram-positive and Gram-negative bacteria and mainly *Bacillus* spp. and *E. asburiae* colonies. On sample day 56, there was a predominance of Gram-positive bacteria in the Panoche soil, particularly for the parental (69%) and transgenic 249 (78%) cotton treatments. The most prevalent species in Panoche soil on sample day 56 in the parental treatment were *Bacillus* spp. (19% of the identified Gram-positive colonies), whereas the transgenic 247 and 249 cotton treatments did not generally have predominant species, in part because a high number (62%) of the Gram-positive colonies in the transgenic 249 cotton treatment could not be identified.

Based on fungal colony morphology, differences were also observed in fungal species composition between the parental and transgenic cotton treatments on a few of the sample days in experiments 1 and 2. For example, in experiment 2 on sample days 21 and 28, all the fungal colonies isolated from the transgenic 249 cotton treatment were *Mucor* sp., whereas only a few *Mucor* sp. were isolated from the parental cotton treatment.

3.4 Bacterial Community Metabolic Substrate Utilization

Differences in bacterial species composition were also observed in the Biolog assays for substrate utilization measured for the diluted soil samples. The principal components analysis of these assay results indicated differences in microbial usage of metabolic substrates among the parental, transgenic 247, and transgenic 249 cotton treatments on sample days 7 and 14. Differences in utilization of the amino acids L-asparagine, L-aspartic acid, and L-glutamic acid on sample days 7 and 14 were important in the separation of the parental and transgenic 247 and 249 treatments, as evidenced by their large contribution to the PCA score. No other single metabolic substrates were as consistently used to separate the treatments. The utilization of these three amino acids was significantly higher in the transgenic 247 and 249 treatments than in the parental treatment on sample days 7 and 14.

3.5 Total DNA Content and Eubacterial DNA Fingerprints

Differences among the treatments in bacterial population levels and species composition were also indicated by the DNA extractions and fingerprints that were performed in experiment 4. As observed with the total bacterial and fungal counts, the addition of plant material caused an increase in microbial levels; two to five times more total DNA was extracted from the treatments containing parental or transgenic plant material than from the treatments containing soil only or soil + purified HD-73 toxin in both the Panoche and the Wasco soils on sample days 7, 14, 28, and 56. For the Wasco soil on sample days 7, 14, and 28, the highest DNA content was measured in the transgenic 247 cotton treatment. For example, on day 28, the A_{260} readings indicated the following DNA recoveries: 20 to 35 μg DNA per g of the soil or soil + purified HD-73 toxin treatments; 50 to 65 μg DNA per g of the soil + parental cotton or soil + parental cotton + purified HD-73 toxin treatments; 50 to 75 μg DNA per g of the soil + transgenic 249 treatment; and 80 to 100 μg DNA per g of the soil + transgenic 247 cotton treatment.

Differences among treatments in the composition of microbial populations were indicated by eubacterial DNA fingerprint patterns (restriction endonuclease digested fragment patterns) of amplified rDNAs from Wasco soil containing parental, transgenic 247, and transgenic 249 cotton on sample days 14 and 56. Fingerprint patterns for the treatments varied in the size or length of fragments, the number, intensity, or absence of bands, and the reoccurring presence of a series of fragments in the fingerprint patterns. for example, on sample day 14 the fragment patterns for the parental treatment showed intense bands at 600–800 bp which were not observed in the fingerprints of either transgenic treatment. On sample days 14 and 56, the transgenic 247 treatment fingerprints contained a unique series of four intense bands between 150–300 bp. In addition, the fingerprint for the transgenic 249 treatment on sample day 56 had bright 650-bp fragment bands present that were not observed in the fingerprints for the other treatments.

4 Discussion

The addition of purified *B.t.k.* endotoxin to natural soil had no measurable effects on the indigenous soil microorganisms. In all of the experiments, no changes were observed in the levels of culturable, aerobic soil bacteria, fungi, or protozoa (effects on protozoa were evaluated only for HD-73) when purified *B.t.k* HD-1 or HD-73 toxin was added to natural soils. In contrast, the increase in culturable, aerobic bacterial and fungal populations that occurred when leaves of cotton were added to soil was significantly higher in the transgenic cotton treatments relative to the parental cotton treatment. This significant stimulation in microbial populations from the transgenic cotton, however, was observed only with the addition of the 247 and 249 lines, but not the 81 line, of the transgenic *B.t.k.*-producing cotton plants. The increase in microbial populations observed with the plate counts from the addition of cotton leaves was also confirmed with the bulk DNA extractions performed in experiment 4; the amounts of DNA extracted in the treatments with parental or transgenic plant leaves increased as total microbial plate counts increased. For example, the greatest amount of DNA extracted and the highest bacterial counts coincided; they both occurred in the transgenic 247 treatment on sample day 28 in the Wasco soil in experiment 4.

Qualitative changes in the soil microbial community were also observed with the addition of the 247 and 249 lines of transgenic *B.t.k.* cotton to Wasco or Panoche soil. The metabolic substrate utilization assays and species identifications of subcultures performed in experiment 4 indicated differences in microbial species composition among the treatments. For example, the community substrate utilization assays indicated major differences on sample days 7 and 14 among the parental, transgenic 247, and transgenic 249 cotton treatments in the type and level of substrates utilized; there was significantly greater utilization of asparagine, aspartic acid, and glutamic acid in the transgenic 247 and 249 cotton treatments than in the parental cotton treatment even though the population levels of the treatments were not significantly different. Interestingly, these substrates are important intermediates, along with glutamine (which is not included as a substrate in the Biolog plates), in nitrogen assimilation reactions. Species identifications of subcultures from plates used for the total bacterial populations in experiment 4 similarly showed the greatest differences between the parental treatment and the transgenic 247 and 249 treatments on sample day 14 (species identifications were not done for sample day 7). DNA fingerprints of amplified rDNAs also reflected qualitative differences among the treatments in eubacterial population. Distinct differences in fragment molecular weight and number and location of bands were observed among the Wasco soil parental, transgenic 247, and transgenic 249 treatments in experiment 4. These differences between the parental treatment and the transgenic treatments were pronounced on sample day 14, as occurred with the substrate utilization measurements and species identifications.

It is somewhat surprising that the line 81 transgenic cotton plant did not cause the same impact on microbial populations that was observed with the line 247 and line 249 transgenic cotton plants. The transgenic cotton lines produce endotoxins (HD-1 for line 81 and HD-73 for lines 247 and 249) that are coded for by genes from closely related, bacterial strains and the HD-1 and HD-73 endotoxins share 86% amino acid identity. It seems unlikely that the small difference in endotoxins between the transgenic plant line 81 and the lines 247 and 249 would account for the difference in effects from the transgenic plants because the purified endotoxins neither differed in their effects nor had any detectable impacts on microbial populations.

One possible explanation for the lack of effects on the total microbial population levels from the addition of the purified HD-1 and HD-73 *B.t.k.* toxins and the transgenic cotton line 81 is that there were characteristics of the transgenic 247 and 249 cotton plants, other than production of the *B.t.k.* toxin, that impacted soil microorganisms. In most of the experiments, the stimulatory effects of the transgenic 247 and 249 cotton plants on bacterial and fungal populations were short-term, suggesting that the transgenic plants may have decomposed faster than the parental plants, and thus more rapidly provided nutrients for microbial growth. This potential increase in decomposition rate of the transgenic 247 and 249 cotton plants may have resulted from genetic manipulation of the plants. Unanticipated changes in plant quality occurring from insertion of genes have been documented. In several laboratory and field studies there have been unintentional changes in plant characteristics due to genetic manipulation and position effects from site(s) of insertion (The Gene Exchange 1992). In addition, there may be genes used to confer a specific trait but which are actually pleiotropic and change several plant traits (Fitzpatrick 1993). Finally, the practice of tissue culturing, used for both traditional and genetically engineered plant breeding, frequently produces plant aberrations, some of which may not be detected during screening for efficacy of the added trait(s).

These sorts of risk assessment studies for transgenic plants need to be performed under a variety of environmental conditions. The greater quantitative and qualitative impacts of the transgenic plant lines in the Wasco soil as compared to the Panoche soil and the higher recovery of DNA from the Wasco soil than the Panoche soil demonstrate the importance of one environmental variable, soil type. The clay and organic matter content of the Panoche soil is higher than that of the Wasco soil and probably resulted in greater binding of the transgenic plant material, making it less available to exert an impact on the soil microorganisms (Stotzky 1986). In past ELISA studies, we obtained much higher recovery of the purified toxin and the transgenic plant toxin from the Wasco soil than from the Panoche soil (Palm et al. 1996). Similarly, target insect bioassays with *Heliothis virescens* have shown higher bioactivity of the same quantity of purified toxin or transgenic plant toxin when measured in the Wasco soil relative to the Panoche soil (Pratt et al. 1993).

In these studies with *B.t.k.* producing transgenic cotton plants, both quantitative and qualitative changes in exposed soil microorganisms are demonstrated. The quantitative effects were generally transient and not what are typically considered detrimental (i.e., population levels were stimulated rather than depressed). The qualitative effects of apparent changes in the microbial species composition, however, have a potential to impact soil processes and may be of ecological significance. These results are valid only for the methods and test conditions used and additional research is necessary to determine their scope, extent, and significance.

5 Conclusion

In the continuing debate about the degree and types of risks posed by the environmental release of transgenic plants, the argument has been made that transgenic plants pose no more risk than the transgenic compounds they produce, and that plants should not be evaluated based on their having originated from genetic engineering. We believe that our results challenge this argument; two of the transgenic *B.t.k.*-producing cotton lines impacted soil microorganisms, yet no effects were observed from the purified *B.t.k.* endotoxins. We suggest that the risk assessment of transgenic plants should address and monitor for potential ecological effects that may result from changes in plant characteristics other than expression of the inserted gene(s).

Acknowledgments. The authors thank Dr. Roy Fuchs, Sharon Berberich, Joel Ream, and Bill Schuler of the Monsanto Company, St. Louis, Missouri, USA, for providing the cotton seeds, purified toxin, and technical information. The authors would like to acknowledge the excellent technical and scientific support provided for this study by Valerie Fieland, Lisa Ganio, Curt Palm, Arlene Porteous, Grahame Pratt, Lynn Rogers, and Debbie Schaller.

References

Ali AUDD, Adellatif MA, Bakry NM, El-Sawaf SK (1973) Studies on biological control of the greater wax moth, *Galleria mellonella*. 1. Susceptibility of wax moth larvae and adult honeybee workers to *Bacillus thuringiensis*. J Agric Res 12:117–123

Biolog MicroStation System Manual, Release 3.01. (1992) Biolog, Inc, Hayward, California

Broder MW, Wagner GH (1988) Microbial colonization and decomposition of corn, wheat, and soybean residue. Soil Sci Soc AM J 52:112–117

Crawley HJ, Hails RS, Rees M, Kohn D, Baxton J (1993) Ecology of transgenic oilseed rape in natural habitats. Nature 363:620–623

Darbyshire JF, Wheatley RE, Greaves MP, Inkson RH (1974) A rapid micromethod for estimating bacterial and protozoan populations in soil. Ecology 61:764–771

Delannay X, LaVallee BJ, Proksch RK, Fuchs RL, Sims SR, Greenplate JT, Marrone PG, Dodson RB, Augustine JJ, Layton JG, Fischhoff DA (1989) Field performance of transgenic tomato plants expressing the *Bacillus thuringiensis* var. *kurstaki* insect control protein. Bio/Technology 7:1265–1269

Donegan KK, Palm CJ, Fieland VJ, Porteous LA, Ganio LM, Schaller DL, Bucao LQ, Seidler RJ (1995) Changes in levels, species and DNA fingerprints of soil microorganisms associated with cotton expressing the *Bacillus thuringiensis* var. *kurstaki* endotoxin. Appl Soil Ecol 2:111–124

Donegan KK, Schaller DL, Stone JK, Ganio LM, Reed G, Hamm PB, Seidler RJ (1996) Microbial populations, fungal species diversity and plant pathogen levels in field plots of potato plants expressing the *Bacillus thuringiensis* var. *tenebrionis* endotoxin. Transgen Res 5:25–35

Donegan KK, Seidler RJ, Fieland VJ, Schaller DL, Palm CJ, Ganio LM, Cardweel DM, Steinberger Y (1997) Decomposition of genetically engineered tobacco under field conditions: persistence of the proteinase inhibitor I product and effects on soil microbial respiration and protozoa, nematode and microarthropod populations. J Appl Ecol 34:767–777

Fitzpatrick T (1993) Pleiotropic gene found in barley plant. Genet Eng News 13(5):1,22

Flexner JL, Lighthart B, Croft BA (1986) The effects of microbial pesticides on non-target, beneficial arthropods. Agric Ecosyst Environ 16:203–254

Fox J (1991) *Bt* resistance prompts early planning. Bio/Technology 9:1319

Garland JL, Mills AL (1991) Classification and characterization of heterotrophic microbial communities on the basis of patterns of community-level sole-carbon-source. Appl Environ Microbiol 57:2351–2359

Gasser CS, Fraley RT (1992) Transgenic crops. Sci Am June 1992:62–69

Hofte H, Whiteley HR (1989) Insecticidal crystal proteins of *Bacillus thuringiensis*. Microbiol Rev 53:242–255

James RJ, Miller JC, Lighthart B (1993) *Bacillus thuringiensis* var. *kurstaki* affects a beneficial insect, the cinnabar moth (Lepidoptera: Arctiidae). J Econ Entom 86:334–339

Jenkins JN, Parrott WL, McCarty JC Jr, Barton KA, Umbeck PF (1991) Field test of transgenic cottons containing a *B.t.* gene. Dept. of Information Services, Division of Agriculture, Forestry and Veterinary Medicine. MAFES Techn Bull 174, Jan 1991, 6 pp

Johnson MT, Gould F (1992) Interaction of genetically engineered host plant resistance and natural enemies of *Heliothis virescens* (Lepidoptera: Noctuidae) in tobacco. Environ Entomol 21:586–597

Kareiva P, Morris W, Jacobi CM (1994) Studying and managing the risk of cross-fertilization between transgenic crops and wild relatives. Mol Ecol 3:15–21

Klinger T, Ellstrand NC (1994) Engineered genes in wild populations: fitness of weed-crop hybrids of *Raphanus sativus*. Ecolog Appl 4:117–120

Koziel MG, Beland GL, Bowman C, Carozzi NB, Crenshaw R, Crossland L, Dawson J, Desai N, Hill M, Kadwell S, Launis K, Lewis K, Maddox D, McPherson K, Meghji MR, Merlin E, Rhodes R, Warren GW, Wright M, Evola SV (1993) Field performance of elite transgenic maize plants expressing an insecticidal protein derived from *Bacillus thuringiensis*. Bio/Technology 11:194–200

Lange P (1990) The German experience gained with field testing of genetically modified plants. Federal Ministry for Research and Technology, Bonn, Germany

Lundstrum L (1992) Monsanto develops beetle-resistant plants – plots show remarkable control. Potato Grower Idaho 21:36–38

MacKenzie D (1990) Jumping genes confound German scientists. New Sci Dec 15:199

Manasse RS (1992) Ecological risks of transgenic plants: effects of spatial dispersion on gene flow. Ecol Appl 2:431–438

Miller JC (1990) Field assessment of the effects of microbial pest control agents on nontarget Lepidoptera. Am Entomol 36:135–139

Molloy D, Jamnback H (1981) Field evaluation of *Bacillus thuringiensis* var. *israelensis* as a black fly biocontrol agent and its effect on nontarget stream insects. J Econ Entomol 74:314–318

Mulla MS, Federichi BA, Darwazeh HA (1982) Larvicidal effect of *Bacillus thuringiensis* serotype H-14 against stagnant-water mosquitoes and its effect on nontarget organisms. Environ Entomol 11:788–795

Palm CJ, Donegan KK, Harris DL, Seidler RJ (1994) Quantitation in soil of *Bacillus thuringiensis* var. *kurstaki* delta-endotoxin from transgenic plants. Mol Ecol 3:145–15

Palm CJ, Schaller DL, Donegan KK, Seidler RJ (1996) Persistence in soil of transgenic plant produced *Bacillus thuringiensis* var. *kurstaki* endotoxin. Can J Microbiol 42:1258–1262

Parr JF, Papendick RI (1978) Factors affecting the decomposition of crop residues by microorganisms. In: Oschwald WR (ed) Crop residue management systems. Amer Soci Agron, Madison, WI, pp 101–129

Perlak FJ, Deaton RW, Armstrong TA, Fuchs RL, Sims SR, Greenplate JT, Fischhoff DA (1990) Insect-resistant cotton plants. Bio/Technology 8:939–943

Porteous LA, Armstrong JA, Seidler RJ, Watrud LS (1994) An effective method to extract DNAs from environmental samples for polymerase chain reaction amplification and DNA fingerprint analysis. Curr Microbiol 29:301–307

Pratt GE, Royce LA, Croft BA (1993) Measurement of toxicity of soils following incorporation of plant residues engineered with *Bacillus thuringiensis* vas. *kurstaki* endotoxin, using a *Heliothis virescens* growth bioassay, Proc 5th Investigators Meeting for EPA's Environmental Release of Biotechnology Research Program, College Park, Maryland 1992

Ream JE, Berberich SA, Sims SR, Rogan GJ, Fuchs RL (1992) In blanta distribution and environmental fate of insect-resistant cotton proteins. Plant Physiol Suppl 99(1):80

SAS Institute (1989) SAS/STAT user's guide, Version 6, vol 2, 4th edn. SAS Institute, Cary, North Carolina, 846 pp

Stotzky G (1986) Influence of soil mineral colloids on metabolic processes, growth, adhesion and ecology of microbes and viruses. In: Huang PM, Schnitzer M (eds) Interactions of soil mineral with natural organics and microbes. Soil Sci Soc America, Madison, Wisconsin, pp 305–428

The Economist (1990) The tomatoes of the free of knowledge. The Economist, 14 July 1990

The Gene Exchange (1992) Unexpected results in transgenic organisms. The Gene Exchange 3(3):6–7

Tolstova YS, Ionova ZA (1976) Toxicity of pesticides to TrichoGramma. Zashch: Rast (Mosc) 9:21

Umbeck PF, Barton KA, Nordheim EV, McCarty JC, Parrot WL, Jenkins JN (1991) Degree of pollen dispersal by insects from a field test of genetically engineered cotton. J Econ Entomol 84:1943–1950

USDA-CRS and USDA-ARS (1992) Scientific evaluation of the potential for pest resistance to the *Bacillus thuringiensis* (Bt) delta-endotoxins. A Conference to Explore Resistance Management Strategies, Beltsville, MD, Jan 21–23, 1992

Vaeck M, Reynaerts A, Hofte H, Jansens S, De Beuckeleer M, Dean C, Zabeau M, Van Montagu M, Leemans J (1987) Transgenic plants protected from insect attack. Nature 328:33–37

Yamada K (1992) Genetic vegomatics splice and dice with weird results. Wall Street J April 13, 1992

V.5 Genetic Engineering of Cotton Fiber

M.E. JOHN[1]

1 Introduction

Fibers from various natural sources have been the exclusive raw materials for textiles and related products for many centuries. However, their dominance was lost with the advent of synthetic polymeric fibers. Breakthroughs in chemical engineering, polymer chemistry and innovations in marketing have enabled man-made fibers to achieve a market share of 62%, while cotton has shrunk to 32% (Textile Fibers: Industry Study 1992). Today, the natural fiber industry, especially cotton, must use newer technologies, such as genetic engineering, to increase yield and quality. Furthermore, they can reduce the cost of production and decrease the adverse impact on our environment.

Plant breeding relies on the exchange of part of the gene pool from one cultivar into another compatible one. Several decades of painstaking plant breeding have resulted in improved cotton varieties for agronomical and fiber traits (Fig. 1; Culp 1992; Meredith 1992). To these varieties, we now can add new agronomical and fiber traits through genetic engineering. For example, insect and herbicide resistance can be added to different elite commercial varieties. Conventional cross-breeding of two transgenic varieties would generate a variety that combines insect- and herbicide-resistant traits in one plant. Further cross-breeding of the insect-herbicide-resistant variety to new fiber trait varieties would produce superior cotton varieties. Improvements in yield and fiber traits are expected to expand cotton marketability, enhance demand, increase profitability, and overall strengthen the cotton industry's competitive position. Theoretical considerations of improved agronomical and fiber traits support the notion of profitability for transgenic cotton (Chiou et al. 1993; Cotton Farming 1996).

2 Fiber Development

Cotton fiber is a differentiated single epidermal cell of the outer integument of the ovule. Fiber cell development is triggered by several plant hormones such

[1] Agracetus, 8520 University Green, Middleton, Wisconsin 53562, USA

Biotechnology in Agriculture and Forestry, Vol. 42
Cotton (ed. by Y.P.S. Bajaj)

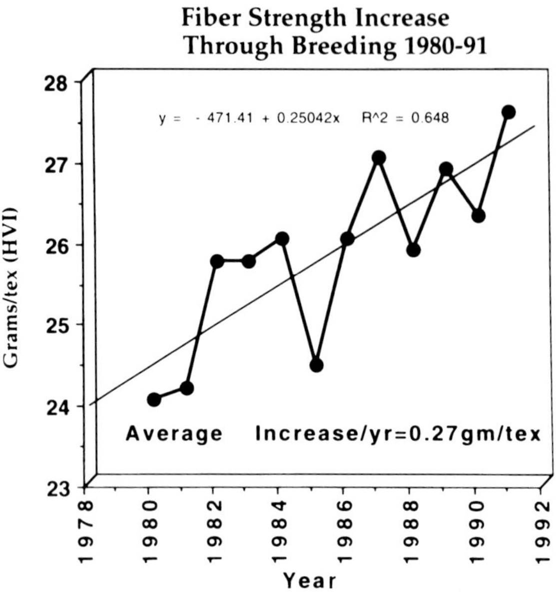

Fig. 1. Trend in national average for high volume instrumentation (HVI) strength for US upland cotton crop. The annual increase in strength since 1980 is 0.25 g/tex per year and is a result of plant breeding and better agronomical practices. (After a Monogram of Cotton Incorporated, 4505 Creedmoor Road, Raleigh, North Carolina, USA Compiled by P.E. Sasser and M.D. Watson)

as gibberellic acid and auxins (Beasley and Ting 1973; Kosmidou-Dimitropoulou 1986). The fiber development spans a period of 45 to 50 days after anthesis, and is delineated into four overlapping phases, initiation, elongation, secondary cell-wall deposition, and maturation (for review see Willison and Brown 1977; Basra and Malik 1984; Ryser 1985; Graves and Stewart 1988). At anthesis, one out of every three epidermal cells of the ovule epidermis undergoes spherical expansion (Stewart 1975). The primary wall of the fiber is formed during the initiation and elongation stages; a large central vacuole is also formed during this time period. The turgor pressure in the vascular compartment, generated by the import and accumulation of potassium and malate, is responsible for fiber expansion. Malate has also been proposed to be an energy and carbon source during secondary cell-wall formation (Cleland 1971). The primary cell wall (0.2 to 0.4 μm thick) is made up of cellulose, neutral (rhamnose, fucose, arabinose, xylose, mannose, galactose, cellulosic and noncellulosic glucose) and acidic polysaccharides, and proteins.

The secondary wall (8 to 10 μm thick) is composed exclusively of cellulose. Secondary wall deposition (16 to 45 days) begins before the cessation of primary wall synthesis. Maturation, the final stage of fiber development (45 to 50 days), is associated with changes in mineral content and protein levels. As the cotton bolls open, dehydration occurs, causing the fibers to dry out and the lumen to collapse, giving the appearance of twisted ribbons. Strength, length, micronaire, uniformity, and maturity of cotton fiber are some of the most important physical properties necessary for textile applications. Some of these properties are being steadily improved upon by classical plant breeding, as shown in Fig. 1. Genetic engineering can complement this effort as well as add new traits, such as dye binding and wrinkle and shrinkage resistance.

3 Critical Tasks in Fiber Modifications

3.1 Selection of Target Properties

A number of factors, such as market demand, volume, technical feasibility, time required to achieve commercialization, profitability, and consumer perceptions, influence the selection of target properties. Some of the specific modifications useful for various applications include superior strength, length, dye binding, dimensional stability, and thermal adaptability (John 1994a,b). Table 1 shows examples of fiber modification objectives and their potential applications.

3.2 Selection of Genes

Theoretical considerations of the structure and function of genes enables the selection of potential genes for fiber modifications. Selected genes must be tested to assess their effect on fiber properties. Lack of a model system to test the genes is a serious drawback in fiber research. Development of an in vitro system (Haigler et al. 1991; Beasley 1992) amenable to testing genes in short periods of time (weeks) would greatly enhance the introduction of new traits into fiber. In vitro ovule culture is a useful system to study the influence of

Table 1. Examples of fiber modification objectives and their potential applications. (John 1994a)

Properties	Applications
Superior strength and length	Textiles
Increased or decreased absorption	Industrial and household absorbents, personal care products, disposable diapers, medical applications
Thermal adaptability	Specialty textiles (winter clothing)
Enzyme immobilization	Food processing, medical and analytical applications, bioremediation
Colored fibers	Textiles
New reactive groups	Textiles, medical and industrial applications
Superior dye binding	Textiles
Surface modifications	Textiles, nonwovens
Better dimensional stability	Textiles

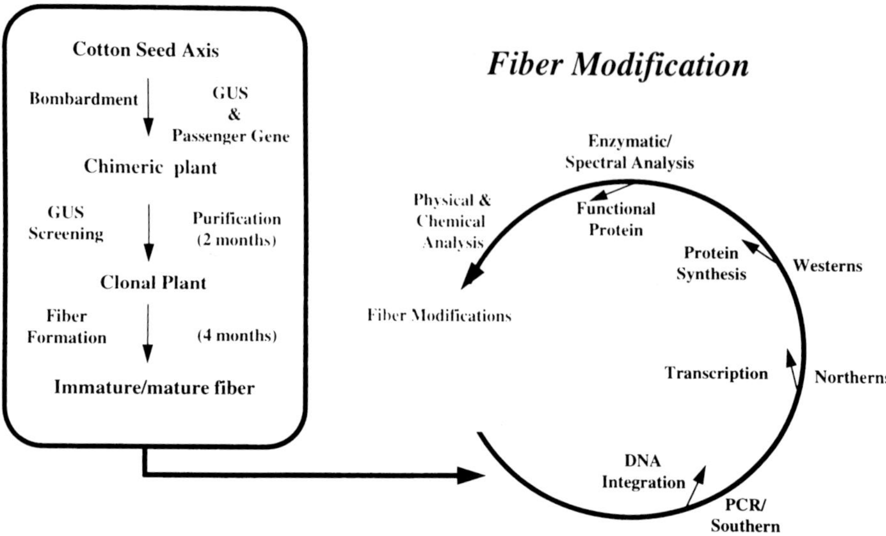

Fig. 2. Steps and timelines for fiber modifications schemes. The process takes 8 to 12 months for completion and requires generation of ten or more plants. (After John 1994b)

nutrition, phytohormones, environmental factors, and genetics on both embryo and fiber development (see Beasley 1977; Davidonis et al. 1993). We investigated in vitro ovule cultures as a model system to test genes. Particle bombardment was used to introduce genes into growing fiber cells in culture. However, gene expression could not be sustained in fiber cells after bombardment (unpubl). Furthermore, lack of well-defined fiber development in in vitro cultures made it difficult to judge the effects of transgene expression. We are currently limited to testing genes in stable transgenic plants, which is a 6-month process. A flow chart describing the introduction of genes into cotton using particle bombardment and steps used in the identification of fiber modifications is shown in Fig. 2. Several transformants expressing the gene are required to assess the effect of gene expression on fiber properties.

3.3 Selection of Transformation Technique

Agrobacterium-mediated transformation is reliable in introducing single gene traits, such as insect and herbicide resistance, into cotton (Umbeck et al. 1987). Particle bombardment technology is recommended for fiber modification strategies, due to its ability to introduce many genes into the same plant. However, this technology requires sophisticated instrumentation and skills (McCabe and Martinell 1993). A review of transformation technology to develop transgenic cotton is presented by McCabe et al. (Chap. V.I, this Vol.).

3.4 Measurement of Fiber Properties

Only limited quantities of fiber are available from each transgenic plant for testing. Current test methods often require larger quantities. Modification of instrumentation and test methodologies which use only a few milligrams of fiber are needed.

3.5 Federal and Local Regulations

Field testing of transgenic cotton requires interstate movement and release notification permits from the United States Department of Agriculture, Animal, and Plant Health Inspection Service (USDA-APHIS). Currently, it takes 30 days to obtain release notification and 10 days to obtain interstate movement permits. Permits from local agencies also may be required. The sale of transgenic fiber is not regulated by any federal or state agencies. However, the sale of seeds, oil, or protein meal require the Environmental Protection Agency (EPA) and the Food and Drug Administration (FDA) approvals.

4 Fiber Modification Strategies

There are two general strategies for fiber modifications. One approach is to increase or decrease levels of fiber proteins or enzymes (John and Stewart 1992; Triplett 1992). This strategy requires that the genes responsible for fiber properties be identified. A second and perhaps more feasible strategy is to select potential genes from sources other than cotton (John 1994a,b). Careful consideration of fiber modification objectives and potential effect of genes in cotton can narrow down the selection to a few candidates.

4.1 Evaluation of Cotton Genes

Cotton genes responsible for fiber strength, length, or other properties are not known. Therefore attempts are being made to isolate and characterize genes that are preferentially expressed in fiber with two objectives in mind. First, a possibility exists that genes that are preferentially expressed in fiber may have some critical role in its development or properties. Hence, identification and characterization of fiber-specific mRNAs may eventually lead to the understanding of their functions in fiber development. A second objective for isolation of fiber-specific mRNAs is the identification of regulatory elements, such as promoters, enhancers, or sequences that determine fiber specificity from corresponding genes. These elements are valuable tools for genetic engineers attempting to express proteins or enzymes for modification of fiber properties. The antisense gene produces transcripts that are complementary to the corresponding mRNA and, therefore, hybridizes with it, forming a double-stranded RNA. This results in reduced expression of the gene through unknown mechanisms. It is possible that the double-stranded RNA may be degraded or its transport into the cytoplasm inhibited (Bird et al. 1991).

Four mRNAs, E6, H6, B6, and FbL2A, that are preferentially expressed in fiber have been characterized (John and Crow 1992; John and Keller 1995; John 1995a, 1996; Rinehart et al. 1996). There are at least two closely related E6 genes in the upland cotton varieties, DP 50 and Coker 312. Sea Island and Pima (*barbadense*) varieties also contain two or more E6 genes. All identified E6 genes are closely related (97 to 100% identity at the nucleotide level). Analyses of various E6 cDNA and genomic clones indicate polymorphism near the N-terminus where a dimer, Ser-Gly, is repeated one to five times (John 1996). E6 is a cytoplasmic protein that is expressed during early fiber development. In order to assess the role of E6 protein in fiber development, its expression was reduced in transgenic cotton through antisense technology. Transgenic plants were identified, where E6 expression was reduced by 98%. However, the fibers of these plants exhibited normal morphology, development, and physical properties. Thus, it appears that the E6 protein does not play a critical role in fiber properties or development (John 1996).

An mRNA (H6), encoding a proline-rich (35 mol%) protein has been identified from fiber (John and Keller 1995). The H6 protein begins to accumulate in fiber during the late primary wall synthesis stage. It contains a repetitive pentameric motif of alanine (or serine)-threonine (or serine)-proline-proline-proline residues. The motif is repeated 17 times in the protein. Many of the attributes of the H6 peptide resemble those of arabinogalactan-type proteins. The H6 protein may function in the development and architecture of the secondary walls (John and Keller 1995).

Another mRNA, FbL2A is expressed during late fiber development (Rinehart et al. 1996). The FbL2A, mRNA appears in 15-day postanthesis (DPA) fibers and maximal levels are found in 20- to 25-DPA fibers. FbL2A encodes a highly hydrophilic protein of 43.5 kDa. A major portion (62%) of the peptide is composed of repeat motifs. A 55-amino acid peptide is repeated four times in a concatenated fashion within the protein. The function of the protein is not known. An mRNA corresponding to lipid transfer protein (LTP) has been identified from fiber. The cotton LTP shows considerable homology to those of spinach, maize, tobacco, tomato, and carrot (Ma et al. 1995).

Several genes involved in the synthesis of cellulose have also been identified from *Acetobacter xylinum* and *Agrobacterium tumefaciens* (Wong et al. 1990; Saxena et al. 1994; Matthysse et al. 1995). Uridine diphosphoglucose (UDPG) is the precursor for cellulose synthesis and is incorporated into the β-(1,4)-glucan polymer by a membrane-bound cellulose synthase. An activator of cellulose synthase, cyclic diguanylic acid, was identified (Ross et al. 1987). Conclusive identification of genes corresponding to cellulose synthase in plants has not been achieved, and many details of plant cellulose synthesis remain to be elucidated (for review see Delmer et al. 1983, 1993; Haigler 1985). Recently, Pear et al. (1996) has identified two cotton cDNA clones that contain conserved amino acid regions that are similar to those of the catalytic subunit of bacterial cellulose synthase. The authors also demonstrated UDP-glucose binding to the peptide fragments and temporal expression that coincides with the onset of secondary cell-wall synthesis, suggesting the genes to be involved in cellulose synthesis.

Thus, a number of genes from cotton have been isolated. Further experimentation would lead to identification of functions of these genes as well as potential applications for fiber modifications. The regulatory elements of some of these genes have already proven to be valuable in cotton genetic engineering (John and Keller 1996).

4.2 Fiber-Specific Promoters

Promoters are DNA sequences that determine the time- and tissue specificity of gene expression. Genetic engineering of cotton for agronomical or product traits requires different promoters. For example, for herbicide or insect resistance a constitutive promoter is required. Whereas fiber modification may require only fiber-specific expression. The most widely used constitutive promoter in plant genetic engineering is the cauliflower mosaic virus 35S promoter (Ow et al. 1986). This promoter directs high levels of gene expression in many plant tissues. A transcriptional enhancer situated upstream of TATA box can be duplicated to increase the gene expression further (Kay et al. 1987).

Targeting the expression of foreign genes to the tissue of interest has a number of advantages. It avoids any detrimental effects of transgene expression on normal growth and morphology of the plant. Large numbers of plant promoters that exhibit predominant expression in certain tissues have been isolated. They are referred to as tissue-specific, although in many instances low levels of expression can be detected in other tissues. Examples include promoters that direct expression to the anthers (Paul et al. 1992), pollen (Twell et al. 1991; John and Petersen 1994), and seeds (Robert et al. 1989).

The corresponding genes for fiber-specific mRNAs discussed earlier (E6, H6, FbL2A, B8) were isolated from genomic libraries. DNA fragments containing putative promoters can be identified from nucleotide sequence comparison between cDNA and genomic clones. A method based on particle bombardment and transient expression of GUS has been developed to identify promoters. In short, the assay consists of linking the *GUS* gene with the putative promoter and bombarding cottonseed axes with the plasmid along with appropriate controls. GUS expression can be detected if the DNA fragment linked to *GUS* contains a promoter (constitutive or tissue-specific). This is most likely due to the large numbers of *GUS* gene copies introduced into each cell by the microparticles (5000 or more copies; 3×10^6 beads/mg; 1 mg DNA [5 kb]/mg bead/1–3 mm). The additive effect of basal transcription from these genes results in detectable levels of GUS. In this assay, the tissue specificity of the promoters could not be assessed. For example, pollen, fiber, or green tissue-specific promoters linked to *GUS* showed enzymatic activity.

This method has been used to identify a number of putative cotton promoters. The results were confirmed by generating stable transformants (John and Crow 1992; John 1996).

Promoters from four E6 genes were identified using the above-described transient assay system. Carrot extensin and a bacterial gene, acetoacetyl CoA reductase, involved in the synthesis of polyhydroxybutyrate, were linked to the E6 promoters, and transgenic plants were generated. Studies of these plants showed that both tissue and temporal specificity of E6 promoters are preserved in transgenic cotton (John and Crow 1992; unpubl.). The E6 promoter is active during early fiber development (5 to 20 DPA). A second cotton promoter, FbL2A, is active during late fiber development (20 to 45 DPA). Based on the limited number of plants studied, it appears that the FbL2A promoter is stronger than the E6 promoter. Two other promoters, H6 and B8, were also tested in transgenic cotton. Both of these promoters appear to be fiber-specific, but are weak promoters (unpubl.). A cotton promoter, Gh-1 linked to the *GUS* gene, was shown to be active in various tissues of transgenic tobacco including trichomes. In transgenic cotton Gh-1 was active in the fiber cells (Dang et al. 1995 and pers. comm.).

Many of these promoters can be used to direct expression of genes from other sources in cotton fiber and assess their potential for fiber modification. An example is the expression of bacterial genes responsible for aliphatic polyester, polyhydroxybutyrate (PHB) in cotton fiber.

5 Evaluation of Genes from Other Sources: Bioplastic Synthesis in Fiber

The synthetic fiber industry developed a value-added product, the bicomponent fibers, which contain a core polymer surrounded by a sheath polymer. They combine properties of two polymers and, in some instances, are a manufacturing necessity (Bach and Knorr 1990). This technical innovation can be duplicated in cotton fiber. If the cotton plant can be genetically engineered to produce a second useful polymer in fiber lumen, such a fiber would have the properties of cellulose along with that of the new polymer.

This hypothesis was tested by genetically engineering cotton to produce an aliphatic polyester compound, polyhydroxybutyrate (PHB) in fiber (John and Keller 1996). PHB is an archetype polyhydroxyalkanoate (PHA) which is a natural biodegradable thermoplastic with chemical and physical properties similar to polypropylene (Fig. 3; Steinbuchel 1991). They are produced by many genera of bacteria as inclusion bodies to serve as carbon sources and electron sink. The formation of PHB in bacteria involves three enzymes, β-ketothiolase, NADPH-dependent acetoacetyl-CoA reductase, and PHA synthase. Two molecules of acetyl-CoA are joined by β-ketothiolase to form acetoacetyl-CoA. Acetoacetyl-CoA is reduced by acetoacetyl-CoA reductase to R-(−)-3-hydroxybutyryl-CoA. This activated monomer is then polymerized

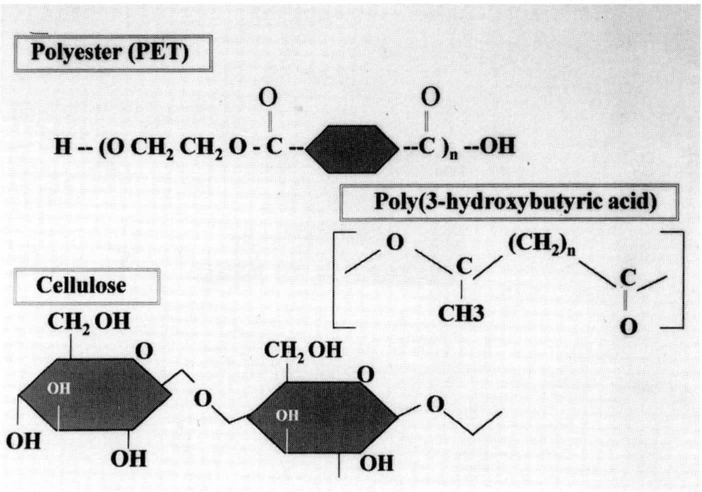

Fig. 3. Structures of cellulose, polyhydroxybutyrate (bioplastic), and polyester poly[ethylene terephthalate] (PET)

by PHA synthase to form PHB. It has been shown that PHB can be synthesized in transgenic plants (Poirier et al. 1992). β-Ketothiolase, involved in the synthesis of mevalonate, is ubiquitous in plants. Therefore, the production of PHB in plants requires only acetoacetyl Co-A reductase and PHA synthase activities. These genes were reengineered for expression in fiber by linking them to fiber-specific promoters, and introduced into cotton by particle bombardment. Transformants were selected based on the expression of the marker gene, *GUS*. Transgenic fibers were examined by epifluorescence microscopy after Nile Blue A staining. Nile Blue A binds to PHB granules and produces a strong orange fluorescence at the excitation wave length of 546 nm. Transgenic fibers exhibited fluorescent granules, whereas control fibers did not. Presence of granules in the cytoplasm of transgenic fibers was also confirmed by transmission electron microscopy. The electron-lucent PHB granules were in the size range of 0.15 to 0.3 μm. Further confirmation of PHB in fibers was obtained by high pressure liquid chromatography (HPLC), gas chromatography (GC), and mass spectra (MS). Figure 4 shows the results of HPLC analysis of PHB isolated from fiber. The PHB is converted to crotonic acid by acid hydrolysis and detected by HPLC (Karr et al. 1983). Duplicate experiments were carried out using extracts of DP50 control fibers. As seen from Fig. 4, transgenic fibers contained crotonic acid, whereas DP50 fibers did not. The chemical nature of fiber-derived PHB was further confirmed by GC-mass spectral analysis. PHB can be converted to ethyl esters and detected by GC-MS (Findlay and White 1983). Fiber extracts were treated with ethanol-chloroform-hydrochloric acid mixtures and subjected to GC analysis. Figure 5 shows the results of GC analysis of transgenic and control fibers along with that of bacterial PHB. The transgenic fiber extract contains a compound with

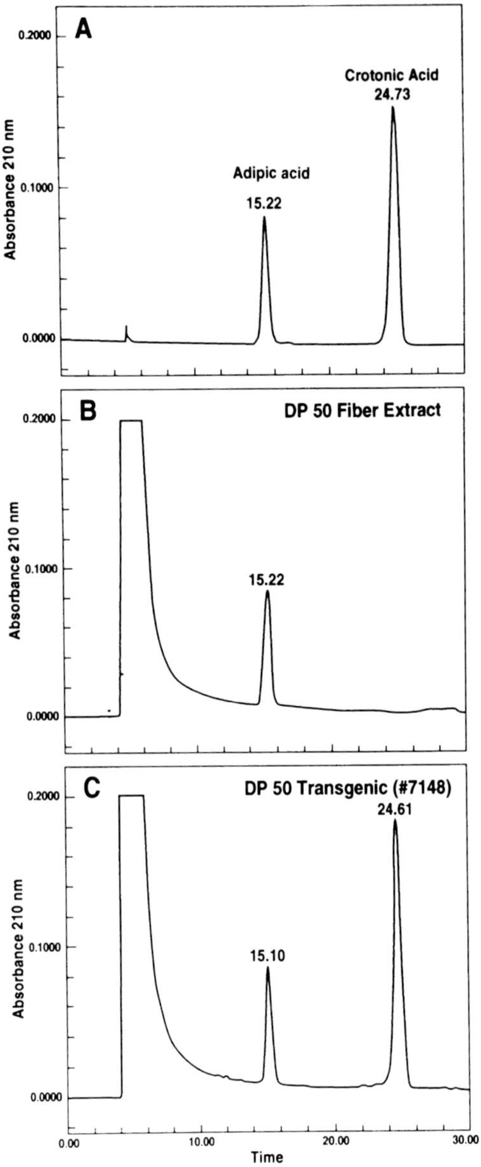

Fig. 4A–C. Detection of PHB by HPLC. Lyophilized immature fibers were homogenized in chloroform using a polytron. The resulting fine suspension was incubated at 65 °C for 18 h. After filtration to separate the chloroform phase, it was dried by N_2 stream at 42 °C and hydrolyzed with sulfuric acid at 90 °C for 45 min. An Aminex HPX-87H column using Beckman System Gold Gradient HPLC system with variable wavelength detector was used to analyze the extract. Standard HPLC retention time for crotonic acid was obtained by subjecting crotonic acid (Sigma) to HPLC analysis. Adipic acid was used as an internal standard. (After John and Keller 1996). **A** Internal standard adipic acid and standard crotonic acid. **B** Control DP50 fiber extract. Note, no peak corresponding to crotonic acid is present. **C** Retention time of compounds in transgenic fiber

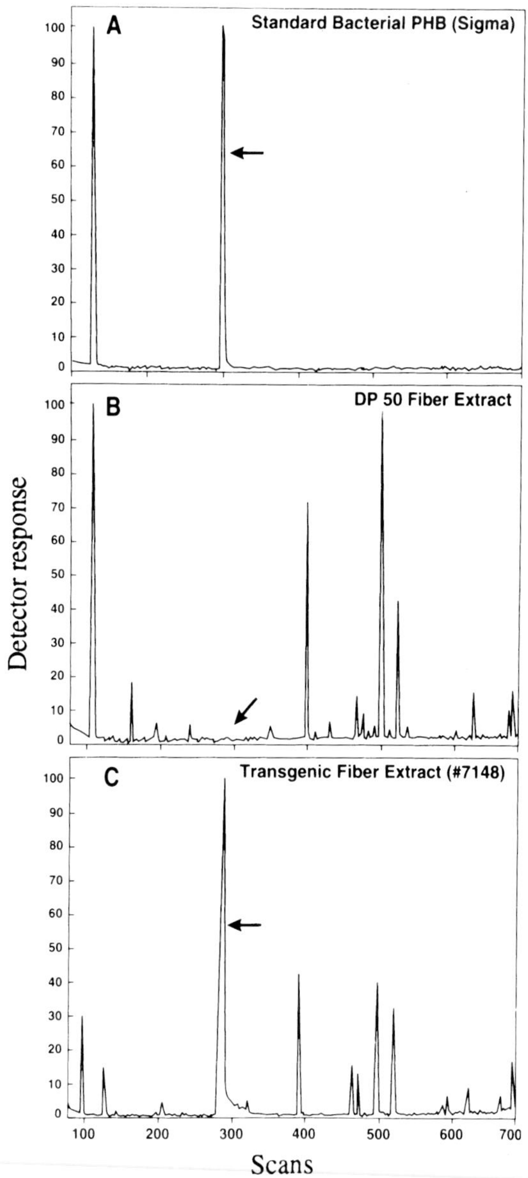

Fig. 5A–C. Gas chromatographic separation of Beta-hydroxy acids from transgenic fibers. **A** Standard bacterial PHB. **B** DP 50 extract was treated with ethanol-chloroform-hydrochloric acid and subjected to GC analysis on a Carlo-Ebra GC with an electron impact detector and a DB 5 column (50 m; Supelco). Scans were taken at 1-s intervals (700 total). Calibrated range was 17–600 mass units. The GC profile was then compared with that of PHB from *A. eutrophus* (Sigma), extracted, and treated under identical conditions. The *arrow* marks the position of the peak present in the profile of PHB from *A eutrophus*. **C** GC analysis of transgenic fiber extract. *Arrow* marks the peak corresponding to the ethyl ester of beta hydroxy acid. (After John and Keller 1996)

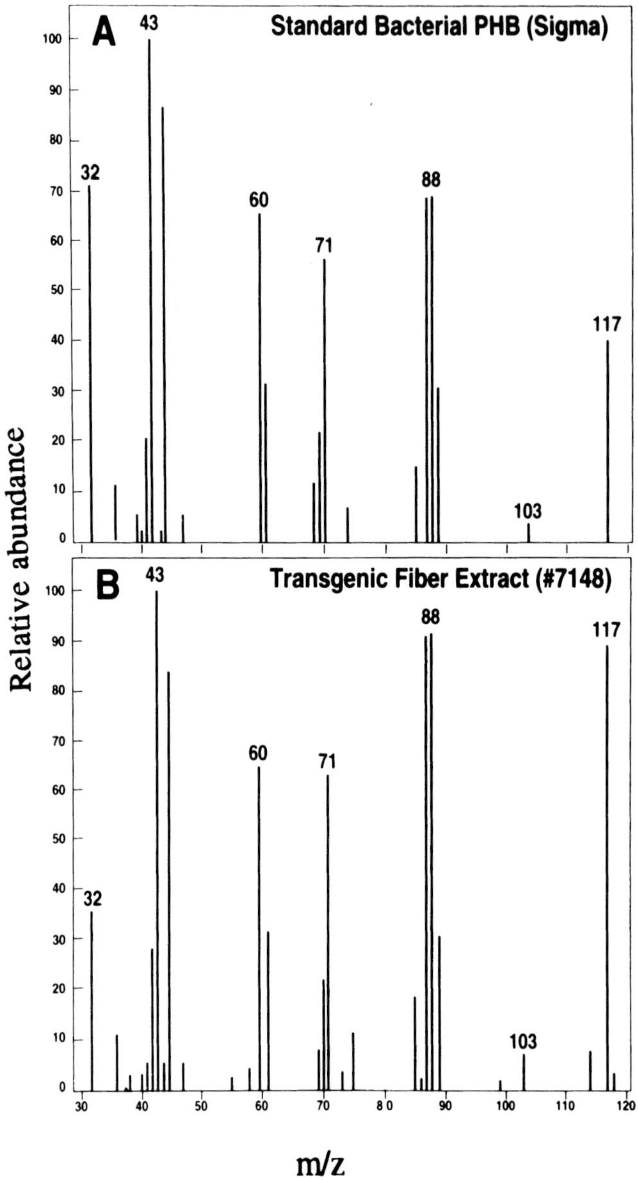

Fig. 6A,B. Confirmation of identity of PHB synthesized in transgenic fibers. The ethyl esters of β-hydroxy acids of bacterial (**A**) and fiber (**B**) origins resolved by GC were subjected to impact electron mass spectroscopy performed on a Kratos MS-25 spectrometer. The pattern of mass fragments of transgenic fiber (**B**) is similar to that of bacterial PHB. (**A**) The mass fragmentation pattern of reference ethyl ester hydroxybutyrate was similar to those of bacterial and fiber-derived materials (not shown). (Adapted from data in John and Keller 1996)

retention time identical to that of bacterial PHB derivative. Mass spectral analysis of ethyl ester derivatives from transgenic fibers, as well as that of bacterial PHB, confirmed that they have similar chemical composition (Fig. 6). From these studies, it was concluded that the PHB produced in fibers is similar to those produced by bacteria. The presence of PHB in fibers resulted in measurable changes in their thermal properties. For example, the rate of heat uptake and cooling was slower and heat uptake capacity was higher for transgenic fibers than controls (Table 2). These results suggest that the transgenic fibers containing PHB have better insulating properties, and therefore may be suitable for winter wear and other textile applications where superior insulating properties are advantageous (John and Keller 1996). These studies demonstrate the potential for fiber modifications through genetic engineering.

Thus, the genes from sources other than cotton hold great promise in enhancing existing or adding new traits to cotton fiber. Success in this regard

Table 2. Thermal properties of transgenic cotton fibers. (After John and Keller 1996)

	TGA (% Mass loss)	TC (BTU/h ft^2/in^3)	DSC (Heat uptake)	SH (J/g °C)
DP 50 Control	Total loss = 82.6%	1.968	619.0 J/g	1.86 (38 °C) 2.692 (60 °C)
	Residue = 17.33%			
No. 7148	Total loss = 85.55%	1.836 (−7%)	690.7 J/g (12%)	2.022 (36 °C) 3.889 (60 °C)
	Residue = 14.45%			945%

TGA 10 °C/min to 700 °C.
DSC 10 °C/min to 320 °C.
TC by steady-state equivalent heat flow (axial/radial).
SH zero baseline of sample sapphire (E 30 °C = 11.053; E 60 °C = 1.063).
Thermal properties of transgenic fibers containing PHB
Thermogravimetric (TGA), thermal conductivity (TC), differential scanning calorimetry (DSC), and specific heat (SH) measurements of cotton fibers were undertaken by MATECH Associates (Scranton, PA) using TA instruments (TA Instruments Inc., New Castle, DE). Thermal analysis protocols confirmed with those recommended by International Confederation for Thermal Analysis. Changes in the mass of trangenic no7148 (16.973 mg) and control DP50 (17.07 mg) were monitored against temperature (10 °C/min to 700 °C) using a thermobalance according to standard operating procedures (handbook for thermal analysis: Therm*al Analysis and Calorimetry* 1991). Mass loss data for each sample is included in the table (column, TGA). A modified differential scanning calorimeter (DSC: TA Instruments, New Castle, DE) was used for the thermal conductivity measurements (TC). All conditions for DP50 were identical to those shown for no. 7148, except dimension = 6.61 mm × 2.36 mm and T2 = 302.18∞k
T1 = temperature at the bottom of the sample; T2 = temperature at the top of the sample; q = heat flow; L = heat flow path; A = total area of the sample). The heat gain/loss characteristics of fibers were measured by DSC. A sample size of 10 mg (no. 7148) and 10.06 mg (DP50) was used. The heat flow rate through samples no. 7148 and control DP50 fibers was monitored against temperature using the steady-state equilibrium heat flow method. Specific heat (SH) values were obtained for 36 and 60 °C each. A sample size of 10 mg was used in the DP50 control, while 10.5 mg was used for no. 7148. Zero baseline was established by measurement of Sapphire. Cell constant values are ^{E}36∞C = (0.7978 × 10 × 60.83)/(60 × 0.8 × 9.6) = 1.053. ^{E}60∞C = (0.8432 × 10 × 60.83)/(60 × 0.8 × 10.05) = 1.063

depends on our ability to test large numbers of genes to identify the few that can introduce changes in fiber properties.

6 Conclusions

Genetic engineering to confer useful agronomic and fiber traits to cotton is likely to lower the cost of production, improve yield and quality, and promote environmentally friendly farm practices. Progress in these areas is likely to be rapid due to new developments in gene identification and transformation technologies. Several genes for insect, herbicide, and stress resistance, as well as fiber modifications, are being tested in various laboratories. A strategy to modify fiber using metabolic pathway engineering to produce aliphatic polyester compounds is under development. Particle bombardment technology has been developed to introduce and test genes in elite varieties of cotton, without the need for regeneration or other tissue culture practices and backcrossing. These developments will lead to improved agronomical and fiber traits in cotton.

7 Experimental Protocols

7.1 cDNA and Genomic Libraries

Cotton plants (*Gossypium hirsutum* L. cv. Coker 312; and DP50; *G. barbadense* L. cv. Sea Island) were grown in the greenhouse. Flowers were tagged on the day of anthesis to determine the age of the fibers. Fibers and other tissues were frozen in liquid nitrogen and homogenized in a buffer (1 : 4 w/v) containing 5 M guanidine isothiocyanate, 0.2 M Tris acetate (pH 8.5), 0.7% β-mercaptoethanol, 0.62% sodium lauroyl sarcosine, and freshly prepared 1% polyvinyl pyrrolidone. RNA and DNA blot analyses were performed as described previously (John and Crow 1992). Complementary DNA libraries were prepared from mRNAs of 10-, 15-, and 23-day fibers in pBR 322 vector. The libraries were screened with single-stranded cDNAs generated from leaf, ovule, fiber, and root mRNAs. Fiber-specific clones were selected based on preferential hybridization of fiber cDNAs (John and Crow 1992). The cDNA inserts were subcloned into a phagemid vector, Bluescript II SK+ (Stratagene, La Jolla, CA) for further manipulations. A genomic library of *G. barbadense* L. cv. Sea Island was prepared by Clonetech (Palo Alto, CA) in phage vector, EMBL 3. Propagation and screening of the genomic library were carried out as described by Ausubel et al. (1987). Sequence analysis of genomic fragments was done by Lofstrand Labs Limited (Gaithersburg, MD). Computer analysis of sequence data was done with the software package of Genetics Computer Group (Madison, WI). Quantitative estimation of marker gene, β-glucuronidase (GUS), was performed by fluorometric assay as follows. Plant tissue was homogenized in 50 mM Na_2PO_4, pH 7; 10 mM EDTA, 0.1% Triton X-100, 0.1% sodium lauroyl sarcosine, 10 mM β-mercaptoethanol. The extract was incubated at 37 °C with 4-methyl umbelliferyl glucuronide (MUG; 1 mg/ml) and the reaction stopped at various time points with 1 ml of stop buffer (0.2 M Na_2CO_3). The fluorometer was calibrated with freshly prepared 4-methyl umbelliferone (MU) standards (Jefferson et al. 1987). Any quenching due to plant materials was accounted for by using control plant extracts during MU standardization. Acetoacetyl-CoA reductase and PHA synthase, enzymes involved in the biosynthesis of polyhydroxyalkonates in bacteria, were cloned by polymerase chain reaction (PCR) amplification (Mullis et al. 1986). The primers for reductase gene containing BamHI sites, (ATTAAGGA-TCCATGACTCAGCGCATTGCG and GGATTAGGAT-CCGCAGGTCAGCCCATATGC) for PCR were based on the sequence of the gene *Alcaligenes eutrophus* (Peoples and Sinskey 1989a). The PCR product (741 bp) was cloned into the Bam HI site of Bluescript SK+ vector. The polyhydroxyalkanoic acid synthase (PHA synthase) gene (1770 bp) was amplified as two fragments and joined (Peoples and Sinskey 1989b). A 580-bp fragment was amplified using two primers (AACATGAATTCATGGC-GACCGGCAAAGG; AATTAG-GATCCGCGAGATCTTGCCGCGTG), and after digestion with Eco RI and Bam HI was cloned into an SK+ vector. This fragment contained an internal Bgl II site. The second fragment (1200 bp) was amplified using primers (CACGCG-GCAAGATCTCGC; TGTAAG GATCCTCATGCCTTGGCTTTGACG) and digested with Bgl II and Bam HI. The SK+ vector containing the 580-bp fragment was then digested with Bgl II and Bam HI and the 1200-bp fragment was ligated to Bgl II/Bam HI sites to generate the PHA synthase gene. Both reductase and PHA synthase genes were sequenced to confirm their nucleotide sequences (Peoples and Sinskey, 1989a,b).

7.2 Plasmid Constructs

The following plasmids were constructed to test cotton promoter activities in transient transformations. Cauliflower mosaic viral 35S promoter was linked to a *GUS* gene at Nco/Xho sites of a promoter-less *GUS* gene, p2117 to generate p35S-GUS. The plasmid 2117 also contained a Nos poly(A) addition signal (280 bp) as an EcoRI fragment. In transient expression studies the 35S-*GUS* was used as a positive control, while p2117 was the negative control. Various cotton putative promoters were linked to p2117 and assayed for GUS expression in transient studies. Following plasmids were used for stable transformation. The acetoacetyl CoA reductase gene was linked to a cotton promoter, E6-3B (John and Crow 1992), Nos poly(A) addition signal and a 35S-*GUS* gene to generate, pE6-Red. The E6 promoter in pE6-Red was then replaced by FbL2A promoter (2.3 kb) to generate pFb-Red. The PHA-synthase gene was linked to FbL2A (2.3 kb) and 35S promoters to generate pFb-Syn and 35s-Syn, respectively.

7.3 Transgenic Cotton

Particle bombardment methodology was used to introduce genes into a commercial cotton variety, Delta & Pine 50 (*G. hirsutum* L. cv. DP50). For this, cottonseed axes were bombarded with gold particles (1.5 to 3 μm; Alfa Chemicals Co.) containing plasmid DNAs. Plasmid DNA was precipitated onto gold beads as described by McCabe and Martinell (1993). Meristems of the seed axes were exposed by dissection for bombardment. An Accell particle delivery device was used to accelerate DNA into meristems as described by Christou et al. (1990). The loading rate for the gold was 0.05 mg gold plus DNA complex per cm^2 of the carrier sheet. The carrier sheets containing gold-DNA complex were accelerated towards the target tissues by the discharge of an 18–20 kV arc.

Transformation of the seed axes was monitored by histochemical detection of GUS activity. Two to 3 weeks postbombardment, leaves of the explants were tested for GUS expression. GUS-positive nodes or auxiliary buds were identified and GUS-negative ones were pruned. This process was repeated until a uniformly transformed plant was obtained. Particle-mediated cotton transformation results in two types of transformants, epidermal and germline (McCabe and Martinell 1993). Stable integration and expression of the transgene occurs in both types of transformants. When only the epidermal cell layer of a plant is transformed, it is designated an epidermal transformant. The progeny of the epidermal transformant will not inherit the transgene, and hence they are vegetatively propagated. Fibers are of epidermal origin and, therefore, epidermal transformants are suitable for assessing transgene expression in fibers. Germline transformants, on the other hand, pass on the transgene into their progeny and are suitable for commercial cultivation.

Acetoacetyl CoA reductase activity in transgenic plant tissues was assayed by the method of Saito et al. (1977). Fiber proteins were extracted from frozen tissue with a buffer containing 20 mM Tris pH 8, 5 mM EDTA, 5 mM β-mercaptoethanol, 5% glycerol, and 0.01% PVP (1 : 4 w/v). The reaction mixture for reductase activity determination contained 100 mM Tris-HCl, pH 8; 0.12 mM NADPH, and fiber extract. The mixture was preincubated at room temperature for 3 min and monitored at 340 nm. Acetoacetyl CoA was added (0.02 mM) and the decrease in NADPH was measured at 340 nm for 5 min. The millimolar extinction coefficient for NADPH is 6.22. One unit of acetoacetyl CoA reductase catalyzes the oxidation of 1 micromol of NADPH in 1 min. Protein concentrations were determined by the method of Bradford (1976). Soluble fiber proteins were extracted with buffer containing 40 mM Tris, pH 6.9, 50 mM β-mercaptoethanol, 10 mM EDTA, 25 μg/ml PMSF, 5% glycerol, and 0.01% PVP (1 : 4 w/v). For Western analysis, proteins were separated on SDS-PAGE (10%) and transferred to nitrocellulose by electroblotting. An alkaline phosphatase-linked immunodetection system (ProtoBlott II AP; Promega, Madison, WI) was used to localize the antibody reaction. PHA synthase (*Alcaligenes eutrophus*) antibody was a gift of Metabolix Inc. (Cambridge, MA).

Polyhydroxybutyrate (PHB) in transgenic cotton fibers was detected by high-pressure liquid chromatography(HPLC), as described by Karr et al. (1983). Quantification was done by Beckman System Gold computer software. Standard HPLC retention time for crotonic acid was obtained by subjecting crotonic acid (Sigma) to HPLC analysis. PHB was converted into ethyl esters of Beta hydroxy acids by treatment with ethanol-chloroform hydrochloric acid mixture and detected by GC.

Acknowledgment. The author thanks Cheryl Scadlock and Peggy Wagner for editorial assistance. This study was partially funded by the Department of Commerce, National Institute of Technology, Advanced Technology Program, grant no. 70NANB5H1061.

References

Ausubel FM, Brent R, Kingston RE, Moore DD, Seidman JG, Smith JA, Struhl K (1987) Current protocols in molecular biology. John Wiley, New York

Bach HC, Knorr RS (1990) Acrylic fibers. In: Kroschwitz JI (ed) Polymers: fibers and textiles, a compendium. John Wiley, New York, pp 1–54

Basra AS, Malik CP (1984) Development of cotton fiber. Int Rev Cytol 89:65–113

Beasley CA (1977) Ovule culture: fundamental and pragmatic research for the cotton industry. In: Reinert J, Bajaj YPS (eds) Applied and fundamental aspects of plant cell, tissue, and organ culture. Springer, Berlin Heidelberg New York, pp 160–178

Beasley CA (1992) In vitro cotton ovule culture: A review. In: Proc Cotton Fiber Cellulose: Structure, Function and Utilization Conf, National Cotton Council, Memphis, TN, pp 65–89

Beasley CA, Egli E (1977) Fiber production in vitro from a conditional fiberless mutant of cotton. Dev Biol 57:234–237

Beasley CA, Ting IP (1973) The effects of plant growth substances on in vitro fiber development from fertilized cotton ovules. Am J Bot 60:130–139

Bird CR, Ray JA, Fletcher JD, Boniwell JM, Bird AS, Teulieres C, Blain I, Bramley PM, Schuch W (1991) Using antisense RNA to study gene function: Inhibition of carotenoid biosynthesis in transgenic tomatoes. Bio/Technology 9:635–639

Braford MM (1976) A rapid and sensitive method for the quantitation of microgram quantities of protein utilizing the principle of protein-dye binding. Anal Biochem 72:248–254

Chiou GT, Chen DT, Capps O (1993) A structural investigation of biotechnological impacts on cotton quality and returns. Am J Agr Econ 75:467–478

Christou P, McCabe DE, Martinell BJ, Swain WF (1990) Soybean genetic engineering-commercial production of transgenic plants. Trends Biotechnol 8:145–151

Cleland R (1971) Cell wall extension. Annu Rev Plant Physiol 22:197–222

Cotton Farming (1996) Genetically altered cotton. Cotton Farming 40:8–12

Culp TW (1992) Simultaneous improvement of lint yield and fiber quality in upland cotton. In: Proc Cotton Fiber Cellulose: Structure, Function and Utilization Conf, National Cotton Council. Memphis, TN, pp 247–288

Dang PM, Heinen JL, Allen RD (1995) Expression of a cotton fiber gene promoter in tobacco. In: Proc Beltwide Cotton Conference, vol 2. National Cotton Council. Memphis, TN, pp 1140–1141

Davidonis GH (1993) Cotton fiber growth and development in vitro: effects of tunicamycin and monensin. Plant Sci 88:229–236

Delmer DP, Benziman M, Klein AS, Bacic A, Mitchell B, Weinhouse H, Aloni Y, Callaghan T (1983) A comparison of the mechanism of cellulose biosynthesis in plants and bacteria. J Appl Polym Sci Appl Polym Symp 37:1–16

Delmer DP, Ohana P, Gonen L, Benziman M (1993) In vitro synthesis of cellulose in plants: Still a long way to go. Plant Physiol 103:307–308

Findlay RH, White DC (1983) Polymeric β-hydroxyalkanoates from environmental samples and *Bacillus megaterium*. Appl Environ Microbiol 45:71–78

Graves DA, Stewart JM (1988) Chronology of the differentiation of cotton (*Gossypium hirsutum* L.) fiber cells. Planta 175:254–258

Haigler CH (1985) The functions and biogenesis of native cellulose. In: Nevell TP, Zeronian SH (eds) Cellulose chemistry and its applications. Ellis Harwood. Chichester, UK, pp 31–83

Haigler CH, Rao NR, Roberts EM, Huang J-Y, Upchurch DR, Trolinder NL (1991) Cultured ovules as models for cotton fiber development under low temperatures. Plant Physiol 95:88–96

Jefferson RA, Kavanagh TA, Bevan MW (1987) GUS fusions: β-glucuronidase as a sensitive and versatile gene fusion marker in higher plants. EMBO J 6:3901–3907

John ME (1994a) Re-engineering cotton fibre. Chem Ind 17:676–679

John ME (1994b) Cotton futures. CHEMTECH 24:27–30

John ME (1995a) Characterization of a cotton (*Gossypium hirsutum* L.) fiber mRNA (Fb-B6). Plant Physiol 107:1477–1478

John ME (1995b) Prospects for modification of fibers through genetic engineering of cotton. In: Gebelein CG, Carraher CE (eds) Industrial biotechnological polymers. Technomic, Inc. Lancaster, PA, pp 69–79

John ME (1996) Structural characterization of genes corresponding to cotton fiber mRNA, E6: reduced E6 protein in transgenic plants by antisense gene. Plant Mol Biol 30:297–306

John ME, Crow LJ (1992) Gene expression in cotton (*Gossypium hirsutum* L.) fiber: cloning of the mRNAs. Proc Natl Acad Sci USA 89:5769–5773

John ME, Keller G (1995) Characterization of mRNA for a proline-rich protein of cotton fiber. Plant Physiol 108:669–676

John ME, Keller G (1996) Metabolic pathway engineering in cotton: biosynthesis of polyhydroxybutyrate in fiber cells. Proc Natl Acad Sci USA 93(23):12768–12773

John ME, Petersen MW (1994) Cotton (*Gossypium hirsutum* L.) pollen-specific polygalacturonase mRNA: tissue and temporal specificity of its promoter in transgenic tobacco. Plant Mol Biol 26:1989–1993

John ME, Stewart JMcD (1992) Genes for jeans: biotechnological advances in cotton. TIBTECH 10:165–169

Karr DB, Waters JK, Emerich DW (1983) Analysis of poly-B-hydroxybutyrate in *Rhizobium japonicum* bacteroids by ion-exclusion high-pressure liquid chromatography and UV detection. Appl Environ Microbiol 46:1339–1344

Kay R, Chan A, Daly M, McPherson J (1987) Duplication of CaMV 35S promoter sequences creates a strong enhancer for plant genes. Science 236:1299–1302

Kosmidou-Dimitropoulou K (1986) Hormonal influences in fiber development. In: Mauney JR, Stewart JMcD (eds) Cotton physiology. The Cotton Foundation, Memphis, TN, pp 361–373

Ma D-P, Tan H, Si Y, Creech RG, Jenkins JN (1995) Differential expression of a lipid transfer protein gene in cotton fiber. Biochim Biophys Acta 1257:81–84

Matthysse A, Thomas DL, White AR (1995) Mechanism of cellulose synthesis in *Agrobacterium tumefaciens*. J Bacteriol 177:1076–1081

McCabe DE, Martinell BJ (1993) Transformation of elite cotton cultivars via particle bombardment of meristems. Bio/Technology 11:596–598

Meredith WR Jr (1986) Fiber quality variation among USA cotton-growing regions. In: Brown JM (ed) Proc Beltwide Cotton Prod Res Conf, Natl Cotton Council, Memphis, TN, pp 105–106

Meredith WR (1992) Improving fiber strength through genetics and breeding. In: Proc Conf Cotton Fiber Cellulose: Structure, Function and Utilization, National Cotton Council, Memphis, TN, pp 289–302

Mullis KB, Faloona FA, Scharf SJ, Saiki RK, Horn GT, Erlich HA (1986) Specific enzymatic amplification of DNA in vivo: the polymerase reaction. Cold Spring Harbor Symp Quant Biol 51:263–273

Ow DW, Wood KV, DeLuca M, de Wet JR, Helinski DR, Howell SH (1986) Transient and stable expression of the firefly luciferase gene in plant cells and transgenic plants. Science 234:856–859

Paul W, Hodge R, Smartt S, Draper J, Scott R (1992) The isolation and characterisation of the tapetum-specific *Arabidopsis thaliana* A9 gene. Plant Mol Biol 19:611–622

Pear JR, Kawagoe Y, Schreckengost WE, Delmer DP, Stalker DM (1996) Higher plants contain homologs of the bacterial celA genes encoding the catalytic subunit of cellulose synthase. Proc Natl Acad Sci USA 93(22):12637–12642

Peoples OP, Sinskey AJ (1989a) Poly-β-hydroxybutyrate biosynthesis in *Alcaligenes eutrophus* H16. Characterization of the genes encoding β-ketothiolase and acetoacetyl-CoA reductase. J Biol Chem 264:15293–15297

Peoples OP, Sinskey AJ (1989b) Poly-β-hydroxybutyrate (PHB) biosynthesis in *Alcaligenes eutrophus* H16. Identification and characterization of the PHB polymerase gene (*phbC*). J Biol Chem 264:15298–15030

Poirier Y, Dennis DE, Klomparens K, Somerville C (1992) Polyhydroxybutyrate, a biodegradable thermoplastic, produced in transgenic plants. Science 256:520–523

Rinehart JA, Petersen MW, John ME (1996) Tissue-specific and developmental regulation of cotton gene, FbL2A. Plant Physiol 112(3):1331–1341

Robert LS, Thompson RD, Flavell RB (1989) Tissue-specific expression of a wheat high molecular weight glutenin gene in transgenic tobacco. Plant Cell 1:569–578

Ross P, Weinhouse H, Aloni Y, Michaeli D, Weinberger-Ohana P, Mayer R, Braun S, de Vroom E, van der Marel GA, van Boom JH, Benziman M (1987) Regulation of cellulose synthesis in *Acetobacter xylinum* by cyclic diguanylic acid. Nature (Lond) 325:279–281

Ryser U (1985) Cell wall biosynthesis in differentiating cotton fibres. Eur J Cell Biol 39:236–256

Saito T, Fukui T, Ikeda F, Tanaka K (1977) An NADP-linked acetoacetyl CoA reductase from *Zoogloea ramigera*. Arch Microbiol 114:211–217

Saxena IM, Kudlicka K, Okuda K, Brown RM (1994) Characterization of genes in the cellulose-synthesizing operon (*acs* operon) of *Acetobacter xylinum*: implications for cellulose crystallization. J Bacteriol 176:5735–5752

Steinbuchel A (1991) Polyhydroxyalkanoic acids. In: Byrom D (ed) Biomaterials: novel materials from biological sources. Stockton Press, New York, pp 124–213.

Stewart JMcD (1975) Fiber initiation on the cotton ovule (*Gossypium hirsutum*). Am J Bot 62:723–730

Textile Fibers: Industry study #420 (1992) The Freedonia Group, 20600 Chagrin Blvd, Cleveland, OH, pp 30–59

Triplett BA (1992) Strategies for improving cotton fiber quality. In: Proc Cotton fiber cellulose: Structure, function and utilization Conf, Natl Cotton Council Memphis, TN, pp 107–114

Trolinder NL, Berlin JD, Goodin JR (1987) Differentiation of cotton fibers from single cells in suspension culture. In Vitro Cellular and Dev Biol 23(1):789–794

Twell D, Yamaguchi J, Wing RA, Ushiba J, McCormick S (1991) Promoter analysis of genes that are coordinately expressed during pollen development reveals pollen-specific enhancer sequences and shared regulatory elements. Genes Dev 5:496–507

Umbeck P, Johnson G, Barton K, Swain W (1987) Genetically transformed cotton (*Gossypium hirsutum* L.) plants. Bio/Technology 5:263–266

Willison JHM, Brown RM (1977) An examination of the developing cotton fiber: wall and plasmalemma. Protoplasma 92:21–41

Wong HC, Fear AL, Calhoon RD, Eichinger GH, Mayer R, Amikam D, Benziman M, Gelfand DH, Meade JH, Emerick AW, Bruner R, Ben-Bassat A, Tal R (1990) Genetic organization of the cellulose synthase operon in *Acetobacter xylinum*. Proc Natl Acad Sci USA 87:8130–8134

Section VI
Secondary Metabolites – Gossypol and Tannins

VI.1 Gossypol – A Unique Biologically Active Compound from the Cotton Plant

J.W. Jaroszewski[1]

1 Introduction

Gossypol (Fig. 1), a yellow pigment produced by the cotton plant, is of profound importance both as a toxic contaminant potentially present in industrial products derived from cottonseed, and because of its unique pharmacological effects. Given the high-volume production of cotton in the world and the high content of the material in the cottonseed (up to several percent), gossypol is among the most important secondary plant metabolites. Much of the early interest in gossypol paralleled the efforts to develop technological processes for production of high-quality cottonseed oil and meal from the cottonseed. More recently, the burst of interest in gossypol has been due to the discovery of the intriguing and potentially useful antifertility and antitumor properties of gossypol. This chapter reviews briefly the importance of this compound, and reports the recent advances in the chemistry and pharmacology of gossypol.

2 Isolation and Elucidation of Structure of Gossypol

The first isolation of a colored matter from the cottonseed oil was reported over a century ago (Longmore 1886). The name gossypol (i.e., *Gossypium* phenol) is due to Marchlewski, who purified the compound in 1899 and ascertained its polyphenolic nature (Marchlewski 1899). He also discovered that gossypol forms a solvate with acetic acid. This complex, considerably less soluble than gossypol itself, can be precipitated from the solutions of gossypol, and has been used to purify and handle gossypol ever since. However, the exact chemical structure of gossypol remained obscure for many years (Carruth 1918; Clark 1927). The long series of degradative and synthetic studies conducted during the 1930s by Adams and coworkers eventually resulted in correct formulation of gossypol, as shown in Fig. 1 (Campbell et al. 1937; Adams et al. 1941, and other papers of this series). Total synthesis of gossypol was reported in 1958 (Edwards 1958). The early chemical studies of

[1] Department of Medicinal Chemistry, Royal Danish School of Pharmacy, Universitetsparken 2, DK-2100 Copenhagen, Denmark

Biotechnology in Agriculture and Forestry, Vol. 42
Cotton (ed. by Y.P.S. Bajaj)

gossypol have been reviewed elsewhere (Adams et al. 1960; Markman and Rzhekhin 1969; Nazarowa et al. 1981).

These early investigations of gossypol were conducted without being aware of the fact that because of the noncoplanarity of the two naphthalene rings, the gossypol molecule is chiral (C_2 symmetry), and hence the compound is amenable to separation into enantiomers (Fig. 2). This stereochemical feature became known only in 1968, when three groups independently reported on the isolation of (+)-gossypol from *Thespesia populnea*, a species related to the cotton plant (Bhakuni et al. 1968; Datta et al. 1968; King and de Silva 1968). The absolute configuration of (+)-gossypol is believed to be (S) (Huang et al. 1988). The chirality of gossypol is important in connection with the biological effects (see below).

Because of the presence of the polyhydroxylated aromatic aldehyde moieties, gossypol exhibits tautomerism, which profoundly influences its chemis-

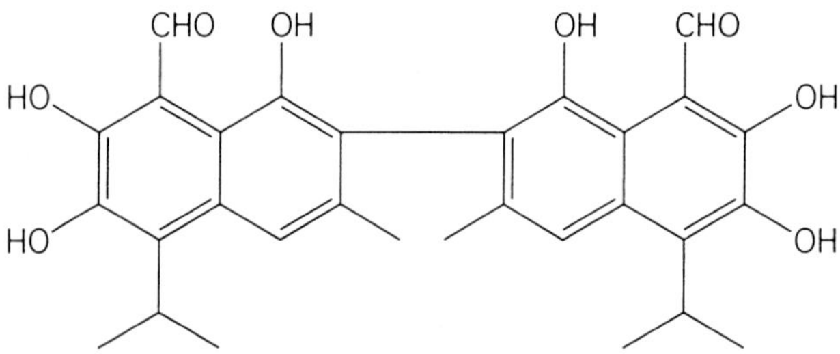

Fig. 1. The structure of gossypol

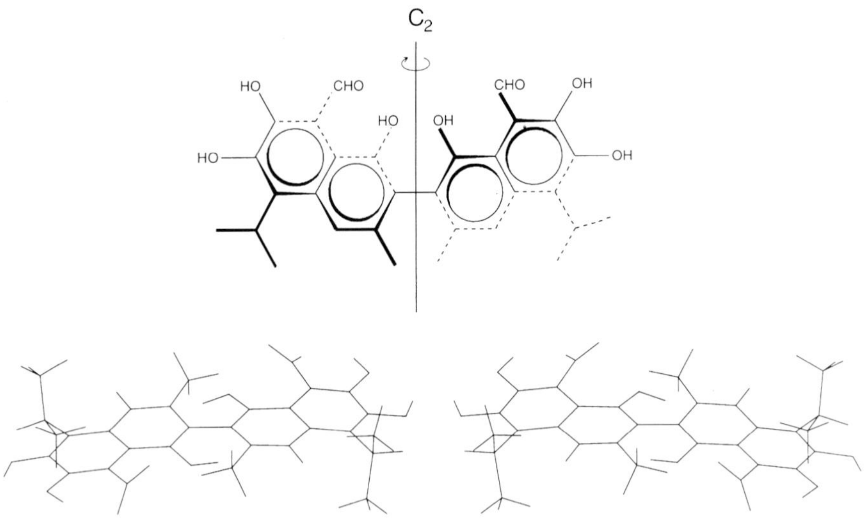

Fig. 2. *Top* Figure shows the C_2 symmetry axis of gossypol molecule; *bottom*-molecular models of the enantiomers of gossypol viewed along the C_2 axis

try (and possibly also biological effects). Thus, the hydroxy group in the peri position to the aldehyde group can participate in the formation of a hemiacetal (lactol form) (Fig. 3), which can be observed in solutions of gossypol by spectroscopic means (Jaroszewski et al. 1988; Golaniec et al. 1996). The formation of two hemiacetal rings results in formation of two asymmetric centers in the molecule (Fig. 3), which, together with the chiral axis present, leads to the formation of complex mixtures of diastereomers. Alkylation or acylation of gossypol results mainly in the formation of stereoisomeric derivatives of the lactol forms, depending on the reaction conditions used. All three possible hexaacetates of the dilactol form (one C_1 isomer and two C_2 isomers) have been isolated and characterized (Jaroszewski et al. 1988).

The third tautomer of gossypol, in which the aldehyde groups are enolized, is frequently encountered in the literature. However, there is no evidence of enolization of the aldehyde groups in the unionized gossypol (J.W. Jaroszewski, unpubl.), and this form appears to be of no significance except in the gossypol anion (Stipanovic et al. 1973), the resonance structure of which bears a significant proportion of the negative charge on the aldehydic oxygen (enolate form). Further complication of the chemistry of gossypol is the easiness with which gossypol eliminates water, giving anhydrogossypol (Fig. 3; Jaroszewski et al. 1992a). In addition to the symmetric forms shown in Fig. 3, mixed structures can be formed.

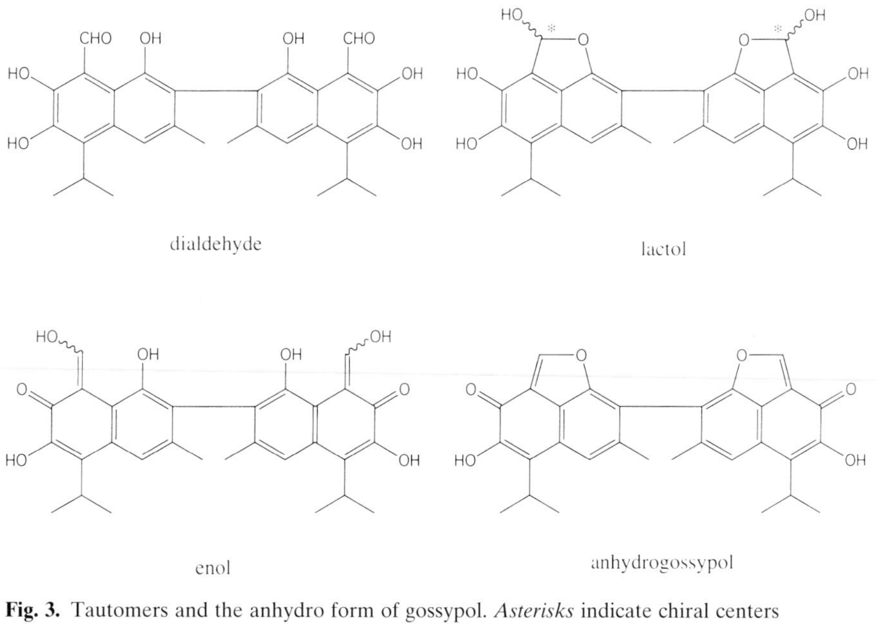

Fig. 3. Tautomers and the anhydro form of gossypol. *Asterisks* indicate chiral centers

3 Biosynthesis, Occurrence, and Function

Gossypol is a bis-sesquiterpene consisting of a total of six isoprene units, originating from mevalonate via (*Z,E*)-farnesyl pyrophosphate or an equivalent (Fig. 4; Masciadri et al. 1985; Stipanovic et al. 1986). The intermediate hemigossypol is oxidatively dimerized to give gossypol. Nothing is known about the stereochemistry of the phenol coupling stage, i.e., whether the coupling step is stereospecific, partially stereospecific, or not stereospecific. In other words, it has yet to be determined whether the natural occurrence of (+)- or (−)-gossypol enantiomers of variable optical purity (see below) represents the (partial) stereospecificity of the coupling step, or an intervention of other processes, such as a (partly) enantioselective turnover or enantioselective binding to proteins.

In accordance with the established biosynthetic scheme for gossypol, the Gossypieae tribe produces a number of other terpenoids (Bell et al. 1975, 1978; Stipanovic et al. 1975, 1977, 1978; Gray et al. 1976), some of which are

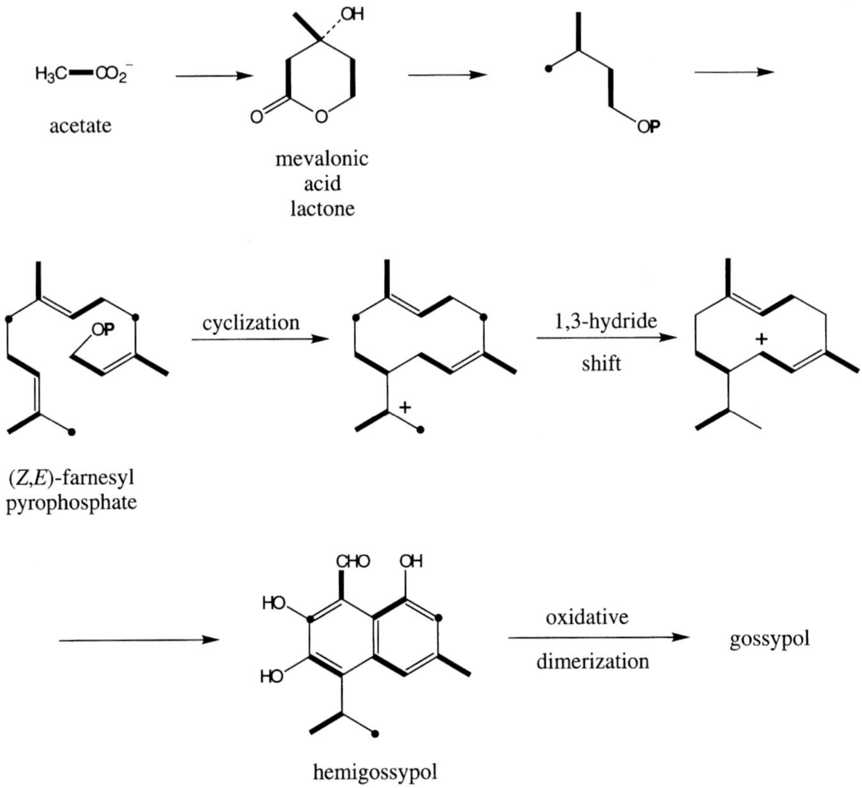

Fig. 4. Biosynthesis of gossypol (Masciadri et al. 1985; Stipanovic et al. 1986). *Solid lines* indicate the acetate units, *P* is the pyrophosphate residue)

shown in Fig. 5. Related sesquiterpenes are also produced by more distant Malvaceae (Ali et al. 1980), but there appears to be no confirmed records about the occurrence of gossypol outside the cotton tribe (Jaroszewski et al. 1992b). In addition to *Gossypium* and the already mentioned *Thespesia*, gossypol has been reported from *Azanza*, *Cienfuegosia*, *Gossypioides*, *Hampea*, *Kokia*, *Lebronnecia*, and *Montezuma* (Lukefahr and Fryxell 1967; Bell et al. 1978; Marcelle et al. 1985; Jaroszewski et al. 1992b).

Gossypol is located mainly in the pigment glands, which occur in all parts of the plant, and which appear to be unique to the Gossypieae (Stanford and Viehoever 1918; Fryxell 1981; Khushk and Vaughan 1985). The content of gossypol in the seeds can be as high as 10% (Carter et al. 1966). Glandless cotton contains only traces of gossypol (Fisher et al. 1988). It is generally recognized that gossypol and other cotton terpenoids have antimicrobial and insecticidal effects, and that they are associated with pest resistance in the cotton plant. Formation of sesquiterpene aldehydes in *Gossypium* cell cultures can be induced by elicitors, supporting their role as phytoalexins (Heinstein 1985; Apostol et al. 1987).

In addition to gossypol and related terpenoids (Fig. 4), cottonseed kernels contain other phenolic compounds: flavonoids derived from quercetol and

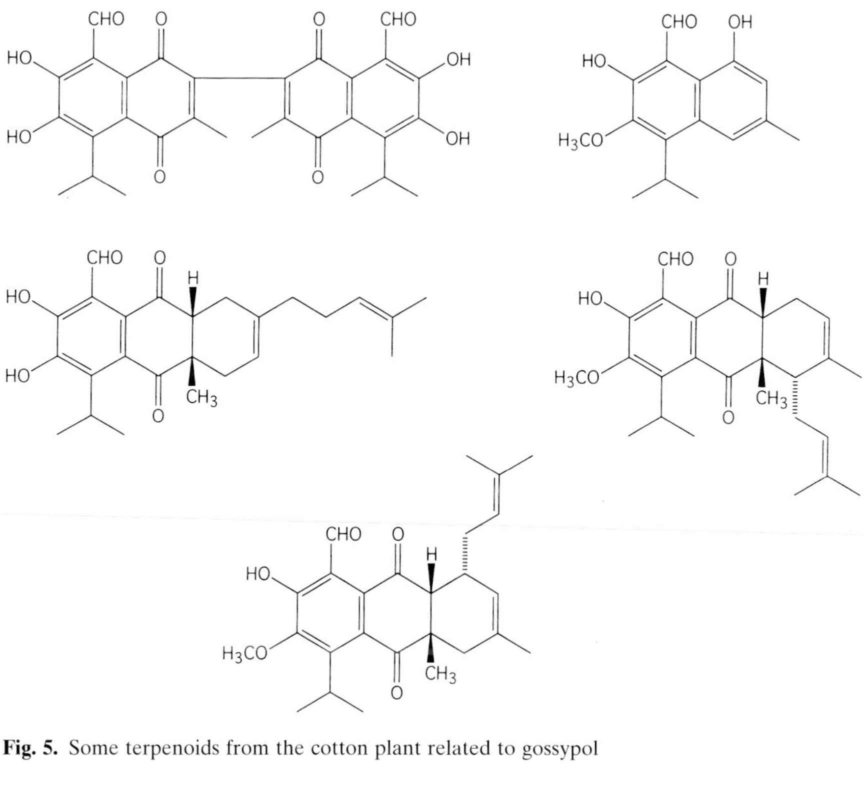

Fig. 5. Some terpenoids from the cotton plant related to gossypol

kaempferol, and phenolic acids (hydroxylated cinnamic acids and derivatives) (Tchiegang and Bourély 1990).

4 Significance of Gossypol

4.1 Toxic Effects of Cottonseed

Originally, the interest in the pharmacology of gossypol was related mainly to the toxic effect of cottonseed. Cottonseed meal is a useful, protein-rich fodder supplement (Alford et al. 1996), whereas cottonseed oil is important for human use. Although the acute toxicity of gossypol is low (LD_{50} orally in rats 2.57 g/kg, El-Nockrashy et al. 1963), there are substantial chronic effects, especially in nonruminant animals, including humans (Abou-Donia 1976). The presence of gossypol residues prohibits or limits the use of cottonseed products as a dietary supplement, and many efforts have been undertaken in order to develop methods of detoxification and analysis of gossypol, and to assess toxic effects and metabolism of gossypol. Authoritative reviews of this subject exist (Abou-Donia 1976; Berardi and Goldblatt 1980).

Because of the presence of the reactive aldehyde groups in gossypol, the compound is able to bind to proteins via the Schiff base formation with terminal amino groups, such as lysine residues (Conkerton and Frampton 1959; Lyman et al. 1959; Finlay et al. 1973; Reddy and Rao 1988; Strøm-Hansen et al. 1989). These and other reactions take place during processing of cottonseed kernels (Markman and Rzhekhin 1969). The binding of gossypol to proteins is a major factor in the toxicity of gossypol and its cumulative effect. The presence of gossypol in cotton protein products not only leads to direct toxic effects of gossypol per se, but also diminishes the quality of proteins by chemical modifications.

4.2 Pharmacological Effects of Gossypol

The turning point in the interest in gossypol was the report by the Chinese National Coordinating Group on Male Antifertility Agents (Anonymous 1978, 1979) that the compound causes infertility in men. In a clinical study, several thousand subjects were exposed to 20 mg of gossypol daily for about 2 months, which resulted in almost complete loss of fertility (assessed as sperm count) in over 99.8% of the subjects. After the onset of infertility, a maintenance dose of 150–220 mg per month could be used. Reportedly, the side effects and general toxicity were low, and complete recovery of fertility was observed in about 3 months (Anonymous 1978, 1979).

This report resulted in intensive investigations of the physiological effects of gossypol, focused on its effects on the biochemical mechanisms of sperm

Table 1. Summary of pharmacological effects of gossypol

Activity	Reference
In vivo male antifertility effect in humans	Anonymous (1978, 1979), Coutinho et al. (1984), Liu et al. (1987), Meng et al. (1988)
In vivo male antifertility effect in animals	Chang et al. (1980), Hahn et al. (1981), Waller et al. (1981), Kalla et al. (1982), Saksena and Salomonsen (1982)
Enantioselective antifertility effect	Waller et al. (1983), Matlin et al. (1985), Lindberg et al. (1987), Wang et al. (1987)
Spermicidal activity in vitro	Waller et al. (1980), Hong et al. (1989)
Antiparasitic effects	Montamat et al. (1982), Eid et al. (1988), Gonzalez-Garza et al. (1997)
Antiviral activity (excl. HIV)	Wichmann et al. (1982), Radloff et al. (1986), Baram et al. (1995)
Anti-HIV activity	Lin et al. (1989, 1993), Royer et al. (1995)
Immunomodulating activity	Baram et al. (1988, 1995)
Antitumor activity	Tso (1984), Jaroszewski et al. (1990), Chang et al. (1993), Flack et al. (1993), Coyle et al. (1994), Gilbert et al. (1995)

production and maturation. The reader is referred to pertinent reviews (Qian and Wang 1984; Segal 1985; Waller et al. 1985; Reyes and Benos 1988; Porat 1990; Randel et al. 1992; Matlin 1994). Although gossypol has now been abandoned as a possible male antifertility drug because of toxic side effects, the pharmacological interest in gossypol resulted in the discovery of other potentially useful activities, which are the subject of current exploration. In particular, gossypol and its analogs show promise in cancer chemotherapy and as inhibitors of HIV. A summary of reported biological effects of gossypol is given in Table 1.

5 Recent Studies on Gossypol

5.1 Synthetic Studies

Newer synthetic studies resulted in more efficient access to the gossypol backbone (Ventuti 1981; Dallacker et al. 1989a,b,c). Synthesized analogs of gossypol include replacement of the isopropyl groups by the ethyl groups (Meltzer et al. 1985), replacement of the aldehyde groups by the methylcarbonyl groups (Wu et al. 1989), and planar analogs obtained by bridging of the substituents ortho to the bond connecting the naphthalene rings (Ognyanov et al. 1989; Zhu et al. 1992). Nitrile analogs of gossypol exhibit antimalarial activity (Royer et al. 1986), whereas dideoxy analogs were synthesized and tested for anti-HIV activity (Royer at al. 1995). Recently, the synthesis of gossypol backbone via stereoselective coupling was achieved (Meyers and Willemsen 1996).

5.2 Optical Stability of Gossypol

Biological effects of gossypol depend on its enantiomeric composition (Table 1), only the (−)-isomer being responsible for the antifertility effect. This selectivity created great interest in the methods of resolution of racemic gossypol (Table 2). In connection with studies involving optically active gossypol, a concern about its optical stability was raised (Ariens 1988).

The existence of gossypol enantiomers is intimately related to the energy barrier for rotation about the single bond connecting the two naphthalene rings: rotation about this bond would lead to racemization of optically active gossypol. This is a thermal process. In order to assess the optical stability of gossypol, thermal racemization experiments and molecular mechanics calculations were carried out (Jaroszewski et al. 1992a). Experimental determination of the susceptibility of gossypol to thermal racemization must take into account its conversion to anhydrogossypol (Fig. 3). Synthesis of optically active anhydrogossypol and thermal racemization experiments with this material, as well as thermal racemization experiments with gossypol in aqueous media (where the formation of anhydrogossypol is suppressed), demonstrated that gossypol is not racemizable under ordinary handling conditions (Jaroszewski et al. 1992a). According to molecular mechanics calculations, formation of gossypol derivatives with five-membered rings between the peri substituents (as in the lactol or anhydrogossypol, Fig. 3) lowers the energy barrier to racemization, but the barrier is still high enough to ensure optical stability of gossypol (Jaroszewski et al. 1992a).

5.3 Gossypol Analysis and the Distribution of Optically Active Forms

As noted above, the cotton plant has been regarded for decades as a source of optically inactive (racemic) gossypol. The discovery of highly enantioselective biological effects of gossypol created interest in natural sources of optically active (enantiomerically pure) gossypol. We carried out a screening of

Table 2. Separation of gossypol enantiomers

Method	Reference
First attempt to separate gossypol enantiomers directly (chromatography on lactose, only scanty separation observed)	Dechary and Pradel (1971)
Chromatography of diastereomeric, peracetylated adducts with 1-methyl-2-phenylethylamine and other chiral amines	Si et al. (1983), Zheng et al. (1985), Si et al. (1987), Zheng et al. (1990a,b)
HPLC of diastereomeric adduct with 1-phenylethylamine on chiral stationary phase	Matlin and Zhou (1984)
Reverse phase HPLC of diastereomeric adducts with β-amino alcohols	Sampath and Balaram (1986a,b)
HPLC of diastereomeric adducts with methyl ester of (+)-phenylalanine (and other amino acids)	Matlin et al. (1987, 1988)

Malvaceae for gossypol, and determined the enantiomeric composition of gossypol samples found (Jaroszewski et al. 1992b). The somewhat surprising result of this investigation was that all plants studied, including *Gossypium* species, contain an excess of one of the enantiomers, although the enantiomeric excess can be quite low (Table 3). Therefore, we have to revise the common understanding (see e.g. Abou-Donia 1976) that the cotton plant produces racemic gossypol (see, however, Dechary and Pradel 1971; Zhou and Lin 1988). The fact that samples of gossypol derived from the cotton plant are usually racemic appears to be related to the isolation procedure, which normally involves precipitation of gossypol as the acetic acid complex. Since optically active gossypol does not form the solvate with acetic acid (King and de Silva 1968), this procedure would yield only a part of the total gossypol present in the extract, i.e., the racemate. The enantiomeric composition of gossypol produced by the cotton plant as a function of species and variety, as well as age, plant part, and conditions of growth, deserves further investigations.

Classically, analysis of cottonseed or other material for gossypol has been carried out using colorimetric methods, usually based on the formation of colored condensation products with aromatic amines (Nazarova et al. 1981). More recently, these methods have been entirely replaced by HPLC methods. The analysis is conveniently carried out using reverse phase systems with detection at 365 nm, which is the high-wavelength absorption maximum of

Table 3. Content, chirality, and optical purity of gossypol isolated from various sources. (Jaroszewski et al. 1992b)

Species	Gossypol content (dry weight) (%)	Chirality of gossypol	Enantiomeric excess (%)
Azanza lampas (Cav.) Alef.	Flowers 0.355	(+)	92
	Leaves 0.07	(+)	80
	Stem 0.03	(+)	83
	Fruits 0.63	(+)	88
	Roots 1.98	(+)	100
Azanza garckeana (F. Hoffm.) Exell & Hillcoat	Seeds 0.04	(+)	95
Gossypium herbaceum L.	Seeds 0.15	(+)	36
	Stem 0.40	(+)	42
Gossypium aboreum L.	Seeds 0.11	(+)	55
	Stem 0.11	(+)	34
	Roots 0.17	(+)	29
Gossypium hirsutum L.	Seeds 0.55	(+)	34
	Roots 0.82	(+)	26
Gossypium barbadense L.	Seeds 0.22	(−)	13
Cienfuegosia drummondi Lewt.	Leaves 0.08	(+)	36
Montezuma speciossisima Sessé & Mos.	Bark 6.1	(+)	96
Thespesia populnea (L.) Soland ex. Correa	Wood 0.81	(+)	100
	Bark 3.26	(+)	100
	Leaves 0.22	(+)	99

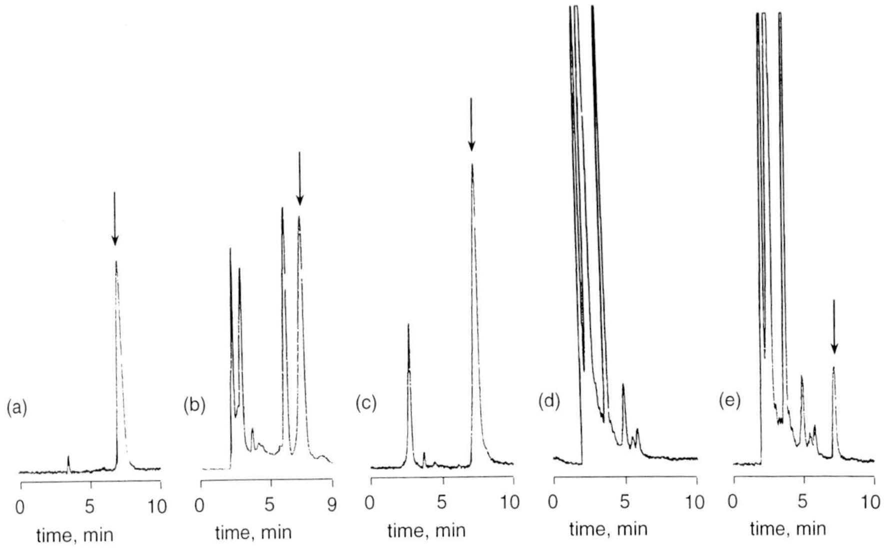

Fig. 6a–e. Examples of HPLC analyses for gossypol (30 × 0.8-cm column of Spherisorb ODS, 7 μm, 2 ml/min of 0.15 M phosphoric acid in water/acetonitrile 4:1, absorbance detection at 365 nm). **a** Gossypol standard. **b** Extract of *Thespesia populnea* leaves. **c** Extract of *Gossypium barbadense* seeds. **d** Extract of *Abelmoschus esculentus* seeds (no gossypol present). **e** The same extract after addition of gossypol standard. *Arrow* indicates gossypol peak. (Jaroszewski et al. 1992b)

gossypol (Fig. 6). The number of published HPLC procedures for gossypol is large. Mass-spectrometric methods offer further improvements in sensitivity and speed (Phillips and Hedin 1990).

5.4 Effect of Gossypol on Energy Metabolism of Mammalian Cells

Among the mechanisms postulated for the antifertility and antitumor action of gossypol is the inhibition of energy production through an interference with mitochondrial function (Reyes and Benos 1988). Since the antifertility effect of gossypol is enantioselective (Table 2), in vitro studies of gossypol should reproduce the differential effect of the enantiomers. In a recent study, the effect of gossypol enantiomers on cell lines derived from murine Leydig cells (the principal source of androgens in male) and Sertoli cells (the major supportive tissue of testes) was studied (Hansen and Jaroszewski 1996). Although (−)-gossypol was more effective in the inhibition of cell growth than (+)-gossypol, little if any difference between the enantiomers on energy production in the cells was observed. Inhibition of energy production in the somatic tissue of the testes is therefore unlikely to be a major mechanism by which gossypol inhibits in vivo production of spermatozoa (Hansen and Jaroszewski 1996).

6 Conclusion

Gossypol, an abundant component of cottonseed, is of great technological and pharmacological interest. During the past decades of investigations of gossypol, dramatic advances have been made in the understanding of its complex chemical structure (tautomeric equilibria), its biosynthesis, botanical distribution, function in plants, stereochemistry, interactions with biological macromolecules, and, most notably, pharmacological effects. Although the expectations that gossypol will find use as a male antifertility drug have not been fulfilled, it continues to attract attention as a pharmacological lead.

References

Abou-Donia MB (1976) Physiological effects and metabolism of gossypol. Residue Rev 61:125–160

Adams R, Geissman TA, Dial WR, Fritzpatrick JT (1941) Structure of gossypol. XXVI. Gossypolic acid. J Am Chem Soc 63:2439–2441

Adams R, Geissman TA, Edwards JD (1960) Gossypol, a pigment of cottonseed. Chem Rev 60:555–574

Alford BB, Liepa GV, Vanbeber AD (1996) Cottonseed protein: what does the future hold? Plant Foods Hum Nutr 49:1–11

Ali S, Singh P, Thomson RH (1980) Naturally occurring quinones. Part 28. Sesquiterpenoid quinones and related compounds from *Hibiscus tiliaceus*. J Chem Soc Perkin 1:257–259

Anonymous (1978) Gossypol – a new antifertility agent for males. Chin Med J 4:417–428

Anonymous (1979) Gossypol – a new antifertility agent for males. Gynecol Obstet Invest 10:163–176

Apostol I, Low PS, Heinstein P, Stipanovic RD, Altman DW (1987) Inhibition of elicitor-induced phytoalexin formation in cotton and soybean cells by citrate. Plant Physiol 84:1276–1280

Ariëns EJ (1988) Stereospecificity of bioactive agents. In: Ariëns EJ, van Rensen JJS, Welling W (eds) Stereoselectivity of pesticides, biological and chemical problems. Elsevier, Amsterdam, pp 39–108

Baram NI, Ziyaev KL, Ismailova GA, Biktimirov L, Ismailov AI, Urazmetov KG (1988) Immunomodulating activity of gossypol derivatives. Khim Prir Soedin: 647–650

Baram NI, Biktimirov L, Ziyaev KL, Paizieva RZ, Iswailov AI (1995) Antivital and interferon-inducing activity of gossypol and its derivatives. Khim Prirod Soedin: 355–361

Bell AA, Stipanovic RD, Howell CR, Fryxell PA (1975) Antimicrobial terpenoids of *Gossypium*: hemigossypol, 6-methoxyhemigossypol and 6-deoxyhemigossypol. Phytochemistry 14:225–231

Bell AA, Stipanovic RD, O'Brien DH, Fryxell PA (1978) Sesquiterpenoid aldehyde quinones and derivatives in pigment glands of *Gossypium*. Phytochemistry 17:1297–1305

Berardi LC, Goldblatt LA (1980) Gossypol. In: Liener IE (ed) Toxic constituents of plant foodstuffs. Academic Press, New York, pp 183–237

Bhakuni DS, Dhar MM, Sharma VN (1968) The chemistry of thespesin. Experientia 24:109–110

Campbell KN, Morris RC, Adams R (1937) The structure of gossypol. I. J Am Chem Soc 59:1723–1728

Carruth FE (1918) Contribution to the chemistry of gossypol, the toxic principle of cottonseed. J Am Chem Soc 40:647–663

Carter FL, Castillo AE, Frampton VL, Kerr T (1966) Chemical composition studies of seeds of the genus *Gossypium*. Phytochemistry 5:1103–1112

Chang MC, Gu Z, Saksena SK (1980) Effects of gossypol on the fertility of male rats, hamsters and rabbits. Contraception 21:461–469

Chang CJG, Ghosh PK, Hu YF, Brueggemeier RW, Lin YC (1993) Antiproliferative and antimetastatic effects of gossypol on Dunning prostate cell-bearing Copenhagen rats. Res Commun Chem Pathol Pharmacol 79:293–312

Clark EP (1927) Studies of gossypol. I. The preparation, purification, and some properties of gossypol, the toxic principle of cottonseed. J Biol Chem 75:725–739

Conkerton EJ, Frampton VL (1959) Reaction of gossypol with free ε-amino groups of lysine and proteins. Arch Biochem Biophys 81:130–134

Coutinho EM, Melo JF, Barbosa I, Segal SJ (1984) Antispermatogenic action of gossypol in men. Fertil Steril 42:424–430

Coyle T, Levante S, Shetler M, Winfield J (1994) In vitro and in vivo cytotoxicity of gossypol against central nervous system tumor cell lines. J Neurooncol 19:25–35

Dallacker F, Schleuter HJ, Wabinski-Westerop K (1989a) Über Gossypol- und Hemigossypol-Derivate – Darstellung von Hydroxy-methyl-naphtho[1,3]dioxolen. Chemiker Z 113:5–11

Dallacker F, Leuther P, Wabinski-Westerop K (1989b) Gossypol und Hemigossypole, II. Darstellung von 8-Hydroxy-4-isopropyl-naphtho[2,3-d][1,3]dioxol-Derivaten. Chemiker Z 113:97–102

Dallacker F, Leuther P, Wabinski-Westerop K (1989c) Gossypol und Hemigossypole, III. Darstellung des Methyl-gossypol-methylenethers. Chemiker Z 113:103–104

Datta SC, Murti VVS, Seshadri TR (1968) A new component of the flowers of *Thespesia populnea*: (+) gossypol. Curr Sci 37:135–136

Dechary JM, Pradel P (1971) The occurrence of (+)gossypol in *Gossypium* species. J Am Oil Chem Soc 48:563–564

Edwards JD Jr (1958) Total synthesis of gossypol. J Am Chem Soc 80:3798–3899

Eid JE, Ueno H, Wang CC, Donelson JE (1988) Gossypol-induced death of African trypanosomas. Exp Parasitol 66:140–142

El-Nockrashy AS, Lyman CM, Dollahite JW (1963) The acute oral toxicity of cottonseed pigment glands and intraglandular pigments. J Am Oil Chem Soc 40:14–21

Finlay TH, Pharmgrongartama ED, Perlmann GE (1973) Mechanism of the gossypol inactivation of pepsinogen. J Biol Chem 248:4827–4833

Fisher GS, Frank AW, Cherry JP (1988) Total gossypol content of glandless cottonseed. J Agric Food Chem 36:42–44

Flack MR, Pyle RG, Mullen NM, Lorenzo B, Wu YW, Knazek RA, Nisula BC, Reidenberg MM (1993) Oral gossypol in the treatment of metastatic adrenal cancer. J Clin Endocrinol Metab 76:1019–1024

Fryxell PA (1981) The natural history of the cotton tribe. Texas A & M University Press, College Station

Gdaniec M, Ibragimov DT, Talipov SA (1996) Gossypol. Compr Supramol Chem 6:117–145

Gilbert NE, O'Reilly JE, Chang, CJG, Lin YC, Brueggemeier RW (1995) Antiproliferative activity of gossypol and gossypolone on human breast cancer cells. Life Sci 57:61–67

Gonzalez-Garza MT, Castro-Garza J, Anaya-Velazguez F, Mata-Cardenos BP, Lozano-Garza G, Carranza-Rosules MD, Vargas-Villareal J, Said-Fernandez S (1997) Gossypol anti-amebic effect in vivo. Arch Med Res 28:298–299

Gray JR, Mabry TJ, Bell AA, Stipanovic RD, Lukefahr MJ (1976) *para*-Hemigossypolone: sesquiterpenoid aldehyde quinone from *Gossypium hirsutum*. J Chem Soc Chem Commun: 109–110

Hahn DW, Rusticus C, Probst A, Homm R, Johnson AN (1981) Antifertility and endocrine activities of gossypol in rodents. Contraception 24:97–105

Hansen LL, Jaroszewski JW (1996) Effect of gossypol on cultured TM3 Leydig and TM4 Sertoli cells: ^{31}P and ^{23}Na NMR study. NMR Biomed 9:72–78

Heinstein P (1985) Stimulation of sesquiterpene aldehyde formation in *Gossypium arboreum* cell suspension cultures by conidia of *Verticillium dahliae* J Nat Prod 48:907–915

Hong CY, Huang JJ, Wo P (1989) The inhibitory effect of gossypol on human sperm motility: relationship with time, temperature and concentration. Human Toxicol 8:49–51

Huang L, Si Y, Snatzke G, Zheng D, Zhou J (1988) Absolute configuration of gossypol. Collect Czech Chem Commun 53:2664–2666

Jaroszewski JW, Cornett C, Strøm-Hansen T (1988) Gossypol. In: Jaroszewski JW, Schaumburg K, Kofod H (eds) NMR spectroscopy in drug research. Munksgaard, Copenhagen, pp 75–98

Jaroszewski JW, Kaplan O, Cohen JS (1990) Action of gossypol and rhodamine 123 on wild-type and multidrug-resistant MCF-7 human breast cancer cells: ^{31}P NMR and toxicity studies. Cancer Res 50:6936–6943

Jaroszewski JW, Strøm-Hansen T, Hansen LL (1992a) Optical stability of gossypol. Chirality 4:216–221

Jaroszewski JW, Strøm-Hansen T, Hansen SH, Thastrup O, Kofod H (1992b) On the botanical distribution of chiral forms of gossypol. Planta Med 58:454–458

Kalla NR, Foo J, Sheth AR (1982) Studies of the male antifertility agent – gossypol acetic acid. Effect of gossypol acetic acid on the fertility of male rats. Andrologia 14:492–500

King TJ, de Silva LB (1968) Optically active gossypol from *Thespesia populnea*. Tetrahedron Lett: 261–263

Khushk MT, Vaughan JG (1985) Seed structure in relation to the taxonomy of the Hibisceae (*Gossypium*, *Lebronnecia* and *Thespesia*). Pak J Bot 17:119–129

Lin T, Schinazi R, Griffith BP, August EM, Eriksson BFH, Zheng D, Huang L, Prusoff WF (1989) Selective inhibition of human immunodeficiency virus type 1 replication by the (−) but not the (+) enantiomer of gossypol. Antimicrob Agents Chemother 33:2149–2151

Lin T, Schinazi RF, Zhu J, Birks E, Carbone R, Si Y, Wu K, Huang L, Prusoff WH (1993) Anti-HIV-1 activity and cellular pharmacology of various analogs of gossypol. Biochem Pharmacol 46:251–255

Lindberg MC, Naqvi RH, Matlin SA, Zhou RH, Bialy G, Blye RP (1987) Comparative anti-fertility effects of gossypol enantiomers in male hamsters. Int J Androl 10:619–623

Liu GZ, Lyle KC, Cao J (1987) Experiences with gossypol as a male pill. Am J Obstet Gynecol 4:1079–1081

Longmore J (1886) Cotton-seed oil: its colouring matter and mucilage, and description of a new method of recovering the loss occurring in the refining process. J Soc Chem Ind (Lond) 5:200–205

Lukefahr MJ, Fryxell PA (1967) Content of gossypol in plants belonging to genera related to cotton. Econ Bot 21:128–131

Lyman CM, Baliga BP, Slay MW (1959) Reactions of proteins with gossypol. Arch Biochem Biophys 84:486–497

Marcelle GB, Cordell GA, Soejarto DD, Fong HHS (1985) Isolation of (+)-gossypol from *Montezuma speciosissima*. J Nat Prod 48:162–163

Marchlewski L (1899) Gossypol, ein Bestandtheil der Baumwollsamen. J Prakt Chem 40:84–90

Markman AL, Rzhekhin VP (1969) Gossypol and its derivatives. Israel Program for Scientific Translations, Jerusalem (translation from Russian)

Masciadri R, Angst W, Arigoni D (1985) A revised scheme for the biosynthesis of gossypol. J Chem Soc Chem Commun: 1573–1574

Matlin SA (1994) Prospects in pharmacological male contraception. Drugs 48:851–863

Matlin SA, Zhou R (1984) Resolution of gossypol: analytical and preparative HPLC. J High Resolut Chromatogr Commun 7:629–631

Matlin SA, Zhou R, Bialy G, Blye RP, Naqvi RH, Lindberg MC (1985) (−)-Gossypol: an active male antifertility agent. Contraception 31:141–149

Matlin SA, Belenguer A, Tyson RG, Brookes AN (1987) Resolution of gossypol: analytical and large-scale preparative HPLC on non-chiral phase. J High Resolut Chromatogr Commun 10:86–91

Matlin SA, Zhou RH, Belenguer A, Tyson RG, Brookes AN (1988) Large-scale resolution of gossypol enantiomers for biological evaluation. Contraception 37:229–237

Meltzer PC, Bickford PH, Lambert GJ (1985) A regioselective route to gossypol analogues: the synthesis of gossypol and 5,5′-didesisopropyl-5,5′-diethylgossypol. J Org Chem 50:3121–3124

Meng GD, Zhu JC, Chen ZW, Wong LT, Zhang GY, Hu YZ, Ding JH, Wang XH, Qian SZ, Wang C, Manchin D, Pinol A, Waites GMH (1988) Recovery of sperm production following the cessation of gossypol treatment: a two-centre study in China. Int J Andro 11:1–11

Meyers AI, Willemsen JJ (1996) An asymmetric synthesis of (+)-apogossypol hexamethyl ether. Tetrahedron Lett 37:791–792

Montamat EE, Burgos C, De Burgos NMG, Rovai LE, Blanco A, Segura EL (1982) Inhibitory action of gossypol on enzymes and growth of *Trypanosoma cruzi*. Science 218:288–289

Nazarova IP, Glushenkova AI, Umarov AU (1981) Gossypol-like compounds of the cotton plant. Methods of determining gossypol. Chem Nat Compd 17:87–102

Ognyanov V, Petrov OS, Tiholov EP, Mollov NM (1989) Synthesis of gossypol analogues. Helv Chim Acta 72:353–360

Phillips VA, Hedin PA (1990) Spectral techniques for structural analysis of the cotton terpenoid aldehydes gossypol and gossypolone. J Agric Food Chem 38:525–538

Porat O (1990) Effects of gossypol on the motility of mammalian sperm. Mol Repr Dev 25:400–408

Qian S, Wang Z (1984) Gossypol: a potential antiferility agent for males. Annu Rev Pharmacol Toxicol 24:329–360

Radloff RJ, Deck LM, Royer RE, Jagt DLV (1986) Antiviral activities of gossypol and its derivatives against herpes simplex virus type II. Pharmacol Res Commun 18:1063–1073

Randel RD, Case CC Jr, Wyse SJ (1992) Effects of gossypol and cottonseed products on reproduction of mammals. J Anim Sci 70:1628–1638

Reddy IM, Rao MSN (1988) Interaction of gossypol with gossypin (11S protein) and congossypin (7S protein) of cottonseed and glycinin (11S protein) of soybean. J Agric Food Chem 36:698–703

Reyes J, Benos DJ (1988) Bioenergetic and membrane actions of gossypol: an uncoupler as a male contraceptive. Comments Mol Cell Biol 5:1–20

Royer RE, Deck LM, Campos NM, Hunsaker LA, Jagt DLV (1986) Biologically active derivatives of gossypol: synthesis and antimalarial activities of peri-acylated gossylic nitriles. J Med Chem 29:1799–1801

Royer RE, Deck LM, Jagt TJV, Martinez FJ, Mills RG, Young SA, Jagt DLV (1995) Synthesis and anti-HIV activity of 1,1′-dideoxygossypol and related compounds. J Med Chem 38:2427–2432

Saksena SK, Salomonsen RA (1982) Antifertility effect of gossypol in male hamsters. Fertil Steril 37:686–680

Sampath DS, Balaram P (1986a) A rapid procedure for the resolution of racemic gossypol. J Chem Soc Chem Commun: 649–650

Sampath DS, Balaram P (1986b) Resolution of racemic gossypol and interaction of individual enantiomers with serum albumin and model peptides. Biochim Biophys Acta 882:183–186

Segal SJ (ed) (1985) Gossypol, a potential contraceptive for men. Plenum Press, New York

Si Y, Zhou J, Huang L (1983) Resolution of racemic gossypol. Kexue Tongbao 28:1574

Si Y, Zhou J, Huang L (1987) Studies on resolution of racemic gossypol. Sci Sin Ser B 30:297–303

Stanford EE, Viehoever A (1918) Chemistry and histology of the glands of the cotton plant, with notes on the occurrence of similar glands in related plants. J Agric Res 13:419–435

Stipanovic RD, Bell AA, Howell CR (1973) Spectral identification of the ketol tautomer of gossypol. J Am Oil Chem Soc 50:462–463

Stipanovic RD, Bell AA, Mace ME, Howell CR (1975) Antimicrobial terpenoids of *Gossypium*: 6-methoxygossypol and 6,6′-dimethoxygossypol. Phytochemistry 14:1077–1081

Stipanovic RD, Bell AA, O'Brien DH, Lukefahr MJ (1977) Heliocide H_2: an insecticidal sesterterpenoid from cotton (*Gossypium*). Tetrahedron Lett 567–570

Stipanovic RD, Bell AA, O'Brien DH, Lukefahr MJ (1978) Heliocide H_1: a new insecticidal sesterterpenoid from cotton (*Gossypium hirsutum*). J Agric Food Chem 26:115

Stipanovic RD, Stoessi A, Stothers JB, Altman DW, Bell AA, Heinstein P (1986) The stereochemistry of the biosynthetic precursor of gossypol. J Chem Soc Chem Commun: 100–101

Strøm-Hansen T, Cornett C, Jaroszewski JW (1989) Interaction of gossypol with amino acids and peptides as a model of enzyme inhibition. Int J Peptide Protein Res 34:306–310

Tchiegang C, Bourély J (1990) Chromatographic analysis of phenolic compounds in cotton seed kernels (*Gossypium hirsutum* L. and *G. barbadense* L.). Coton Fibres Trop 45:27–43

Tso W (1984) Gossypol inhibits *Ehrlich ascites* tumor cell proliferation. Cancer Lett 24:257–261

Venuti MC (1981) Efficient synthesis of the gossypol binaphthyl backbone. J Org Chem 46:3124–3127

Waller DP, Zaneweld LJD, Fong HHS (1980) In vitro spermicidal activity of gossypol. Contraception 22:183–187

Waller DP, Fong HHS, Cordell GA, Soejarto DD (1981) Antifertility effects of gossypol and its impurities on male hamsters. Contraception 23:653–660

Waller DP, Bunyapraphatsara N, Martin A, Vournazos CJ, Ahmed MS, Soejarto DD, Cordell GA, Fong HHS, Russell LD, Malone JP (1983) Effect of (+)-gossypol on fertility in male hamsters. J Androl 4:276–279

Waller DP, Zaneweld LJD, Farnsworth NR (1985) Gossypol: pharmacology and current status as a male contraceptive. Econ Med Plant Res 1:87–112

Wang N, Zhou L, Guan M, Lei H (1987) Effects of (−)- and (+)-gossypol on fertility in male rats. J Ethnopharmacol 20:21–24

Wichmann K, Vaheri A, Luukkainen T (1982) Inhibiting of herpes simplex virus type 2 infection in human epithelial cells by gossypol, a potent spermicidal and contraceptive agent. Am J Obstet Gynecol 142:593–594

Wu GP, Ying HQ, Yan ZL, Guo ZM, Hou JY, Zhu CQ, Shi DX, Bian J (1989) Synthesis and antifertility actions of gossypol derivatives and phenol aldehydes. Acta Pharm Sin 24:502–511

Zheng DK, Si YK, Meng JK, Zhou J, Huang L (1985) Resolution of racemic gossypol. J Chem Soc Chem Commun: 168–169

Zheng DK, Si YK, Meng JK, Huang L (1990a) Studies on the resolution of racemic gossypol. II. By chiral α-methylphenethylamine and β-methylbenzylamine. Acta Pharm Sin 25:417–422

Zheng DK, Meng JK, Si YK, Ma SY, Huang L (1990b) Studies on the resolution of racemic gossypol. IV. Use of threo (−) or (+)-1-(p-nitrophenyl)-1,3-dihydroxypropylamine-2 as the resolving agent. Acta Pharm Sin 25:430–434

Zhou R, Lin X (1988) Isolation of (−)-gossypol from natural plants. Cintraception 37:239–245

Zhu G, Chen D, Huang J, Chi C, Liu F (1992) Regioselective bromination and fluorination of apogossypol hexamethyl ether. J Org Chem 57:2316–2320

VI.2 Tannins in Cotton

K.E. LEGÉ

1 Introduction

Profitable cotton production in most areas of the world requires some measure of plant protection against insect and disease pest organisms. Since numerous insect pests (Aston and Winfield 1972) and disease organisms (Ridgeway et al. 1984) have an economic impact on cotton production worldwide, chemical control tactics have been and will continue to be a predominant method of plant protection. However, host plant resistance breeding measures using plant morphology characteristics, allelochemicals, and other resistance or tolerance mechanisms will continue to play an important role in decreasing losses due to pests, as well as decreasing the number and amount of pesticides applied to the crop.

Several allelochemicals, such as condensed tannins (Legé et al. 1992, 1993; Smith et al. 1992), phenolic acids (Wakelyn et al. 1974; Legé et al. 1995), gossypol (Dilday and Shaver 1980; Shaver et al. 1980; Hedin et al. 1984; Jaroszewshi et al. 1992) and most recently, bacterial endotoxins (Hofte and Whiteley 1989; Gill et al. 1992), have been investigated and utilized in host resistance-breeding programs in cotton. This chapter will focus on condensed tannins and related compounds.

2 Chemistry of Tannins

Tannins are polyphenolic compounds with molecular weights ranging between 500 and 3000 that possess the property of precipitating proteins. Numerous phenolic hydroxy groups allow the compounds to form multiple hydrogen bonds with the NH^-, NH_2, or OH^- groups of proteins and other macromolecules. This reaction produces complexes that are dissociable at normal pH and that are not readily degraded by enzymes. The term tannins originated from the leather industry and was used to describe substances that were capable of converting animal hides into leather (Swain 1979). Tannins can be classified into two groups on the basis of biological origin. Hydrolyzable

Sure Grow Seed, Inc., Route 1, Box 310, Centre, Alabama 35960, USA

Biotechnology in Agriculture and Forestry, Vol. 42
Cotton (ed. by Y.P.S. Bajaj)

tannins are complex polyphenols that are degradable by hydrolysis to form either gallic or ellagic acid, in addition to a sugar unit, usually glucose. Nonhydrolyzable, or condensed, tannins arise from the polymerization of hydroxy flavans to give dimers, trimers, or oligomers. Some condensed tannins may be insoluble due to strong bonding with proteins, nucleic acids, or polysaccharides during extraction to form precipitates. Condensed tannins are also known as procyanidins, since anthocyanidins are formed upon heating with acids (Swain 1979).

3 Tannins as Host Plant Resistance Mechanisms in Cotton

Although the biological role of tannins is not known and no overall generalization regarding the effects of tannins on herbivores can be made (Bernays et al. 1989), it is thought that their primary role is most likely that of a plant-defense mechanism (Swain 1979). Two characteristic properties are thought to contribute to the host plant resistance action of tannins. Tannins bind to various chemical groups in proteins, especially saliva and glycoproteins in the mouth, to produce a characteristic astringent taste (Feeny 1968), which translates into a repellent effect, reducing palatability (Lipke and Fraenkel 1976). Condensed tannins cause the availability of dietary protein (Feeny 1970) and some individual amio acids (Delort-Laval and Viroben 1969) to be reduced. Additionally, enzyme activity is greatly inhibited in tissues with high tannin levels (Goldstein and Swain 1965). A direct toxic action has also been proposed for condensed tannins (Feeny 1970). However, a more recent study indicated that the ultimate cause for the growth-inhibiting action of condensed tannins seemed to be a reduction in the rate of food consumption by *Helicoverpa zea* larvae (Klocke and Chan 1982). The report indicated no effect on assimilation or on efficiency of the conversion of digested food in larvae treated with condensed tannins, although larvae fed the condensed tannin diets did exhibit decreased digestive enzyme activities (Klocke and Chan 1982). Regardless, each or all of these characteristics decreases the attractiveness of these tissues to certain pests, and as a result, the invasion of an insect or microbe into the host tissue is impeded (Bate-Smith 1973).

Although tannins are considered to be the most important secondary plant compounds involved in plant defense against insects and diseases (Swain 1979), the most extensively researched secondary plant compound related to host plant resistance in cotton has been gossypol, a terpenoid aldehyde compound (see Chap. VI.1, this Vol.). High concentrations of gossypol have been associated with insect resistance in cotton (Bottger and Patana 1966; Lukefahr and Martin 1966; Lukefahr and Houghtaling 1969; Bell and Stipanovic 1977). Similar host plant resistance responses have been reported for tannins. Consequently, condensed tannin has been one of the major allelochemicals selected in cotton breeding programs (White et al. 1982).

3.1 Arthropod Resistance

Condensed tannins have been reported to exhibit resistance characteristics to many species of arthropods, especially lepidopterous species. One of the earliest reports concerning tannins as plant-defense mechanisms involved the oak winter moth (*Operophtera bruata*) larvae (Feeny 1968). The larvae were fed an artificial diet containing at least 1% tannin. Tannins were found to increase as the season progressed, and since the winter moth larvae were found to feed only early in the season, it was thought that tannins may play a significant role in resistance to this pest in the latter portion of the season.

A 45% reduction in cotton seed damage caused by pink bollworm (*Pectinophora gossypiella* Saunders) was attributed to high condensed tannin concentrations (Wilson and Wilson 1977). Chakravorty and Basu (1981) suggested that the astringent character of tannins resulted in reduced palatability of cotton bolls to pink bollworm larvae, since bolls containing higher levels of tannins were less preferred than those with lower levels of tannins.

Condensed tannins have been documented as host plant resistnace mechanisms for cotton bollworm (*Helicoverpa zea* Boddie) and tobacco budworm (*Heliothis virescens* F.). A diet containing freeze-dried squares of the cotton race accession Texas 254, which had a level of up to 3.4% condensed tannins, was fed to tobacco budworm larvae. At 1% of the diet, the square powder reduced larval growth to about a third of the control; at 2.5% of the diet, larval growth was severely inhibited. When condensed tannin levels were at only 0.2% of the diet, tobacco budworm larvae growth was retarded by 84% (Fig. 1; Chan et al. 1978). Similar results have been found with cotton bollworm larvae (Klocke and Chan 1982). The tannin levels of the first fully expanded leaf in the terminal of cotton was negatively correlated with the number of cotton bollworm-damaged fruit at one-third-grown square and at second week of bloom stages (Zummo et al. 1983). This particular leaf was sampled because the cotton bollworm most often lays eggs on or near this leaf, and therefore the leaf is likely to be the first plant tissue consumed by the emerging larvae (Wilson et al. 1980). Other studies have reported similar associations between condensed tannin levels and cotton fruit insect damage (Schuster and Lane 1980; Zummo et al. 1984b). Resistance to the spotted bollworm (*Earias vittella* F.) in *Gossypium arboreum* has been associated with condensed tannin levels (Sharma et al. 1982; Sharma and Agarwal 1982a,b, 1983, 1984). The resistance was related to the feeding-deterrent activity of condensed tannins, which ultimately increased the development period of the spotted bollworm (Sharma and Agarwal 1982b).

Condensed tannins also act as a host plant resistance mechanism to two-spotted spider mites (*Tetranychus urticae* Koch; Schuster 1980). Several race stocks of cotton were found to exhibit resistance to the two-spotted spider mite, and condensed tannins were subsequently determined to serve as the mechanism of the observed antibiosis (Schuster 1979; Lane and Schuster 1979, 1981). These race stocks have since been used to transfer two-spotted spider mite resistance, primarily based on condensed tannin levels, into cotton with commercial-type backgrounds (Schuster 1980).

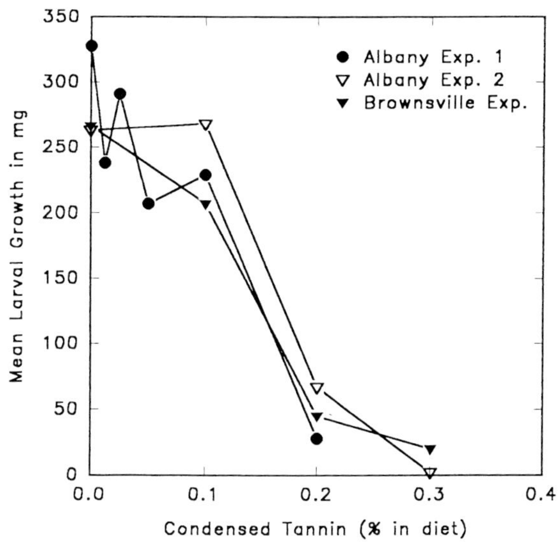

Fig. 1. Relationship between condensed tannins in a diet of cotton square powder and growth of tobacco budworm (*Heliothis virescens*) larvae. Means of 20 larvae are presented for the Albany, CA, experiments; means of 16 larvae are presented for the Brownsville, TX, experiment. (Data from Chan et al. 1978)

Condensed tannins have also been documented to convey resistance to boll weevil (*Anthonomous grandis* Boh) (Maxwell et al. 1967; Benedict and Bird 1981) by acting as a feeding deterrent. Adult boll weevil weight was reduced by as much as 20% when fed a diet containing 0.2 to 1% of tannin (Maxwell et al. 1967). Condensed tannins do not impart resistance to the boll weevil by reducing oviposition (Hedin and McCarty 1990).

Several complicating factors have been suggested recently involving the use of condensed tannins and related chemicals as mechanisms of resistance to various pests. First, the resistance associated with condensed tannins may not be absolute; for example, despite other reports of effective resistance against tobacco budworm, there is evidence that condensed tannin levels are not associated with resistance to this pest (Hedin et al. 1983). Negative correlations have been reported between flavanol (including condensed tannins) levels and jassid (*Amrasca biguttula biguttula* Ishida) feeding; the report further indicated that greater feeding on susceptible cultivars most likely stimulated higher flavanol synthesis relative to resistant cultivars. Similarly, high tannin concentrations were negatively correlated with cigarette beetle (*Lasioderma serricorne* F.) growth, but it was speculated that the high tannin levels were associated with tissues that exhibited high protein content, which is most likely a more important determinant of growth of the insect (Yokoyama and Mackey 1987; Yokoyama et al. 1987). These findings were further complicated by the fact that applications of methyl parathion decreased tannin levels and increased protein levels in cotton foliage, making affected plants more suitable as hosts of the cigarette beetle (Yokoyama et al. 1987).

The most recent complication involving condensed tannins as an effective means of conveying resistance is related to transgenic cotton varieties containing the gene-controlling synthesis of the endotoxin derived from *Bacillus thuringiensis*, or *B.t.* A report by Navon et al. (1993) echoes previous studies' findings regarding the ability of condensed tannins to serve as a feeding deterrent to tobacco budworm larvae, but condensed tannins were also found to antagonize the effects of the *B.t.* toxin. The resistance imparted by condensed tannins deterred feeding by the larvae, which ultimately reduced the amount of *B.t.* toxin ingested by the insect; as a result, larvae feeding on high-tannin, *B.t.*-toxin-containing diets had lower mortaility rates than insects feeding on diets containinig condensed tannins or *B.t.* toxin alone. Condensed tannins could have also directly reduced the availability and activity of the *B.t.* toxin, since tannins can precipitate proteins. Therefore, the report suggests that insect control tactics employing the use of the *B.t.* toxin, either as a topically applied microbial insecticide or in transgenic cotton plants, may not be compatible when used in conjunction with plants containing high tannin concentrations.

3.2 Disease Resistance

Control strategies for diseases, in contrast to those applied for arthropods, has been more successfully approached by modifying cultural practices, including the development and implementation of host plant resistance. While some chemical control tactics have been developed for some seedling diseases and nematodes, most other disease-causing organisms in cotton are not routinely controlled or suppressed with chemical tools. Therefore, the development of resistant or tolerant cultivars has played an important role in maintaining cotton production in parts of the world that historically possess economically limiting levels of inoculum for various diseases. Condensed tannins and related polyphenolic compounds have been attributable to resistance to numerous disease organisms (Bell et al. 1992).

Cotton seedlings are very susceptible to infection by several fungal diseases, such as *Rhizoctonia*, *Pythium*, *Colletotrichum*, *Alternaria*, *Rhizopus*, and *Fusarium*. As the seedling ages, resistance is progressively enhanced until little or no infection is observed in a 2- to 3-week-old seedling (Bell et al. 1992). Flavanol (catechins plus tannins) concentrations are reportedly correlated with resistance to *Rhizoctonia solani* at any age in the cotton seedling (Hunter 1978). Additionally, flavanol levels in cotton seedlings are nearly doubled 24 h following infection with *R. solani*; the rate of increase is greater for older (12-day-old) than for younger (6-day-old) seedlings (Hunter 1974). The mechanism of resistance to *R. solani* in cotton seedlings is related to the inactivation of the fungal enzyme endopolygalacturonase by flavanols. The enzyme is responsible for host-plant tissue degradation and subsequent hypocotyl lesion development. Condensed tannins, as well as oxidation products of catechin, are thought to inhibit the enzyme and impart resistance to *R. solani* (Hunter 1974).

Host-plant resistance to *Verticillium* wilt caused by *Verticillium dahliae* has been attributible to condensed tannins in apricot (*Prunus armeniaca*; Somers and Harrison 1967), potato (*Solanum tuberosum*; Patil and Dimond 1967), strawberry (*Fragaria* spp.; Okasha et al. 1968), and cotton (Bell 1968; Bell and Stipanovic 1978). Tannins have an antibiotic effect (Bell and Stipanovic 1978), and they inhibit pectinase enzymes of the fungus (Patil and Dimond 1967; Okasha et al. 1968). Cotton strains that are more indeterminate have been reported to have higher tannin levels and exhibit higher degrees of resistance to *Verticillium* wilt (Bell et al. 1992). Additionally, both constitutive and induced condensed tannins appear to be involved in cotton host-plant resistance to *V. dahliae*; resistant cultivars synthesize new flavanols (catechins plus tannins) about 24h sooner and in greater concentrations than susceptible cultivars (Bell et al. 1992). Tannin quality as well as quantity is important for resistance to *Verticillium* wilt. Resistant cultivars have smaller and/or less oxidized tannin polymers compared to susceptible cultivars (Bell et al. 1992).

Cottonseed resistance to infection by various diseases is primarily derived from characters that restrict rates of water imbibition, which maintains low seed moisture levels in open bolls in the field during alternating periods of wetting and drying (Bell et al. 1992). The cells that form the pigment layers of the seed coat are filled with flavanols in the immature cotton seed (Halloin 1982). Subsequently, the flavanols are oxidized to form melanin when the seed ripens (Chakravorty et al. 1982). Immature seed removed from bolls prior to dehiscence exposed to nitrogen during ripening do not form melanin from flavanols; the resulting mature seed are highly permeable to water (Halloin 1982) and the seed coats are fragile, so that the epidermal layer is readily removed during the ginning process. Therefore, flavanols, which may occur at concentrations up to 20% (Sadikov 1965; Chakravorty et al. 1982; Chan 1985), are suggested to protect the immature seed from microbial invasion, while melanins (oxidation products of flavanols) serve to restrict water imbibition and to act as a mechanical barrier to prevent deterioration of mature cotton seed (Bell et al. 1992).

Several species of nematodes can invade the root tissues of cotton and cause a wound through which other disease organisms can enter and further infect the cotton plant. While absolute resistance to any species of nematode does not exist in any commercial cultivar, a degree of resistance has been reported to root-knot nematode (*Meloidogyne incognita*) (Veech 1977). The mechanism of the resistance reaction involves hypersensitivity in the cells of the head of the nematode, which causes those cells to become brown and necrotic. In a resistant cultivar, Auburn 623, a 35% decrease in tannin concentrations occurred in response to nematode infestation, whereas the susceptible cultivar, Deltapine 16, did not change in tannin concentration when nematodes were invading (Veech 1977). It was suggested, therefore, that the oxidation of flavanols around the head of the nematode (as suggested by the brown color) could be part of a resistance response. However, toxicity of tannins and oxidized flavanols to nematodes has not been investigated.

Tannins are implicated in resistance reactions to various fungal species (e.g., *Diplodia*, *Glomerella*, *Fusarium*, *Alternaria*) causing boll rots, and may convey some resistance to bacterial blight caused by *Xanthomonas campestris* pv. *malvacearum* (Bell et al. 1992, and references within).

4 Variation in Tannin Content

Numerous studies have documented environmental influences on tannin levels in various plant species (Swain 1979; Haslam 1979; Krischik and Denno 1983; Zummo et al. 1984b). Additionally, significant variation in tannin content exists in many plant organs sampled at different times of the year or growing season. This inherent variation in tannin levels has complicated breeding efforts in cotton aimed at improving resistance to pests by enhancement of tannin levels. A detailed discussion of variation of tannins in different tissues of the cotton plant can be found in Bell et al. (1992). The following discussion will focus on seasonal variability in tannin content at the plant organ level.

4.1 Seasonal Variation

Early reports of seasonal variation in tannin content were aimed at characterizing tannin levels in relation to insect pest infestation periods, as well as oviposition and feeding sites within the plant. Zummo et al. (1984b) observed that tannin increased in concentration in the terminal leaf of cotton as the season progressed from the cotyledonary stage to one-third-grown square stage, then decreased significantly; tannin content subsequently recovered rapidly through first-cracked boll stage, which was the final sampling (Fig. 2). The terminal leaf was sampled because cotton bollworm often lays eggs on or near this leaf; hence, it is most likely the first plant tissue consumed by an emerging larvae (Wilson et al. 1980). In an earlier study (Guinn and Eidenbock 1982), a similar drop in tannin content occurred at 12 weeks after planting, or approximately 2 weeks prior to first-cracked boll stage. Tannin content, however, then increased to a peak at 18 weeks after planting, or 50% open boll. This increase in tannin content associated with plant age has been found in many perennial plant species (Krischik and Denno 1983). A more recent study suggests that the tannin levels in cotton plants peak at first bloom stage (Fig. 3), based on averages of condensed tannin concentrations of six plant parts sampled from six genotypes varying in spider mite resistance (Legé et al. 1992).

4.2 Within-Plant Variability

The condensed tannin concentration of the terminal leaf (Zummo et al. 1983, 1984a,b; Legé et al. 1992, 1993), bolls (Chakravorty and Basu 1981) and

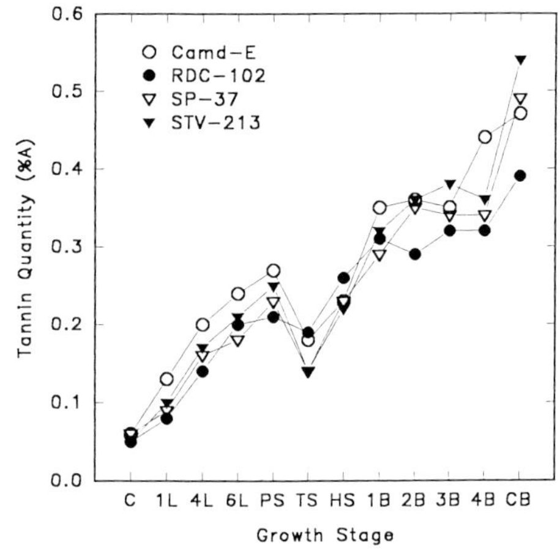

Fig. 2. Tannin quantity, in % spectrophotometric absorbance (%A), from the first terminal leaf over 12 weeks for four cotton cultivars. Growth stages are: *C* cotyledonary; *1L* first true leaf; *4L* fourth true leaf; *6L* sixth true leaf; *PS* pinhead square; *TS* one-third-grown square; *HS* one-half-grown square; *1B* first week of bloom; *2B* second week of bloom; *3B* third week of bloom; *4B* fourth week of bloom; *CB* cracked boll. (Data from Zummo et al. 1984b)

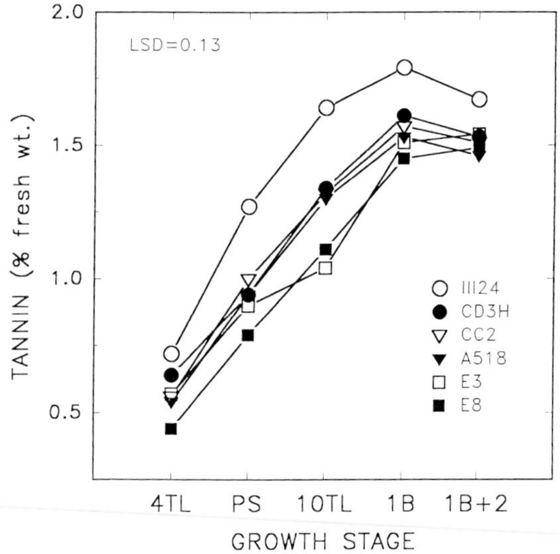

Fig. 3. Condensed tannin concentration of six cotton genotypes at fourth true leaf (*4TL*), pinhead square (*PS*), tenth true leaf (*10TL*), first bloom (*1B*), and 2 weeks post first bloom (*1B* + *2*) growth stages. Values are means from six plant parts and three planting dates over 2 years. (Data from Legé et al. 1992)

squares (Chan et al. 1978) of cotton have been reported. Regardless of the plant organ, younger plant parts are consistently higher in condensed tannins than older or mature plant parts. In some primitive cotton strains, tannin concentration was low in the cotyledons and first true leaves, but increased with each successive leaf until the tenth true leaf (Lane and Schuster 1981). Terminal leaves, bracts from 6-day-old squares, and sepals from 10-day-old squares of cotton (i.e. young tissue) were significantly higher in condensed tannins than their mature counterparts (i.e., leaves from the fourth mainstem node from the terminal, bracts from white blooms, and sepals from white blooms, respectively) throughout the season (Fig. 4; Legé et al. 1993). The distribution of condensed tannins within the cotton plant further substantiates its possible role as a natural plant-defense mechanism. Many insect species often feed on terminal leaves and young squares; since condensed tannins are higher in concentration in these younger tissues, some suggest that natural selection for higher tannin levels in these tissues occurred in the primitive strains from which recent high-tannin lines (Smith et al. 1990; Schuster et al. 1990) have been developed. Since variability of condensed tannin levels in plant parts does exist for cotton, breeding for high-tannin levels should involve sampling for more than one plant part to obtain a clearer indication of tannin levels throughout the plant. If one or just a few plant parts are sampled, breeders should consider carefully which plant tissue to sample and meticulously examine the same age tissue each time. Also, breeders should sample the tissue normally attacked by the pest in question. For example, young leaves and squares should be sampled if high-tannin levels are to be used for bollworm-budworm resistance breeding; for spider mite resistance, select for high tannins in the more mature leaves (Legé et al. 1993).

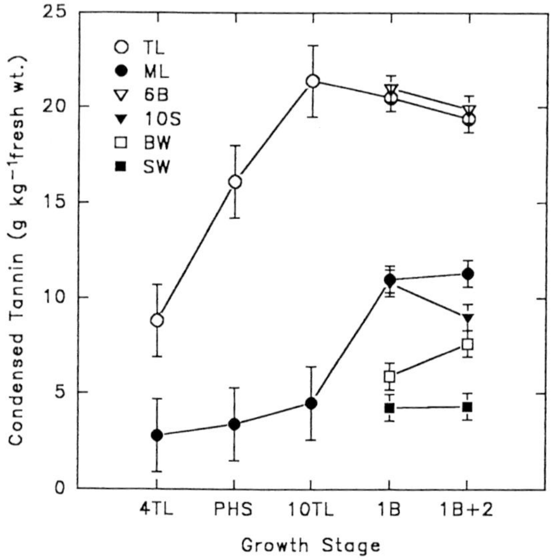

Fig. 4. Condensed tannin concentration of the topmost unfolded leaf (*TL*), fourth mainstem leaf (*ML*), 6-day-old bract (*6B*), 10-day-old sepal (*10S*), bract from a white bloom (*BW*), and sepal from a white bloom (*SW*) at fourth true leaf (*4TL*), pinhead square (*PHS*), tenth true leaf (*10TL*), first bloom (*1B*), and 2 weeks post first bloom (*1B + 2*) growth stages. Values are means of six genotypes and three planting dates over 2 years. (Portions of data from Legé et al. 1993)

5 Interaction of Tannins and Phenolic Acids

A recent study has suggested that a relationship exists between condensed tannin levels and some phenolic acid levels in genotypes selected for spider mite resistance (Legé et al. 1995). Thirteen phenolic acids (gallic, gentisic, protocatechuic, *p*-hydroxybenzoic, vanillic, caffeic, salicylic, syringic, *p*-coumaric, benzoic, ferulic, sinapic, and cinnamic acids) and condensed tannins were analyzed in four gentoypes selected for spider mite resistance and high tannins. The total phenolic acid concentration, as well as those of five individual phenolic acids, were significantly correlated with condensed tannin levels. The genotypic rankings of caffeic, syringic, and *p*-coumaric acids closely followed the ranking of condensed tannins for those genotypes (Legé et al. 1995). These results suggest that selecting for elevated levels of condensed tannins could concomitantly increase levels of some phenolic acids, which are reported to convey resistance to various pests of cotton (Weise 1968; Wakelyn et al. 1974; Bell and Stipanovic 1978; Guerra et al. 1990). However, much more research is needed to determine if the relationship between condensed tannins and phenolic acids and other phenolic compounds has any biological significance.

6 Human Health Aspects of Tannins

Tannins can potentially have adverse effects on humans who are regularly exposed for prolonged periods to cotton dusts associated with ginning and milling processes. Symptoms of byssinosis, a chronic respiratory disease, have been observed when purified tannins are added to isolated animal cells or tissues. Since much of the dust associated with the processing of raw cotton originates from dried bract tissue, many investigations dealing with cotton dusts and human respiratory problems have focused on determining the tannin concentration of the bract. However, it is uncertain how much tannin is actually inhaled by gin and mill workers and whether the levels in dusts are at biologically active concentrations; additionally, bacterial endotoxins, fungal cell walls, and cotton terpenoids have been suggested as causes of symptoms of byssinosis (Bell et al. 1992). Accordingly, researchers have had difficulty distinguishing among the potential causes of byssinosis. The toxicity of cotton tannins to animal cells has been reviewed earlier in detail (Bell and Stipanovic 1983; Rohrbach et al. 1992).

7 Summary, and Future of Tannins

Tannins serve as an effective host-plant resistance mechanism to many species of arthropods and disease organisms that cause economic damage to cotton.

Research cocerning resistance to tobacco budworm, cotton bollworm, pink bollworm, spotted bollworm, two-spotted spider mite, and boll weevil has suggested that the mode of action of tannins is primarily that of a feeding deterrent to these arthropod pests. Tannins impart resistance or tolerance to disease organisms mainly through critical enzyme inhibition. Much work has been conducted to give insight into what tissues possess the highest tannin concentrations and at what point during the growing season tannins can be expected to increase or decrease in these tissues. This has given researchers valuable information regarding the most appropriate tissue to sample when breeding and screening lines for resistance to these various pests. Accordingly, tannins have been a significant part of cotton breeding programs involved in developing resistant lines to any of these pests of cotton.

Since the genetic control of tannin synthesis is not understood, and since the advent of transgenic varieties with near-absolute resistance to some pests, breeding for elevated levels of tannins to increase pest resistance has not progressed to the degree seen in the late 1980s and early 1990s. Regardless, tannins still serve as a valuable tool to provide some level of resistance or tolerance to important economic pests of cotton, and still have potential as a critical part of host-plant resistance breeding in the crop.

Many areas associated with pest resistance and condensed tannins have remained virtually univestigated. The biological activity of tannins on nematodes is very poorly understood; the effectiveness of tannins in resisting infection by organisms causing boll rots is not well documented; and while breeding for high-tannin levels has realized some success, the genetic control of tannin synthesis is not understood. The potential of utilizing condensed tannins and related compounds is dependent upon documenting these relationships. However, any antagonistic effects of tannins, such as the interaction of tannins and endotoxins from *Bacillus thuringiensis*, will need to be strongly considered in efforts to utilize these polyphenolic allelochemicals in future pest resistance breeding endeavors.

References

Aston JL, Winfield B (1972) Insect section introduction. In: Cotton. CIBA-Geigy Agrochemicals Tech Monograph No 3, Greensboro, NC, pp 44–45

Bate-Smith EC (1973) Haemanalysis of tannins: the concept of relative astringency. Phytochemistry 12:907–912

Bell AA (1968) Gossypol and tannin synthesis in *Verticillium*-infected cotton tissues. Proc Beltwide Cotton Prod Res Conf, Memphis, TN, p 29

Bell AA, Stipanovic RD (1977) The chemical composition, biological activity, and genetics of pigment glands in cotton. Proc Beltwide Cotton Prod Res Conf, Memphis, TN, p 97

Bell AA, Stipanovic RD (1978) Biochemistry of disease and pest resistance in cotton. Mycopathologia 65:91

Bell AA, Stipanovic RD (1983) Biologically active compounds in cotton: an overview. Proc Cotton Dust Res Conf, Memphis, TN, pp 77–80

Bell AA, El-Zik KM, Thaxton PM (1992) Chemistry, biological significance and genetic control of proanthocyanidins in cotton (*Gossypium* spp.). In: Hemingway RW, Laks PE (eds) Plant polyphenols. Synthesis, properties, signficance. Plenum Press, New York, pp 571–595

Benedict JH, Brd LS (1981) Relationship of microorganisms within the plant and resistance to insects and diseases. Proc Beltwide Cotton Prod Res Conf, Memphis, TN, pp 149–150
Benays EA, Cooper-Driver GC, Bilgener M (1989) Herbivores and plant tannins – a review. Adv Ecol Res 19:263–302
Blasingame DW (1995) Cotton disease loss estimate committee report. Proc Beltwide Cotton Conf, San Antonio, TX, pp 195–196
Bottger GT, Patana P (1966) Growth, development and survival of certain *Lepidoptera* fed gossypol in the diet. J Econ Entomol 59:1166–1168
Chakravorty SC, Basu AK (1981) Changes in tannin during growth of bolls in three cotton cultivars known for their differential susceptibilities to pink bollworm (*Pectinophora gossypiella*) incidence. Sci Cult 47:373–376
Chakravorty SC, Basu AK, Sahni VM (1982) Variability in tannin and protein contents in fruiting parts in relation to the incidence of pink bollworm in cotton. Indian J Agric Sci 52:767–773
Chan BG (1985) From L-phenylalanine to condensed tannins: a brief review of tannin chemistry and biochemistry and a report on its distribution in cotton plant parts. In: Wakelyn PJ, Jacobs RR (eds) Proc 9th Cotton Dust Res Conf, Memphis, TN, pp 49–52
Chan BG, Waiss AC, Lukefahr M (1978) Condensed tannins, an antibiotic chemical from *Gossypium hirsutum*. Insect Physiol 24:113–118
Delort-Laval J, Viroben G (1969) Nutrition. CR Hebd Seances Acad Sci Paris 269:1558–1561
Dilday RH, Shaver TN (1980) Variability in flower-bud gossypol content and agronomic and fiber properties within the primitive race collection of cotton. Crop Sci 20:91–95
Feeny PP (1968) Effect of oak leaf tannins on larval growth of the winter moth *Operophtera brumata*. J Insect Physiol 14:805–817
Feeny PP (1970) Oak leaf tannins and winter moth feeding. Ecology 51:556–581
Gill SS, Cowles EA, Pietrantonio PV (1992) The mode of action of *Bacillus thuringiensis* endotoxins. Annu Rev Entomol 37:615–636
Goldstein JL, Swain T (1965) The inhibition of enzymes by tannins. Phytochemistry 4:185–192
Guerra DJ, Cothren JT, Phillips JR (1990) Influence of selected phenolic compounds on development of bollworm (Lepidoptera: Noctuidae) larvae. J Econ Entomol 83:2115–2118
Guinn G, Eidenbock MP (1982) Catechin and condensed tannin contents of leaves and bolls of cotton in relation to irrigation and boll load. Crop Sci 22:614–616
Halloin JM (1982) Localization and changes in catechin and tannins during development and ripening of cottonseed. New Phytol 90:651–657
Haslam E (1979) Vegetable tannins. In: Swain T, Harborne JB, Van Sumere CF (eds) Recent advances in phytochemistry, vol 12. Plenum Press, New York, pp 475–517
Hedin PA, McCarty JC (1990) Possible roles of cotton bud sugars and terpenoids in oviposition by the boll weevil. J Chem Ecol 16(3):757–772
Hedin PA, Jenkins JN, Collum DH, White WH, Parrott WL (1983) Multiple factors in cotton contributing to resistance to the tobacco budworm, *Heliothis virescens* F. In: Hedin PA (ed) Plant resistance to insects. Am Chem Soci Symp Ser 208. Washington, DC, pp 80–101
Hedin PA, Shepherd RL, Kappelman AJ Jr (1984) Evaluation of cotton polyphenols as factors of resistance to root-knot nematode and fusarium wilt. J Agric Food Chem 32:635–638
Hofte H, Whiteley HR (1989) Insecticidal crystal proteins of *Bacillus thuringiensis*. Microbiol Rev 53:242–255
Hunter RE (1974) Inactivation of pectic enzymes by polyphenols in cotton seedlings of different ages infected with *Rhizoctonia solani*. Physiol Plant Pathol 4:151–159
Hunter RE (1978) Effects of catechin in culture and in cotton seedlings on the growth and polygalacturonase activity of *Rhizoctonia solani*. Phytopathology 68:1032–1036
Jaroszewski JW, Strom-Hansen T, Hansen LL (1992) Optical stability of gossypol. Chirality 4:216–221
Klocke JA, Chan BG (1982) Effects of cotton condensed tannin on feeding and digestion in the cotton pest *Heliothis zea*. J Insect Physiol 28:911–915
Krischik VA, Denno RF (1983) Individual population and geographic patterns in plant defense. In: Denno RF, McClure MS (eds) Variable plants and herbivores in natural and managed systems. Academic Press, New York, pp 463–512
Lane HC, Schuster MF (1979) Tannins in cotton. Proc Beltwide Cotton Prod Res Conf, Memphis, TN, p 43

Lane HC, Schuster MF (1981) Condensed tannins of cotton leaves. Phytochemistry 20:425–427

Legé KE, Smith CW, Cothren JT (1992) Genotypic and cultural effects of condensed tannin concentration of cotton leaves. Crop Sci 32:1024–1028

Legé KE, Smith CW, Cothren JT (1993) Planting date and plant density effects on condensed tannin concentration of cotton. Crop Sci 33:320–324

Legé KE, Cothren JT, Smith CW (1995) Phenolic acid and condensed tannin contentrations of six cotton genotypes. Environ Exp Bot 35(2):241–249

Lipke H, Fraenkel G (1976) Insect nutrition. Annu Rev Entomol 1:17–44

Lukefahr MJ, Houghtaling JE (1969) Resistance of cotton strain with high gossypol content to *Heliothis* spp. J Econ Entomol 62:588–591

Lukefahr MJ, Martin DF (1966) Cotton plant pigments as a source of resistance to bollworm and tobacco budworm. J Econ Entomol 59:176–179

Maxwell FG, Jenkins JN, Parrott WL (1967) Influence of constituents of cotton plant on feeding, oviposition and development of the boll weevil. J Econ Entomol 60:1294–1296

Navon A, Hare JD, Federici BA (1993) Interactions among *Heliothis virescens* larvae, cotton condensed tannin and the CryIA(c) ‰-endotoxin of *Bacillus thuringiensis*. J Chem Ecol 19:2485–2499

Okasha KA, Ryugo K, Wilhelm S, Bringhurst RS (1968) Inhibition of growth of *Verticillium albo-atrum* sporelings by tannins and polyphenols from infected crowns of *Verticillium*-resistant and -susceptible strawberry cultivars. Phytopathology 58:1114

Patil SS, Dimond AE (1967) Inhibition of *Verticillium* polygalacturonase by oxidation products of polyphenols. Phytopathology 57:492

Ridgeway RL, Bell AA, Veech JA, Chandler JM (1984) Cotton protection practices in the USA and the world. In: Kohel RJ, Lewis CF (eds) Cotton. ASA, CSSA, SSSA, Madison, Wisconsin, pp 265–365

Rohrbach MS, Kreofsky TJ, Vuk-pavlovic' Z, Lauque D (1992) Cotton condensed tannin: a potent modulator of alveolar macrophage host-defense function. In: Hemingway RW, Laks PE (eds) Plant polyphenols. Synthesis, properties, significance. Plenum Press, New York

Sadikov AS (1965) Recent advances in the chemistry of the cotton plant. J Sci Ind Res 24:77

Schuster MF (1979) New sources of high tannin resistance to *Heliothis* in upland cotton resulting in feeding deterrence. Proc Beltwide Cotton Prod Res Conf, Memphis, TN, pp 85–87

Schuster MF (1980) Insect resistance in cotton. In: Harris MK (ed) Biology and breeding for resistance to arthropods and pathogens in agricultural plants. Tex Agric Exp Sta Misc Publ 1451:101–112

Schuster MF, Lane HC (1980) Evaluation of high-tannin cotton lines for resistance to bollworms. Proc Beltwide Cotton Prod Res Conf, Memphis, TN, pp 83–84

Schuster MF, Smith CW, Niles GA (1990) Registration of nine upland cotton germplasm lines having elevated level of condensed tannin. Crop Sci 30:1375

Sharma HC, Agarwal RA (1982a) Consumption and utilization of bolls of different cotton genotypes by larvae of *Earia vittella* F. and effect of gossypol and tannins on food utilization. Z Angew Zool 68:13–16

Sharma HC, Agarwal RA (1982b) Effect of some antibiotic compounds in *Gossypium* on the post-embryonic development of spotted bollworm (*Earias vittella*). Entomol Exp Appl 31:225–228

Sharma HC, Agarwal RA (1983) Factors affecting genotypic susceptibility to spotted bollworm *Earias vittella* Fab. in cotton. Insect Sci Appl 4:363–365

Sharma HC, Agarwal RA (1984) Factors imparting resistance to stem damage by *Earias vittella* F. (Lepidoptera: Noctuidae) in some cotton phenotypes. Prot Ecol 6:35–36

Sharma HC, Agarwal RA, Singh M (1982) Effect of some antibiotic compounds in cotton on post-embryonic development of spotted bollworm (*Earias vittella* F.) and the mechanism of resistance in *Gossypium arboreum*. Proc Indian Acad Sci (Anim Sci) 91:67–68

Shaver TN, Dilday RH, Wilson FD (1980) Use of glandless breeding stocks to evaluate unknown *Heliothis* growth inhibitors (X-factors) in cotton. Crop Sci 20:545–548

Smith CW, Schuster MF, Niles GA (1990) Registration of 17 upland cotton germplasm lines having elevated levels of condensed tannins. Crop Sci 30:1374–1375

Smith CW, McCarty JC, Altamarino TP, Legé KE, Schuster MF, Phillips JR, Lopez JD (1992) Condensed tannins in cotton and bollworm-budworm (Lepidoptera: Noctuidae) resistance. J Econ Entomol 85(6):2211–2217

Somers TC, Harrison AF (1967) Wood tannins – isolation and significance in host resistance to *Verticillium* wilt disease. Aust J Biol Sci 20:475

Swain T (1979) Phenolics in the environment. In: Swain T, Harborne JB, Van Sumere CF (eds) Recent advances in phytochemistry, vol 12. Plenum Press, New York, pp 624–637

Veech JA (1977) Terpenoid aldehydes and condensed tannins in the roots of five cotton varieties with differing responses to the root-knot nematode. Proc Beltwide Cotton Prod Res Conf, Memphis, TN, p 37

Wakelyn PJ, Stipanovic RD, Bell AA (1974) Identification of scopoletin in dried bract of the cotton plant. J Agric Food Chem 22:567–568

Weise MV (1968) Alteration in phenolics and growth regulators in cotton infected with *Verticillium albo-atrum*. PhD Thesis. University of California, Davis, CA, USA

White WH, Jenkins JN, Parrott WL, McCarty JC, Collum DH, Hedin PA (1982) Strain and within-season variability of various allelochemics within a diverse group of cottons. Crop Sci 22:1235–1238

Wilson LT, Gutierrez AP, Leigh TF (1980) Within-plant distribution of the immatures of *Heliothis zea* (Boddie) on cotton. Hilgardia 48:12–23

Wilson RL, Wilson FD (1977) Effects of cotton differing in pubescence and other characters on pink bollworms in Arizona. J Econ Entomol 70:196–198

Yokoyama VY, Mackey BE (1987) Protein and tannin in upper, middle, and lower cotton plant strata and cigarette beetle (Coleoptera: Anobiidae) growth on the foliage. J Econ Entomol 80:843–847

Yokoyama VY, Mackey BE, Leigh TF (1987) Relation of cotton protein and tannin to methyl parathion treatment and cigarette beetle (Coleoptera: Anobiidae) growth on foliage. J Econ Entomol 80:834–839

Zummo GR, Benedict JH, Segers JC (1983) No-choice study of plant-insect interactions for *Heliothis zea* (Boddie) (Lepidoptera: Noctuidae) on selected cottons. Environ Entomol 12:1833–1836

Zummo GR, Benedict JH, Segers JC (1984a) Effect of the plant growth regulator mepiquat chloride on host plant resistance in cotton to bollworm (Lepidoptera: Noctuidae). J Econ Entomol 77:922–924

Zummo GR, Segers JC, Benedict JH (1984b) Seasonal phenology of allelochemicals in cotton and resistance to bollworm (Lepidoptera: Noctuidae). Environ Entomol 13:1287–1290

Subject Index

Springer and the environment

At Springer we firmly believe that an international science publisher has a special obligation to the environment, and our corporate policies consistently reflect this conviction.
We also expect our business partners – paper mills, printers, packaging manufacturers, etc. – to commit themselves to using materials and production processes that do not harm the environment. The paper in this book is made from low- or no-chlorine pulp and is acid free, in conformance with international standards for paper permanency.

Printing: Saladruck, Berlin
Binding: Buchbinderei Lüderitz & Bauer, Berlin